P9-EEF-394

Understanding Wood

Understanding Wood

A craftsman's guide to wood technology

R. Bruce Hoadley

Frontispiece—Boards cup during drying because wood shrinks more around the annual rings than it shrinks across the annual rings. The amount of cup is thus a function of where the board was when it was inside the tree. To illustrate, Hoadley drew straight lines across this red oak disc when it was freshly cut. Upon drying to 4% moisture content, the unequal shrinkage caused the disc to crack and the lines to bend. To control the experiment, Hoadley bandsawed a narrow stress-relief slit from the bark to the pith; as the wood dried, the slit opened to the wedge-shaped cutout you see. Left to itself, a ragged crack would have opened at the weakest point in the wood's structure.

First printing: November 1980

The Taunton Press, Inc.
52 Church Hill Road
Box 355
Newtown, Connecticut 06470

International Standard Book Number 0-918804-05-1
Library of Congress Catalog Card Number 78-65177
Printed in the United States of America

Contents

Tables

Foreword

The properties and characteristic qualities of the timber available are so numerous and important, and yet so little understood generally, that I am induced by solicitations of many friends to give, in these pages, information respecting them.

A handy-book on timber is, in the opinion of many, much required. The botanical treatises which are accessible are too strictly scientific in their form and treatment to interest the general reader, and they lack that practical application of knowledge to the wants of the shipwright and carpenter, which is one of the aims of this book to give.

I wish I had written those words, for they summarize perfectly the reasons why I have written this book. But they were, in fact, written in London in the year 1875 by Thomas Laslett, timber inspector to the Admiralty of the British Empire, in his book, *Timber and Timber Trees.*

Like Laslett, I have written my book for woodworkers, but this is not a book about woodworking. Rather, it is about wood itself, surely mankind's first workable material, and an ever-present part of our ever-changing world. We are not likely to run out of wood, because unlike most other materials, we can always grow more. And the more we learn about it, the more there is to know. A look at what else Laslett wrote in his book will make this clear.

Laslett believed, for example, that sap collected between the bark and the wood and eventually congealed to form a new growth ring. He could not know about cellular reproduction and the additive formation of new wood cell by cell. Laslett deduced that trees grew taller because the bark squeezed the sap, forcing it upward. Today, with a microscope, we can see cells building sideways in the cambium layer, and twigs growing longer by cellular division. Lest we be tempted to smugness, however, we should imagine how primitive our scientific knowledge might look a hundred years from now.

The main reason I have attempted this book is the realization that a wealth of knowledge about wood has been accumulated by scientists, but almost none of it has been translated and interpreted for the individual craftsman. Working from scientific principles, technology is routinely developed by commercial and academic agencies but mainly shared among themselves in highly technical textbooks and obscure journals. The technology of industrial-scale woodworking has been well developed and widely published. But the same scientific principles have yet to be applied to the small-scale woodworking shop, whether that of the serious amateur, the independent cabinetmaker or the artist/craftsman. For example, volumes have been written on how to dry lumber in carload quantities, yet it is almost impossible to find guidance when drying the boards cut from a single tree, or when drying a single board.

Much of what has been written about the craft of woodworking is reduced to sets of instructions or directions relative to tools and procedures, without any supporting information about the material itself. These how-to books commonly assume that all conditions relative to the wood are under control, or else they dismiss the point with an airy instruction such as, "Get some suitable hardwood of appropriate quality and dryness." Yet, for a person to pursue a craft with success, knowledge and understanding of the material must develop along with manual skill.

I also feel compelled to attack the mountains of misinformation available and commonly accepted by woodworkers. Most of it comes innocently from the misinterpretation of observations. For example, one book on sculpture states that bright light makes wood check, a conclusion reached when cracks appeared in wood brought up from a dark (and probably damp) cellar into daylight (and drier air). Another says that wood cups because its annual rings try to straighten out as it dries—a correct observation of the direction of cupping, but pure guesswork as to its cause. Tradition carries along such misleading terms as "dry rot" and such misconceptions as "wood has to breathe." Dry wood will not decay, nor does wood breathe in the animal sense. Wood doesn't eat either, and it doesn't require feeding with furniture polish.

My strategy has been to begin with the tree, to examine the wood as the cellular product of the tree's growth. I have given special attention to wood/moisture relationships and dimensional change before going on to physical properties, strength in particular. Then I have tried to analyze such everyday woodworking operations as machining, bending, joining and finishing in terms of the wood's physical and biological nature. It is my hope that the examples I have given can serve as models for readers in analyzing problems that arise in their own woodworking endeavors.

I also hope that this book will encourage craftsmen to delve further into the literature of wood science and technology, and to help them do so I have included an annotated list of reference books. Frequently, what makes technical literature opaque to the layman is its terminology. Therefore, against the wishes of some who would have me avoid scientific terms and "say it in simple words," I have tried to present and explain the standard terminology throughout the book, and I have included a detailed glossary. The serious woodworker will find that it is important to know that *rake angle* and *hook angle* have nothing to do with gardening or fishing, and that *terminal parenchyma* is not a horrible disease.

I suppose every author wonders when a book actually began—for me, it goes back more than 30 years. I grew up in the Connecticut countryside where the surrounding woodlands were both playground and the source of material for "making stuff." The cellar of our house had an old workbench and chests of grandfather's tools, worn from years of use but begging for the chance once again to work miracles in wood. My earliest memories include climbing a wobbly stool to get on the workbench to turn on the light, the screech of the huge square-threaded screw of the vise, and lifting the heavy lid of the toolchest to stare at the mysteries within. I recall more trouble than triumph from my early years of woodworking experiments. I remember nails that bent over when driven into oak, saws that bound up tight in green wood, screws twisted off when driven without pilot holes, and planed surfaces ridged by nicks in the iron. But I still "made stuff"—my frustrations were nothing compared to my fascination with wood.

I'll never forget the first time I saw a chain saw in operation. It was at a late-summer farm-equipment demonstration, when I was in my early teens. The farmers and loggers all watched in amazement as the saw bar melted through a 12-in. oak log in a matter of seconds, effortlessly taking slice after slice. It was quite a machine.

But I was not watching the raucous machine. I stood transfixed by those marvelous discs of wood, a dozen or more, that lay in the grass and sawdust. The demonstration over, the entire crowd followed the saw operator back to his table to learn more about the machine.

Except me. I was excitedly stacking up as many of the wooden discs as I could carry. They were red oak, creamy sapwood and medium brown heartwood, just tinged with peach. As I staggered through the goldenrod toward my father's car, they were unbelievably heavy. I still recall the vivid pattern of the rings, their pie-crust of bark, the cool dampness of the top disc under my chin, the pungent odor of the wood. The aching in my arms was a tiny price to pay for the unending array of things I would make from such a magnificent product of nature: lamp bases, clock faces, desk sets, picture frames. . . . I could not believe that woodworkers had not already put this beautiful natural log to better use.

My next recollection is of having the discs safely home and proudly lined up along the wall shelf in our cellar workshop. I probably realized that some sort of drying would be necessary, but that could wait. I was content just to admire my treasure.

I am sure you already know the sad ending. By the following morning, the brilliant end grain had faded to a lifeless sandy color. In the days that followed, my castle of hope crumbled as the first few hairline cracks in the sapwood grew and reached toward the pith. Soon each disc had a gaping radial crack. In final mockery, even the bark fell off.

But why?

It is easy for me to believe that this one incident was the turning point that eventually led me to pursue my career as a wood technologist, for ever since I have wanted to know why the wood does what it does. I have been lucky enough to make some progress in my quest, and I have been able to share in the classroom some of what I have learned. Now I hope my book helps craftsmen understand wood, too.

—R. Bruce Hoadley
Amherst, Mass., April 1980

Photo and art credits—Richard Starr, cover; *Wood Structure and Identification,* Harold A. Core, Wilfred A. Cote and Arnold C. Day (Syracuse University Press, 1979), 11, 17; Max Wilson, 18, 25; Fine Hardwoods American Walnut Association, 22, 23, 25; Tim Savard, 35; Stephen Smulski, 182, 183; *Workbench* magazine, 196; American Plywood Association, 202; American Forest Institute, 204; Capital Machine Co., Inc., 217, 218; Dick C. Wittenberg, 226.

Acknowledgments

The idea for this book came from the many students I have worked with over the years, especially those in craft workshops, short courses and seminars. Their many questions, discussions, reactions and frequent suggestions to put it all down on paper were the impetus I needed to get me started. For his encouragement to begin the writing and for his support throughout, I owe special thanks to Dr. Donald R. Progulske, head of the department of forestry and wildlife management at the University of Massachusetts, Amherst.

During the course of publication, many people became involved with this book. My deepest gratitude goes to Dee Ann Civello, who, with the help of my daughter Susan, transformed my longhand into typed manuscript with efficiency and accuracy, adding an encouraging word when I needed it most. Much of the early editing was done by Stanley N. Wellborn, and my good friend and colleague, Dr. Alan A. Marra, read the galleys thoroughly, clarified the text and refined it technically. Technical details in the section on seasoning were provided by Dr. William W. Rice. Ethan V. Howard located and machined many of the wood samples for photography, and the microtome surfacing of the wood samples for the chapter on wood identification was the work of Philip L. Westover and Susan M. Smith.

Those many handsome photographs that stand apart in the text are the painstaking work of Richard Starr, a woodworker and teacher himself. His dedication and patience made it possible to capture precise technical details that are also visually striking. The scanning electron microphotographs on page 11 were provided through the courtesy of Dr. Wilfred Cote.

Much of the technical data on physical and mechanical properties of wood as well as some of the illustrations were reproduced or adapted from the *Wood Handbook* and other publications of the U.S. Forest Products Laboratory at Madison, Wis. During my visits there and in correspondence, members of the staff were always helpful and cooperative.

* * *

Although writing this book was more fun than I had expected, it also turned out to be a far larger project than I had ever dreamed. I was indeed fortunate to have the constant support of a wonderful family who accepted the late suppers, the aborted holidays and the interrupted vacations. To my wife, Barbara, and to my daughters, Susan and Lindsay, for their patience and understanding, I dedicate this book.

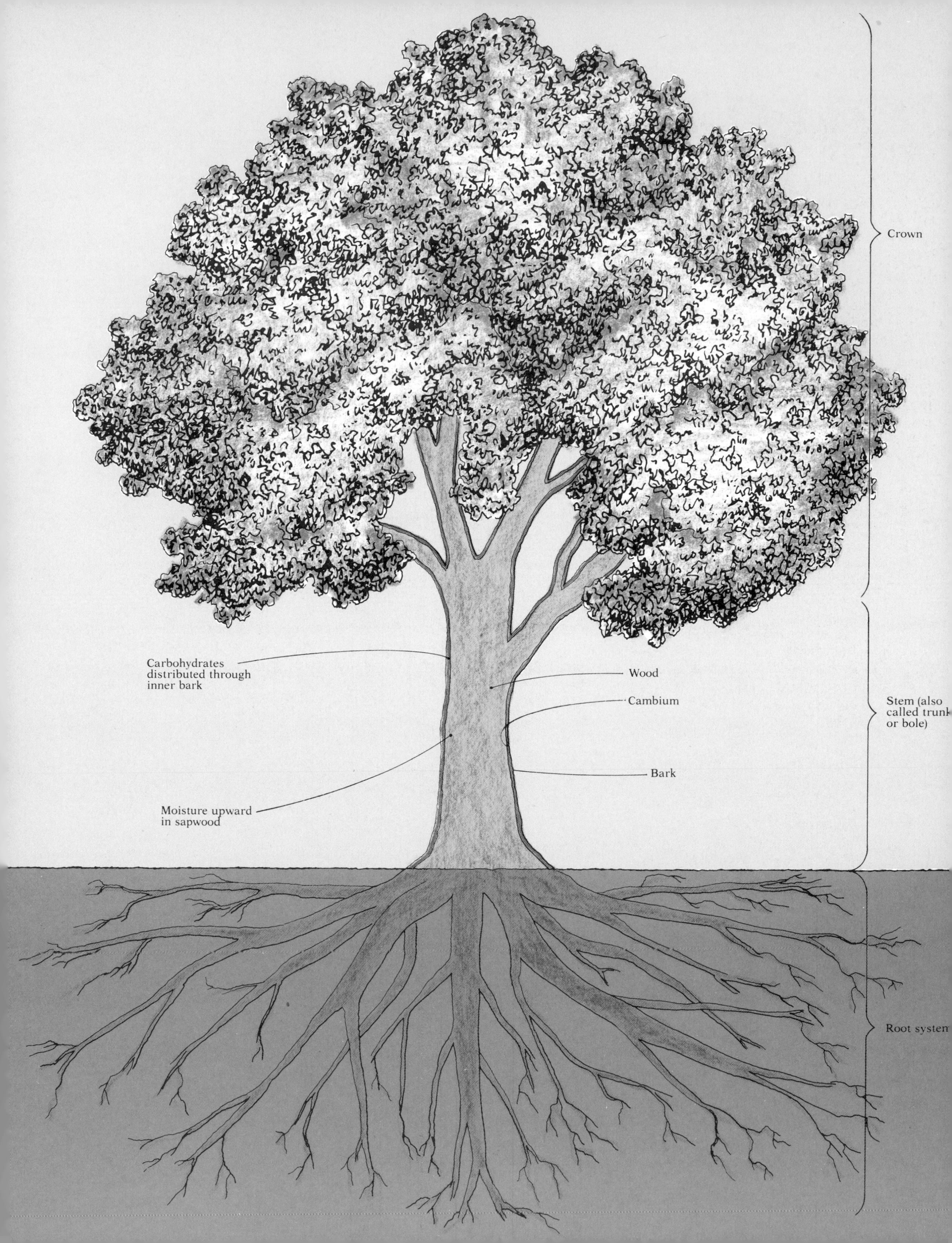

Crown
Carbohydrates
distributed through
inner bark
Wood
Cambium
Stem (also
called trunk
or bole)
Bark
Moisture upward
in sapwood
Root system

The Nature of Wood

Wood comes from trees. This is the most important fact to remember in understanding the nature of wood. Whatever qualities or shortcomings wood possesses are traceable to the tree whence it came. Wood evolved as a functional tissue of plants rather than as a material designed to satisfy the needs of woodworkers. Thus, knowing wood as it grows in nature is basic to working successfully with it.

Since prehistoric times, man has used the beauty and economic value of trees in commerce and art, for shelter and furnishings, and one is tempted to discuss at length the virtues of these noble representatives of the plant kingdom. But since our goal is understanding wood, we will concentrate on the functional and physical aspects of trees rather than on their aesthetic aspects. We must consider the tree at all scales—from the entire plant down to the individual cell.

At the smallest level, understanding cell structure is the key to appreciating what happens when wood is sanded across the grain, or why stain penetrates unevenly, or why adhesives bleed through some veneers but not others. But to understand where feather grain is to be found, or to visualize a knot's internal structure based on its surface appearance, or to anticipate which boards are susceptible to decay, it is necessary to examine the structure of the entire tree as a living organism. So this is a logical starting point.

Despite their wide diversity, all trees have certain common characteristics *(1)*. All are vascular, perennial plants capable of secondary thickening, that is, of adding yearly growth to previous growth. The visible portion of the tree has a main supporting **stem** or **trunk**. If large enough for conversion into sawtimber or veneer, the trunk is often termed the **bole**. The trunk is the principal source of wood used by woodworkers, although pieces having unusual beauty and utility also come from other parts of the tree. The trunk has limbs, which in turn branch and eventually subdivide into twigs. This subdivision, carrying the leaves or foliage, is collectively termed the **crown**. But a casual glance hardly reveals the awesome complexity of the internal structure of these impressive plants. One can begin an acquaintance by exposing a crosswise surface of a tree trunk *(2)*. At the periphery of the log surface, the **bark** layer can be recognized easily. Within the bark, and comprising the bulk of the stem, is the **wood**, which is characterized by its many **growth rings** concentrically arranged around the central **pith**. Between the bark and the wood is the **cambium**, a microscopically thin layer of living cells (that is, having protoplasm in their cavities).

The tree stem parts are accumulations of countless **cells**. The cell is the basic structural unit of plant material. Each cell consists of an outer **cell wall** surrounding an inner **cell cavity**. Within some cells, there is living

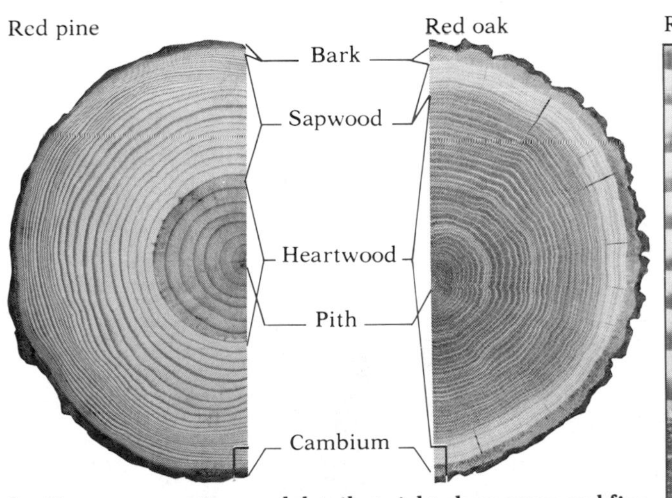

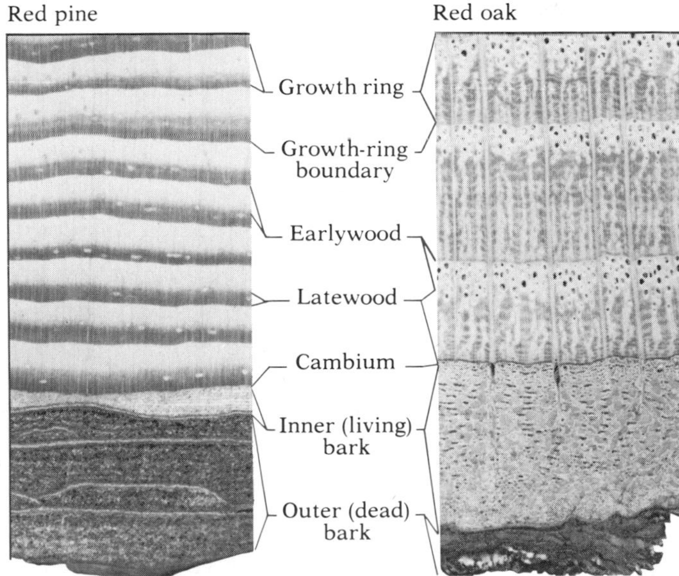

2—Stem cross sections and detail at right show gross and fine structures of a typical softwood, red pine *(Pinus resinosa)*, and a typical hardwood, northern red oak *(Quercus rubra)*. Narrow rays are invisible in red pine, while portions of broad rays in red oak are visible to the naked eye.

1—A tree is a living, functioning organism, not just an inert source of material for the woodworker.

protoplasm, while others are nonliving and contain only sap or even some air space.

Wood cells typically are elongated. The proportion of length to diameter varies widely among cell types, however, from short barrel shapes to long needlelike cells. The majority of wood is composed of **longitudinal cells**, whose axes are oriented vertically in the tree. The largest cells, as in woods such as chestnut and mahogany, are visible to the unaided eye, but most are too small to be seen without magnification.

Scattered through the wood are groups of cells whose axes are horizontal. These groups extend radially outward from the pith and are therefore called **rays**. Rays are like flattened ribbons of cells stretched horizontally in the tree, with the plane of the ribbon vertical. The size of rays varies according to the number of cells contained within them. Although individual ray cells and even the smaller rays are not visible to the eye, the largest rays of certain species, as in beech and oak, can be seen easily.

Cells of similar type or function are collectively referred to as **tissue**, so we may refer to **wood tissue** or **bark tissue**. Bark and wood are both **permanent tissues**, for once formed and matured, their cells retain their shape and size. The cells of certain reproductive tissues (termed **meristems** or **meristematic tissue**) can divide to form new cells. The growing tips of twigs (**apical meristems**) reproduce, and their cell division is responsible for elongation of the tree's growing points. The pith cells at the center of the stem are left in the path traveled by the apical meristem. All further growth of the tree is the result of thickness growth produced by division of the cambium, a **lateral meristem**.

Though apparently static, the tree has an amazingly dynamic internal system during the growing season. Water from the soil, carrying nutrients, enters the roots and moves upward through the wood cells to the leaves. Much of this water evaporates through the leaf surfaces; this **transpiration** helps cool the foliage. In addition, the leaves perform the supreme miracle of plants, **photosynthesis**, whereby water from the soil is combined with carbon dioxide from the atmosphere, catalyzed by chlorophyll and energized by sunlight *(1)*. This process has a

1—Carbon dioxide (CO_2) from the atmosphere combines with water (H_2O) in the leaves in photosynthesis, a process catalyzed by chlorophyll and energized by sunlight. This produces a basic sugar ($C_6H_{12}O_6$) and releases oxygen (O_2) to the air.

double dividend. First, it adds oxygen to the air while removing carbon dioxide, and it produces a basic sugar ($C_6H_{12}O_6$) for the tree's own use. The **sap** (i.e., the water in the tree plus any dissolved nutrients or other materials) carries this sugar down through the inner living bark to where it is used to build new cells in the cambium layer.

The cambial cells have extremely thin **primary walls**, and during the growing season their contents are quite fluid. As a result, this layer is unusually fragile, and bark will often peel away readily from a freshly cut log *(2)*. During the winter dormant season, however, the cell contents thicken. The cambial layer stiffens and becomes less vulnerable to mechanical damage, and if the tree is cut the bark usually remains firmly attached.

This fact may be quite important to woodworkers. If stem cross sections are being prepared for coasters or clock faces, for example, where retaining the bark is desired, the tree should be harvested during dormancy. If, on the other hand, a product such as peeled poles for a log cabin is desired, they had best be cut during the growing season to permit easy separation of the bark. The bark might actually have to be carved off if the tree were cut during the dormant season.

2—Bark peels easily (with a screwdriver) from summer-cut eastern white pine because the cambial layer is fragile during the growing season.

In the cambium "mother" cells, division takes place lengthwise. Of the two "daughter" cells formed, one enlarges to become another cambial mother cell. The other begins to mature into either a bark cell (if toward the outside of the cambium) or a new wood cell (if formed toward the inside of the cambium). Soon after formation—perhaps within a week—the new wood cell wall undergoes a sequence of changes. Its shape may change, either elongating, enlarging or both, to the shape of the wood cell type it will eventually become.

When the developing daughter cells have attained their ultimate size and shape, a **secondary wall** is added to the inner surface of the fragile primary wall. This wall, which is permanent, becomes the dominant layer of the cell, and its formation fixes forever the cell size and shape: There will never be any change in cell type or dimension, or in thickness of the cell walls. The secondary wall is built mainly of **cellulose***, long-chain molecules that are strong and stable. The cellulosic structure of the cell wall is further fortified by **lignin****, the material that characterizes woody plants.

During the last stages of this maturing process, most of the wood cells lose their living protoplasm and have only sap left in them. Some may still perform the function of conducting sap, even though they are no longer living. These cells, whose function is simply to support the tree and in some cases to conduct sap, are sometimes termed **prosenchyma**. In this newly formed **sapwood** a small percentage of the cells, principally those found in the rays, retain their living protoplasm and can assimilate and store food. Such cells are termed **parenchyma**.

*Cellulose molecules are long-chain polymers, that is, multiples of simpler molecules (monomers) linked end-to-end. Cellulose has the general formula $(C_6H_{10}O_5)_n$, the subscript *n* indicating the number of monomer units linked together to form the polymer. In wood cellulose, *n* may be as high as 10,000, which would produce polymer lengths up to 1/100 mm. The orientation of the long-chain cellulose molecules with the long axis of the cells accounts for many of the basic properties of wood. Comparison of chemical formulas suggests that the sugar produced by photosynthesis ($C_6H_{12}O_6$) is converted to the basic monomer of cellulose by the elimination of H_2O per monomer.

**Lignin comprises about 25% of the wood's composition, but its exact chemical nature is extremely complex and not fully understood, principally because isolating it from wood changes it chemically.

Growth rings

Activity of the cambium (i.e., growth) continues as long as environmental conditions are suitable and the tree is healthy. In the temperate climate that prevails in most of the United States, the characteristic annual cycle includes a growing and a dormant season. In most trees, the nature of wood cell formation is similarly cyclic, resulting in visible **growth layers**. These increments are also called **growth rings**, or **annual rings** when formed in association with yearly growth. These rings may vary in width as a characteristic of the species and as a result of growing conditions *(1)*. The distinctiveness of the growth ring in a particular species is determined by its cell structure, usually by the variation of cell diameter or cell-wall thickness, or by the distribution of different types of cells.

Where there is visible contrast within a single growth ring, the first formed layer is termed **earlywood**, the remainder **latewood**. The terms **springwood** and **summerwood**, respectively, are also used to indicate these layers but they are misleading in suggesting a correlation with the calendar seasons of the year. In some species no separable earlywood and latewood portions may occur, or the ring may have indistinct or gradual transition from earlywood to latewood. In some tropical areas growth may continue with little interruption, although intermittent rainfall may cause erratic layering of wood. Transition is usually judged by the visual appearance of the cell structure in a cross-sectional surface. There is also usually a difference in density between earlywood and latewood. Transition should not be judged by the degree of unevenness of the grain.

1—Growth rings.

Grain

No discussion of wood can proceed very far without encountering the word **grain**. There are well over 50 ways in which this word can be used in some ten different categories *(Table 1)*. Each term is defined in the glossary, and each category will be dealt with as appropriate through the text. The term grain alone often describes the direction of the dominant longitudinal cells in the tree. Substituting the term **grain direction** for this meaning adds clarification, because the word grain alone can be meaningless without an accompanying context. For example, the meaning is hardly clear in the statement, "The grain of this piece of wood is unsatisfactory."

It is appropriate here to introduce the word grain as used to indicate the degrees of contrast between earlywood and latewood. If a pronounced difference exists, the wood is said to have **uneven grain** (e.g., southern yellow pine) *(2)*. If little contrast is evident the wood is **even-grained**, as in basswood. Intermediates can be indicated by compound modifiers such as fairly even (e.g., eastern white pine) or moderately uneven (e.g., eastern hemlock).

Table 1—Usages of the word grain in woodworking.

Category			
1. Planes and surfaces	END GRAIN	LONGITUDINAL GRAIN	SIDE GRAIN
	FACE GRAIN	RADIAL GRAIN	TANGENTIAL GRAIN
	LONG GRAIN		
2. Growth-ring placement	BASTARD GRAIN	PLAIN GRAIN	SIDE GRAIN
	EDGE GRAIN	QUARTER GRAIN	SLASH GRAIN
	FLAT GRAIN	RADIAL GRAIN	TANGENTIAL GRAIN
	MIXED GRAIN	RIFT GRAIN	VERTICAL GRAIN
3. Growth-ring width	CLOSE GRAIN	DENSE GRAIN	NARROW GRAIN
	COARSE GRAIN	FINE GRAIN	OPEN GRAIN
4. Earlywood/latewood contrast	EVEN GRAIN	UNEVEN GRAIN	
5. Alignment of longitudinal cells	ACROSS-THE-GRAIN	DIP GRAIN	SPIRAL GRAIN
	ALONG-THE-GRAIN	GRAIN DIRECTION	STEEP GRAIN
	AGAINST-THE-GRAIN	INTERLOCKED GRAIN	STRAIGHT GRAIN
	CROSS GRAIN	SHORT GRAIN	WAVY GRAIN
	CURLY GRAIN	SLOPE-OF-GRAIN	WITH-THE-GRAIN
6. Relative pore size	CLOSED GRAIN	FINE GRAIN	OPEN GRAIN
	COARSE GRAIN		
7. Figure types	BIRD'S-EYE GRAIN	FIDDLEBACK GRAIN	RIFT GRAIN
	BLISTER GRAIN	FLAME GRAIN	ROEY GRAIN
	COMB GRAIN	LEAF GRAIN	SILVER GRAIN
	CROTCH GRAIN	NEEDLE-POINT GRAIN	STRIPE GRAIN
	CURLY GRAIN	QUILTED GRAIN	TIGER GRAIN
	FEATHER GRAIN		
8. Machining defects	CHIPPED GRAIN	RAISED GRAIN	TORN GRAIN
	FUZZY GRAIN	SHELLED GRAIN	WOOLLY GRAIN
	LOOSENED GRAIN		
9. Figure imitation	GRAINING	WOODGRAIN DESIGN	
10. Surface failure	SHORT IN THE GRAIN		

2—Uneven-grained woods like southern yellow pine (left) have visually prominent growth rings. In fairly even-grained woods like eastern white pine (right), the contrast between earlywood and latewood is not pronounced.

Sapwood and heartwood

In twigs and small saplings, the entire wood portion of the stem is involved in sap conduction upward in the tree and is thus termed **sapwood**. Some nonliving prosenchyma cells are active in conduction, and the living parenchyma cells also store food. As the tree develops, the entire trunk is no longer needed to satisfy the leaves' requirements for sap. In the center of the stem, nearest the pith, the prosenchyma cells cease to conduct sap and the parenchyma cells die. The sapwood is thus transformed into **heartwood**. The transition to heartwood is also accompanied by the formation in the cell wall of material called **extractives**. To the woodworker, the most significant aspect of heartwood extractives is color, for the sapwood of all species ranges from whitish or cream to perhaps yellowish or light tan. The dark, distinctive colors we associate with various woods—the rich brown of black walnut, the deep purple red of eastern redcedar or the reddish black striping of rosewood—are the result of heartwood extractives.

Sapwood is not generally resistant to fungi, so any noteworthy decay resistance of a species is due to extractives that are toxic to fungi. Fungal resistance is therefore restricted to the heartwood portion of the tree. Some trees, such as spruces, do not have pigmented extractives associated with heartwood formation. Colorless or nearly colorless extractives may nevertheless provide decay resistance, as in northern white cedar.

Heartwood extractives may change the properties of the wood in other ways as well. In some species, they reduce the permeability of the wood tissue, making the heartwood slower to dry and difficult or impossible to impregnate with chemical preservatives. Extractives often make the heartwood a little denser than the sapwood, and also a little more stable in changing moisture conditions. (Though when green, sapwood sometimes contains five times as much moisture as heartwood, it shrinks more than heartwood because of its lack of extractives.) Extractive materials in the heartwood of some species may be so abrasive that they dull cutting tools, and they may contribute to the wood's surface hardness. But as sapwood becomes heartwood, no cells are added or taken away, nor do any cells change shape. The basic strength of the wood is not affected by sapwood cells changing into heartwood cells.

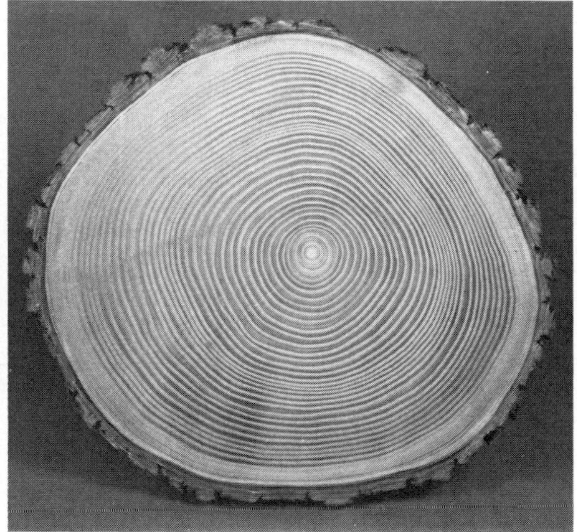

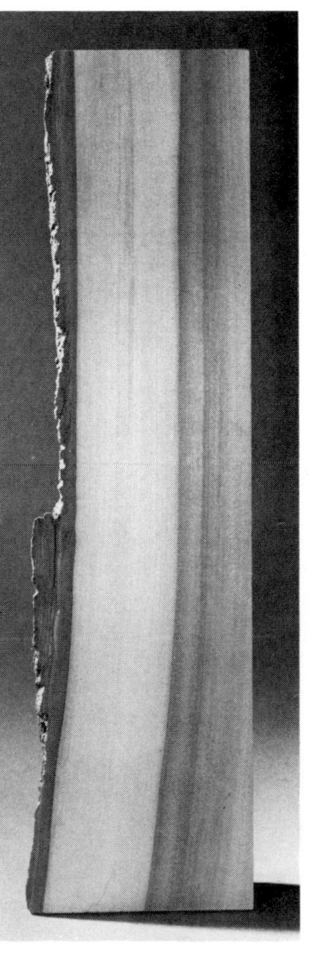

1—Relative width of heartwood and sapwood varies according to species; heartwood (dark area) is small in white ash, above left, and wide in catalpa, above right. Heartwood/sapwood growth may also vary tremendously within a species, according to growing conditions. Both discs in the photo below were cut from eastern redcedar trees. In the one at right, growth was stunted because the tree was shaded by larger trees; the one at left grew normally, in an open meadow. At right, definite color change marks heartwood/sapwood division in yellow poplar.

As the girth of the tree increases with the addition of new sapwood, the diameter of the heartwood zone also expands proportionately *(1)*. The width of the sapwood is characteristic for some species, or at least relatively consistent *(Table 2)*. For example, catalpa retains only one or two growth rings of sapwood. Black locust, on the other hand, retains two or three. Sapwood is commonly 1½ in. to 2 in. wide in mature trees of eastern white pine and cherry but may be up to 6 in. wide in aspen, birch and maple.

Table 2—Number of rings in the sapwood of some hardwoods.

Common name of tree	Scientific name of tree	Number of rings in sapwood
Northern catalpa	*Catalpa speciosa*	1-2
Black locust	*Robinia pseudo-acacia*	2-3
American chestnut	*Castanea dentata*	3-4
Black cherry	*Prunus serotina*	10-12
Honey locust	*Gleditsia triacanthos*	10-12
Black walnut	*Juglans nigra*	10-20
American beech	*Fagus grandifolia*	20-30
Sugar maple	*Acer saccharum*	30-40
Dogwood	*Cornus florida*	30-40
Silver maple	*Acer saccharinum*	40-50
Sweet birch	*Betula lenta*	60-80
Magnolia	*Magnolia grandiflora*	60-80
Black tupelo	*Nyssa sylvatica*	80-100

Structural arrangement

Because of the arrangement of the layers of growth in the tree, as well as the vertical or horizontal orientation of the individual cells, it is appropriate to consider the structure of wood in three-dimensional terms *(2)*.

One plane is perpendicular to the stem axis and is termed the **transverse plane**, or **cross-sectional plane**, typically observed at the end of a log or stump.

Because the tree cross section is analogous to a circle, a plane passing through the pith of the wood (as a radius of the circle) is called a **radial plane** or surface.

A plane parallel to the pith, but not passing through it, forms a tangent to the circular growth-ring structure and is termed a **tangential plane** or surface. The curvature of the growth ring is not geometrically regular, and the surface in question is most ideally tangential where the plane is perpendicular to a radial plane. However, any slabbed log surface is usually accepted as a tangential surface. In a small cube of wood the curvature of the rings is negligible, so the cube can be oriented to contain quite accurate transverse (cross-sectional), radial and tangential faces.

Thin slices or sections of wood tissue, as commonly removed from the surfaces for study, are termed transverse, radial and tangential sections. These planes or sections are often designated simply by the letters **X, R** and **T**, respectively.

2—Block of Douglas fir is cut to show three planes: the transverse or cross-sectional (X), the radial (R) and the tangential (T) surfaces.

In describing lumber or pieces of wood, the term **end-grain surface** (or simply **end grain**) refers to the transverse surfaces. By contrast, any plane running parallel to the pith is either **side grain** or **longitudinal surface**.

With lumber or veneer, additional terminology is used for various longitudinal surfaces *(1)*. For example, pieces whose broad face is more or less in the radial plane are termed **radial grain, edge grain** or **vertical grain**, since the edges of the growth rings emerge at the surface, with the growth rings vertical when viewed from the end-grain surface. One method of producing such pieces is first to saw the log into longitudinal quarters and then to saw each quarter radially. Pieces so produced are said to be **quartersawn** and their surfaces are **quarter-grain**. Other terms for such pieces are **comb grain** and **rift grain**. These terms are flexible and may be applied to pieces in which the growth rings form angles of anywhere from 45° to 90° with the surface.

Lumber and veneer whose face orientation is approximately tangential are said to be **flatsawn, flat-grained** or **tangential-grained. Plain grain** and **slash grain** are sometimes used synonymously with flatsawn. The term **side grain** sometimes means flat-grained. All these terms can include growth-ring orientations from 0° to 45° with the surface. The term **mixed grain** refers to quantities of lumber having both edge-grain and flatsawn pieces. **Bastard grain** usually refers to growth rings oriented between 30° and 60° to the surface.

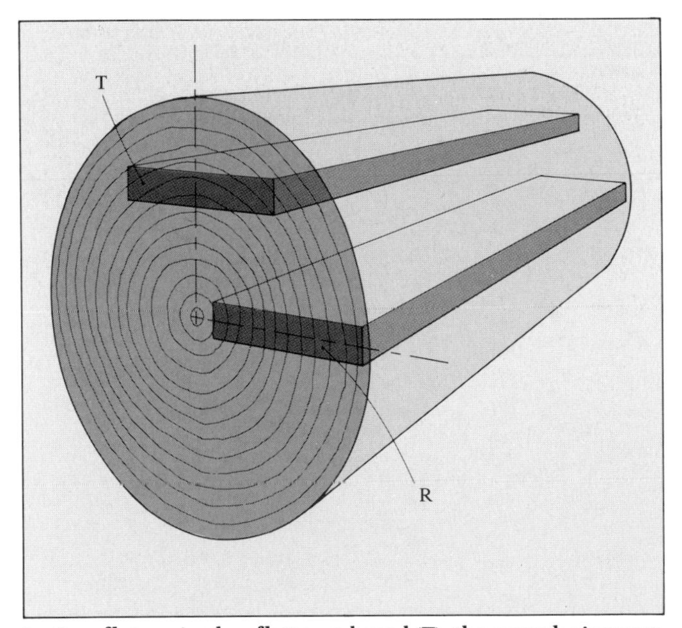

1—In a flat-grained or flatsawn board (T), the growth rings are approximately parallel to the wide faces. In an edge-grained or quartersawn board (R), the rings are approximately perpendicular to the wide faces.

Density and specific gravity

Density (weight per unit volume) is the single most important indicator of strength in wood and may therefore predict such characteristics as hardness, ease of machining and nailing resistance. Dense woods generally shrink and swell more, and usually present greater problems in drying. The densest woods also make the best fuel.

Specific gravity is the ratio of the density of a substance to the density of a standard substance (water, in the case of wood and other solids). Specific gravity is often called the **density index**. In measuring density and specific gravity, it is customary to use oven-dry weight and current volume. Because of volumetric shrinkage

Table 3—Specific gravity of some common species.

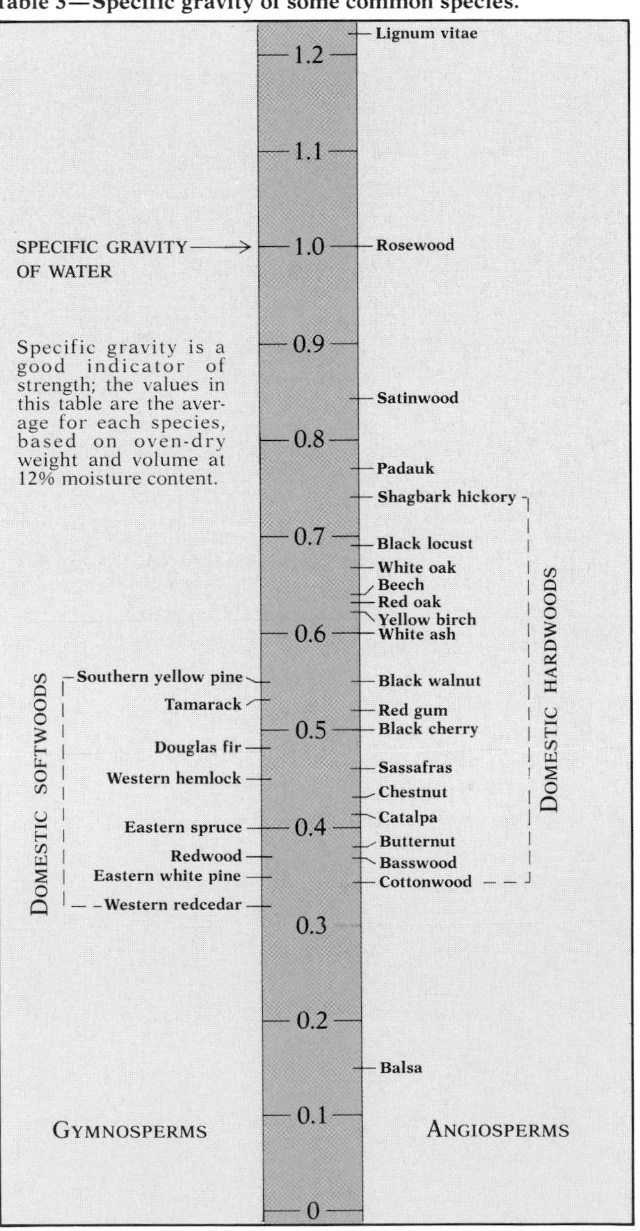

and swelling, the volume of wood may vary slightly with moisture content.

Density is expressed as weight per unit volume, customarily as pounds per cubic foot (English) or grams per cubic centimeter (metric). Water has a density of 62.4 lb./ft.^{3} or 1 g/cm^3. A wood weighing 37.44 lb./ft.^{3} or 0.6 g/cm^3 is thus six-tenths as heavy as water and has a specific gravity of 0.6.

Table 3 gives the average specific gravity of some familiar woods *(2)*. There is a twelvefold range from balsa to lignum vitae. *The Guinness Book of World Records* lists black ironwood, also known as South African ironwood *(Olea laurifolia)* as the heaviest, (93 lb./ft.^{3}) and acknowledges *Aeschynomene hispida* of Cuba as the lightest, weighing only 2¾ lb./ft.^{3} The respective specific gravities are 1.49 and 0.044.

A most important message conveyed by the density chart is how misleading the familiar terms **hardwood** and **softwood** really are—they have little relevance to the softness or hardness of wood. There are indeed differences between hardwoods and softwoods, but, as will be revealed in the next section, the terms hardwood and softwood simply distinguish the two broad groups of related trees within the plant kingdom.

2—Specific gravity (relative density) is an important guide to wood properties—these blocks of domestic and imported species all weigh the same, despite their great differences in size. At the extremities of the range, a big block of balsa wood (left scale pan) precisely balances a little block of rosewood.

1. Beech
2. Tulipwood
3. Spanish cedar
4. White pine
5. Hickory
6. Sumac
7. Purpleheart
8. Black locust
9. Redwood
10. Balsa
11. Walnut
12. Butternut
13. Ebony
14. Southern yellow pine
15. Lignum vitae
16. Barberry
17. Teak
18. Eastern redcedar
19. Rosewood
20. Western redcedar
21. Eastern spruce
22. Douglas fir
23. Red oak
24. White ash
25. Catalpa
26. Obeche
27. Basswood
28. Yellow-wood
29. Myrtle
30. Sugar maple
31. Goncalo alves

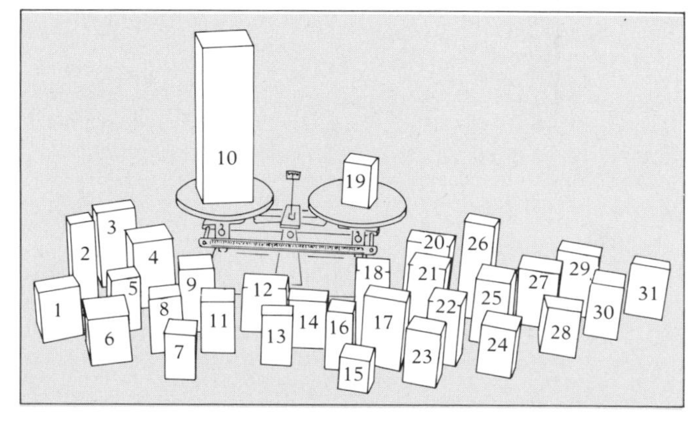

Systematic classification

The woodworker typically reacts to the idea of scientific classification and naming of woods with distaste and discouragement, because gaining a mastery of the subject seems impossible. Such need not be the case, for the woodworker does not need to become an expert in taxonomy (plant classification according to natural relationships) to become familiar with the field.

The **plant kingdom** comprises several major divisions. The one that includes all seed plants, **spermatophytes**, is further divided into two broad groups (separated according to how the seeds are borne). The **gymnosperms** (naked seeds) include all trees producing softwood lumber, and the **angiosperms** (covered seeds) include all trees yielding hardwood lumber. These groups are subdivided into orders, families, genera (singular genus) and species (singular species).

Within the gymnosperms (softwoods), trees are classified in four families of the order *Coniferales*. The term **conifer** thus indicates softwood trees. These trees are characterized by needlelike or scalelike foliage (usually evergreen) and have an **excurrent** tree form (a dominant main stem with lateral side branching) *(1)*. Familiar examples are pines, spruces, firs, hemlocks and cedars.

Conifers range from about 0.30 to 0.55 average specific gravity. Were they lighter, like balsa, they would be too soft for many building uses. If they were heavier, like oak or hickory, they could not be worked easily with hand tools, nor could they be nailed easily without preboring. This density range, the wide distribution of conifers in our temperate zone and the excurrent stem form explain the dominant use of softwoods for structural lumber. (Hardwoods can also be used structurally, if they are of suitable form and workable. Chestnut and yellow poplar have both attributes and have been widely used.)

1—Dendritic form characteristic of hardwoods, left, contrasts with excurrent form typical of softwoods, right.

Another order of gymnosperms, *Ginkgoales*, has but a single species, *Ginkgo biloba*, the maidenhair tree of China and Japan. It is mainly an ornamental tree, and its wood has had little use.

The angiosperms (hardwoods) include some 22 families in the United States. Hardwood trees are mostly **deciduous**, that is, their leaves drop in autumn. Their form tends to be **dendritic** or **deliquescent** (characterized by branching and rebranching of the main stem). The higher density range of some species and the attractive heartwood color and figure have earned them a favored place among woodworkers.

Using this system, eastern white pine would be classified as follows: Kingdom: plant; division: spermatophytes; subdivision: gymnosperms; order: *Coniferales*; family: *Pinaceae*; genus: *Pinus*; species: *strobus*. A particular species of wood is designated by the combination of the **genus** (generic name) and **species** (specific epithet)—*Pinus strobus* for eastern white pine. A full scientific designation includes the name (or abbreviated name if well known) of the botanist who classified the species. Eastern white pine would be written *Pinus strobus L.*, indicating the Swedish botanist Linnaeus. (However, except for scientific writing, it is customary to omit the author designation.) Italics are used for the genus and species because they are Latin names. The genus is capitalized, but the species is not. Where the genus is obvious, as in a discussion exclusively about pines, it is sometimes abbreviated—*P. strobus* for eastern white pine, *P. resinosa* for red pine, *P. ponderosa* for ponderosa pine. Where a piece of wood can be identified as to genus, but not to exact species, it is designated by the genus name followed by the abbreviation sp., as in, for example, southern yellow pine, *Pinus* sp., or red oak, *Quercus* sp. The notation spp. indicates plural.

Familiarity with scientific names is important for two reasons. First, there is often a difference between the common name used for a tree and the lumber that comes from it. Second, there is a great deal of inconsistency among common names. A single species may have several common names, especially in different localities. For example, the *Liriodendron tulipifera* tree is called tuliptree or tulip poplar; the preferred common name for its wood is yellow poplar, but locally it is known as whitewood, tulipwood, tulip poplar, hickory poplar, white poplar or simply poplar—and even popple. But poplar and popple commonly describe cottonwoods and aspens in the genus *Populus*, and whitewood is used for several other species. Ironwood is another common name that keeps popping up. Among tropical species, designation by scientific names is especially important. Several excellent reference books of scientific names for both native and tropical species are listed in the bibliography. Also see Appendix 1.

Cellular structure

Although the average woodworker need not have a detailed familiarity with the cell structure of every species, an acquaintance with anatomical detail will increase understanding of the broad differences between softwoods and hardwoods, and the reasons for differences among species *(2)*. Within a species, anatomical structure also reveals the extent of variability and helps pinpoint locations of relative hardness and softness. It may also show sites of potential splitting, critical to every phase of woodworking from carving and finishing to drilling, nailing and gluing. Understanding cell structure is also vital to appreciating the permeability of wood, which affects, among other things, drying and stain absorption.

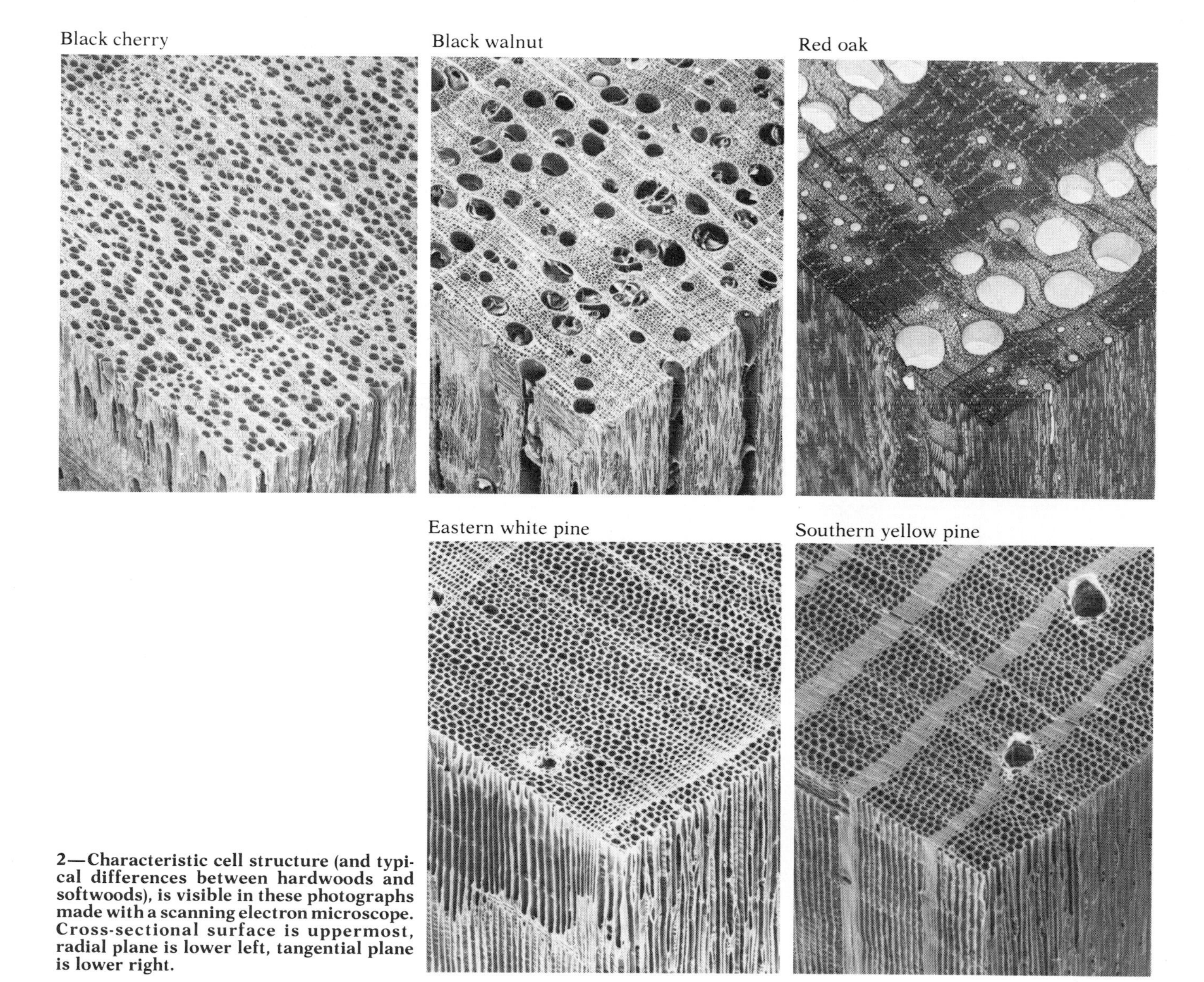

2—Characteristic cell structure (and typical differences between hardwoods and softwoods), is visible in these photographs made with a scanning electron microscope. Cross-sectional surface is uppermost, radial plane is lower left, tangential plane is lower right.

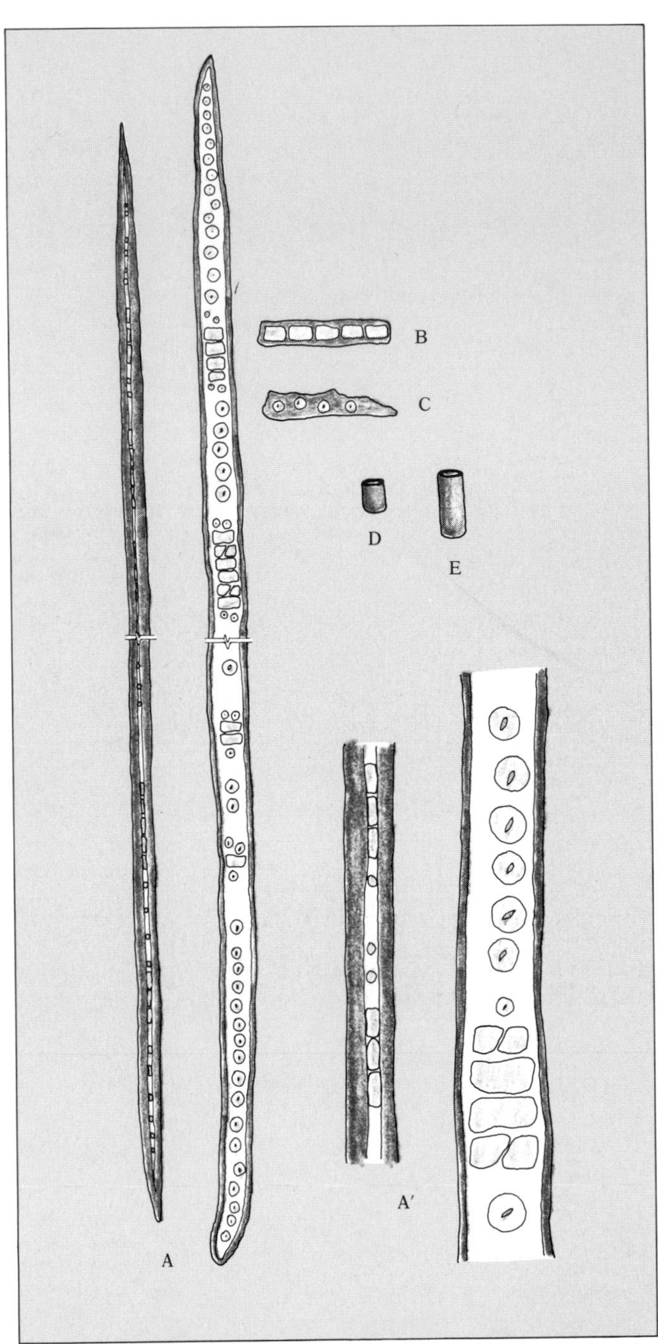

1—Some coniferous cell types: Tracheids (A, enlarged view, A′) comprise more than 90% of the wood volume. The remainder is mostly ray tissue, either ray parenchyma cells (B) or ray tracheids (C). Some species also have a very small percentage of epithelial cells (D), which line resin canals, or longitudinal parenchyma cells (E).

Softwoods

Though it seems complicated at first, the structure of softwoods *(1)* is relatively simple compared to that of hardwoods. Most of the cells found in coniferous wood are **tracheids**, which comprise 90% to 95% of the volume of the wood. Tracheids are fiberlike cells about 100 times longer than they are in diameter. Among different species their average lengths range from about 2 mm to about 6 mm. In an average coniferous species, a cubic inch of wood contains some four million tracheids. These fibers are excellent for making paper.

Average tracheid diameters range from 20 to 60 microns (1 **micron** equals 0.001 mm). This gives a basis for classifying **texture** in wood. Redwood has the largest-diameter tracheids and is termed coarse-textured; eastern redcedar has the smallest-diameter tracheids and is fine-textured. With a 10-power hand lens, texture can be estimated as coarse, medium or fine on the basis of how clearly individual tracheids can be seen in cross section. Texture, a valuable aid in wood identification, is also related to surface smoothness and finishing qualities.

The functions of conduction and support of the tree are carried out by the nonliving (prosenchymatous) tracheids. A trade-off between these two functions seems to take place from earlywood to latewood: In earlywood the tracheids are larger in diameter and thin-walled, well suited for conduction; in latewood the thicker-walled, smaller-diameter tracheids are less suited to conduction but give more support to the tree. This overall variation in tracheids from earlywood to latewood determines the evenness of grain. In white pine the grain is quite even, for example, but in southern yellow pine there may be as much as a threefold difference in density from earlywood to latewood (specific gravity 0.3 to 0.9).

One typical consequence of uneven grain in conifers is the color reversal that results from staining *(2)*. Another common problem develops when flatsawn pieces are used for stair treads or flooring. The earlywood wears away at a faster rate, leaving raised ridges of latewood.

Some coniferous species also have **resin canals**, tubular passageways lined with living parenchyma (epithelial) cells, which exude resin or "pitch" into the canals. Resin canals—normally not found among North American hardwoods—are found in four genera of the conifers, all within the family *Pinaceae: Pinus* (pine), *Picea* (spruce), *Larix* (larch) and *Pseudotsuga* (Douglas fir). Resin canals are largest and most numerous in pines, usually distinct to the naked eye. In other species, magnification may be necessary to locate them.

In sapwood the resin in the canals is quite fluid, and it may be years before it eventually solidifies in the wood. On new millwork the resin often flows to the surface of the wood, forming a droplet where each resin canal intersects the wood surface *(3)*. Kiln-drying schedules that include final temperatures above 175°F are necessary to set the resin adequately and to minimize bleed-out problems. Tiny resin spots tend to bleed through paint films, resulting in a yellowish speckling of the surface. This is why white pines and sugar pine are not the best for painted woodcarvings.

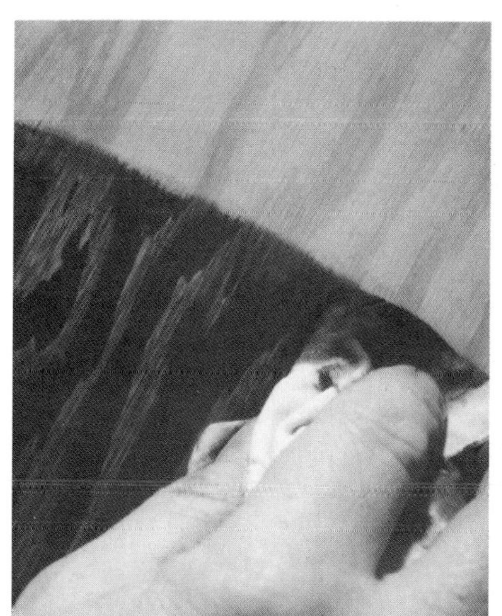

2—Earlywood of Douglas fir, far left, is lighter in color than latewood, but more porous. Thus it retains stain more readily and finishes darker in color. Large earlywood pores of red oak, left, already appear darker than latewood, and the effect is accentuated as stain accumulates in them.

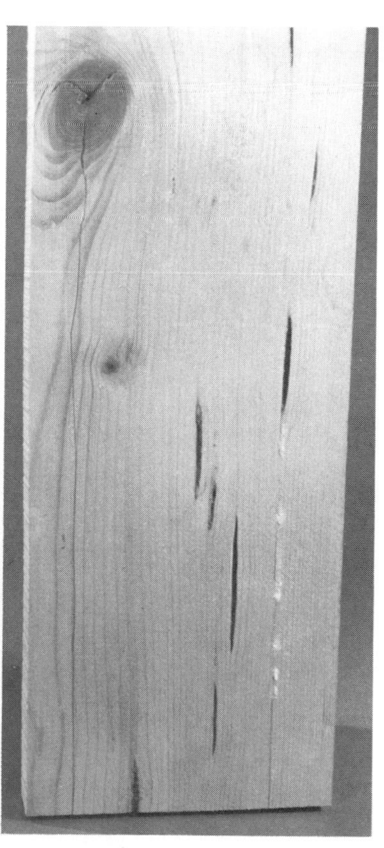

3—Unless resins are solidified, or 'set,' by high temperatures in kiln-drying, droplets of resin may exude from the canals, as on the surface of this turned baluster of eastern white pine. Resin droplets remain sticky for a time but then harden. They can be removed by careful scraping or sanding. Right, pitch pockets in spruce board.

Pitch from resin canals has been traditionally harvested from southern yellow pine by gashing the trees and collecting the resin that oozes out. Processing the pitch yields such products as turpentine, rosin and pine tar oil, which are often called **naval stores** because of their original importance in shipbuilding.

The rays in softwoods are narrow, usually one cell wide (except for occasional rays with horizontal resin canals in some species). They may be 40 or more cells high in some species, but they are essentially invisible even with the use of a 10-power hand lens. Microscopic examination is needed to see them *(1)*. Most of the living parenchyma cells in coniferous woods are found in the rays. In addition to those in the rays, a few parenchyma cells occur elsewhere in some species, but a microscope is necessary to detect them.

Hardwoods

In comparing the anatomy of the hardwoods with that of the softwoods, several general differences are immediately apparent. Most hardwoods lack resin canals (although some tropical hardwoods have gum ducts). Rays in hardwoods vary widely in size, well into the range of visibility with the unaided eye. Many more types of cells are present in hardwoods, and there is more variation in their arrangement. It is believed that the hardwoods evolved much later than the softwoods. The noteworthy difference in this respect is the degree to which the functions of conduction and support are accomplished by the evolution of specialized cells; **vessel elements** represent the ultimate in functional specializations *(2)*.

Vessel elements are extremely large in diameter but have relatively thin walls. They form in the tree in end-to-end arrangement and their end walls have been lost, thus

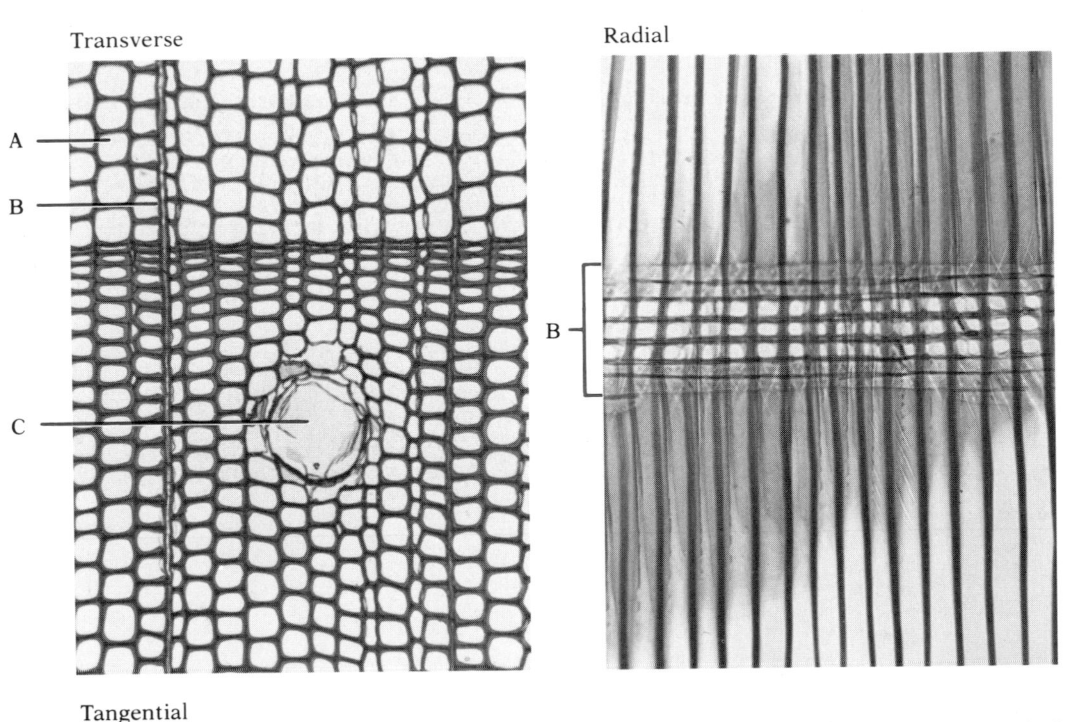

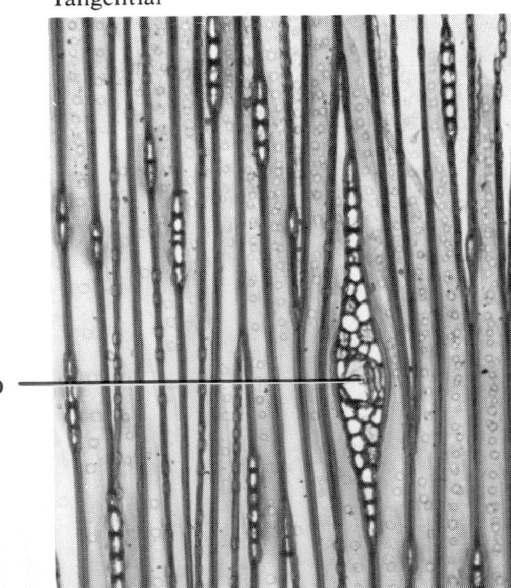

1—Thin sections of eastern white pine magnified about 100 times. The cross section (transverse view) is dominated by rows of tracheids (A); also visible are the narrow rays (B) and a single resin canal (C). In the radial view, elongation of the tracheids is apparent, and the rays (B) appear as brick-like bands of cells. In tangential section the rays are seen in end view. One ray, called a fusiform ray (D), has a horizontal resin canal through its center.

forming continuous pipelines ideal for sap conduction. The open nature of the vessels can be demonstrated by blowing smoke through a piece of wood or using it as a bubble pipe *(3)*. Vessels vary in size among and within species. Some are seen easily with the naked eye. Others require hand-lens magnification but are distinguishable from other cell types because they are slightly larger. **Fibers**, by contrast, are smallest in diameter, with closed ends and thick walls, and therefore poorly suited for conduction. They make the wood strong.

When vessels are cut transversely (across the end grain), the exposed open end is referred to as a **pore**. Because all hardwoods possess vessels they are sometimes called **porous** woods (softwoods are therefore called **nonporous** woods).

The size, number and distribution of vessels and fibers are determining factors in the appearance and uniformity of hardness of a particular wood. In some species (oak, ash, elm, chestnut, catalpa) the pores are concentrated in the earlywood. Such woods are said to be **ring-porous** *(4)*. This structure usually causes pronounced uneven grain.

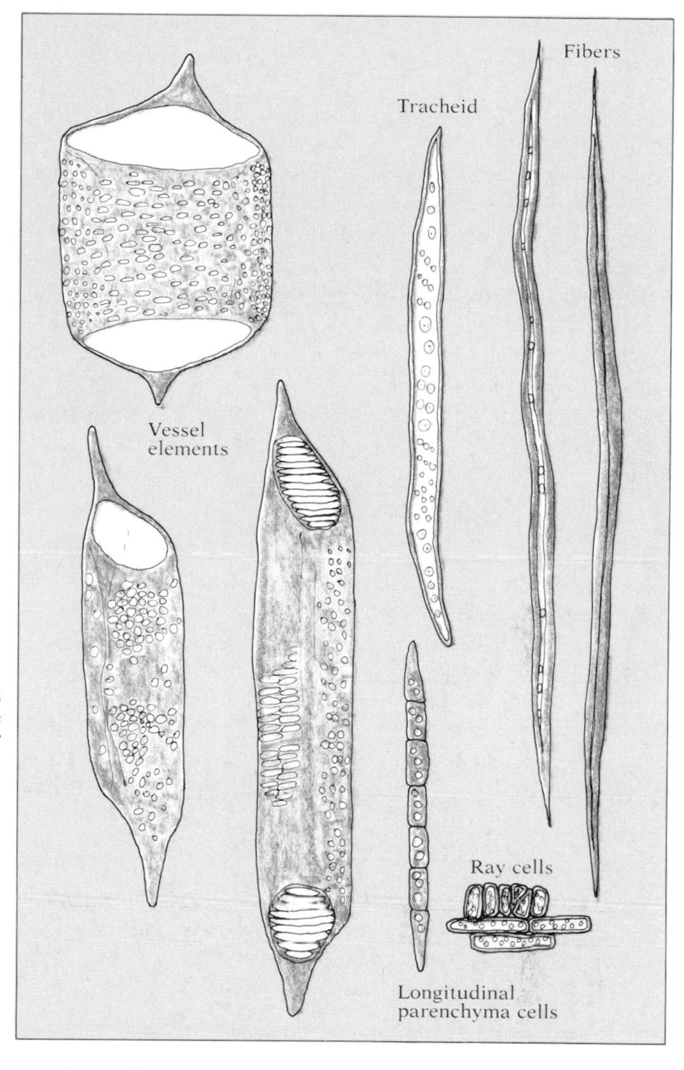

2—Hardwood cell types are extremely varied. The drawing indicates their relative size and shape.

3—Air can be blown through a straight-grained piece of red oak because of the open vessel elements. This piece had its end dipped in liquid dish detergent.

4—Red oak, a ring-porous hardwood, magnified about 100 times. Large, thin-walled pores (A) are concentrated in the earlywood and are many times the diameter of the thick-walled fibers (B) abundant in the latewood. The pores are surrounded by small-diameter, thin-walled tracheids (C). Large, multiseriate rays (D), easily visible to the naked eye, are a distinctive feature of oaks. Uniseriate rays (E), seen only with a microscope, are also numerous.

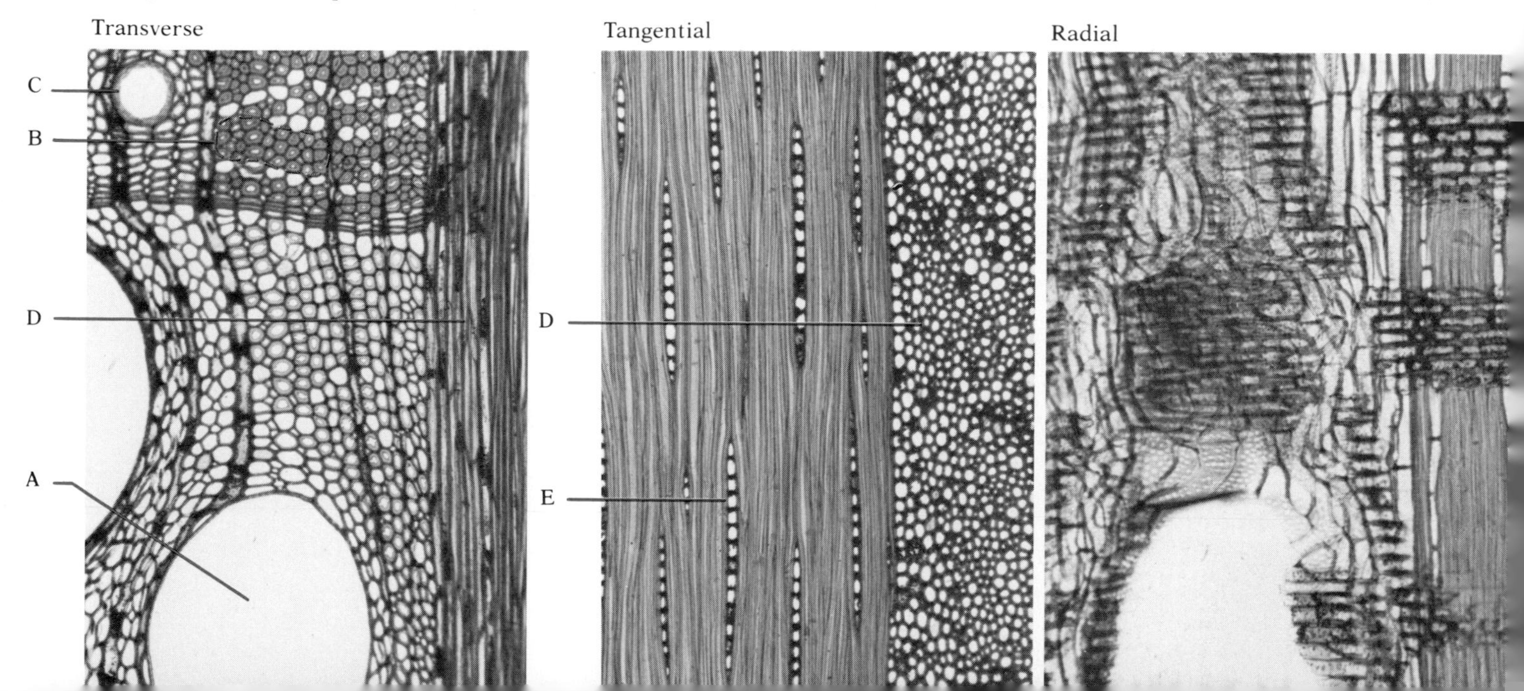

Usually these woods have distinct figures and patterns *(1)*. The uneven uptake of stain makes these ring-porous layers even more pronounced. But these planes of weakness may be used advantageously, as in ash where the layers can be separated for basket strips *(2)*. Some species, such as hickory, have an obvious earlywood concentration of pores, and are classed as ring-porous, although there is not much noticeable variation in wood density. Other hardwoods (cherry, maple, birch, basswood, yellow poplar) have pores distributed fairly evenly and are termed **diffuse-porous** woods *(3)*. Most domestic diffuse-porous woods have relatively small-diameter pores, but among tropical woods some diffuse-porous species (e.g., mahogany) have rather large pores. In some woods the pores are large in the earlywood and get smaller toward the latewood, but with no distinct zoning. These woods are called **semi-ring-porous** or **semi-diffuse-porous**. Examples are black walnut and butternut. Some species that are usually diffuse-porous may sometimes tend toward semi-ring-porous structure—black willow and cottonwood are two examples.

In hardwoods, pore size is used as a measure of texture. Red oak, with its large pores, is coarse-textured; red gum is fine-textured because of its small-diameter pores.

1—Catalpa carving by the author displays the pronounced figure typical of ring-porous hardwoods.

2—If freshly cut black ash is pounded on the tangential surface, the ring-porous earlywood buckles, allowing the wood to be peeled apart.

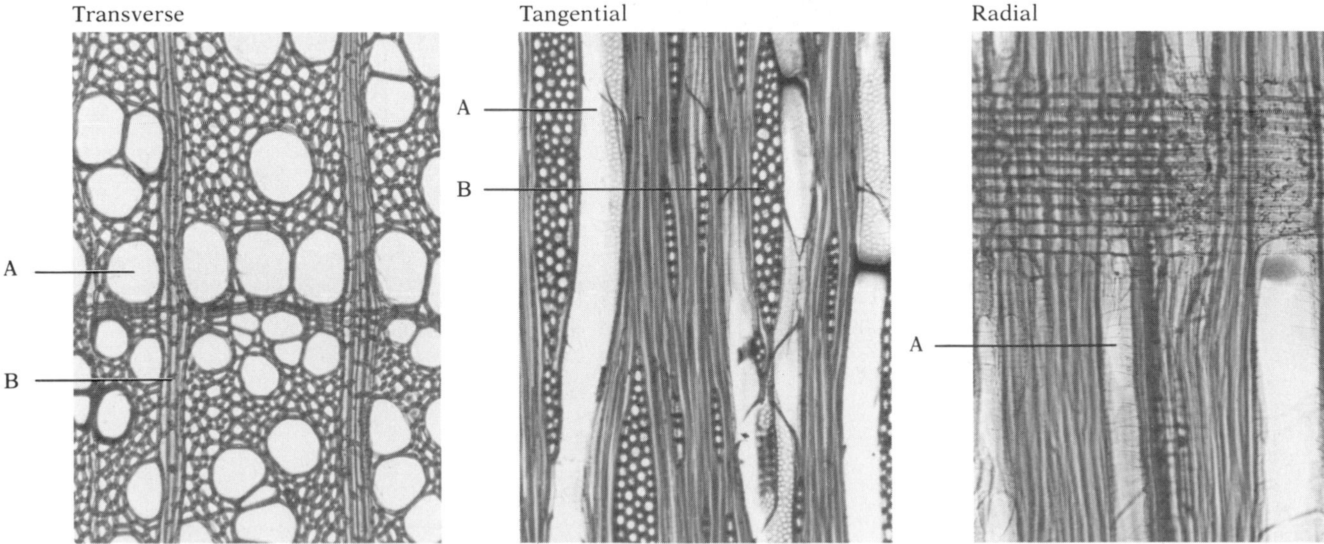

3—Black cherry, a typical diffuse-porous wood, has uniform sized vessels (A) evenly distributed among the fibers. The rays (B), up to six cells wide, are distinct with a lens on the transverse surface.

In some hardwoods, as the transition from sapwood to heartwood takes place, bubble-like structures appear in the cavities of the vessel elements. These structures are known as **tyloses** and usually can be seen in the pores with a hand lens *(4)*. In some species they are sparse or absent (red oaks) or unevenly distributed (hickory, chestnut). In others they are numerous (white oak). In some species (black locust, Osage-orange) the vessel elements are densely packed with tyloses. You could not blow bubbles through a piece of white oak as you can with red oak. Because liquid passage is difficult in white oak, it is the obvious choice over red oak for making barrels and other tight cooperage.

Besides the vessels, the other types of longitudinal cells in hardwoods (fibers, tracheids and parenchyma cells) are of uniformly small diameter, so even with a hand lens they cannot be seen individually. But because of their difference in cell-wall thickness, masses of them are distinguishable. Fiber masses usually appear darker; tracheids and parenchyma cells are lighter. Such characteristic arrangements are valuable in identification.

Hardwood rays vary widely in size and therefore in appearance. The smallest rays are only one cell wide and may be visible only with a microscope, as in chestnut or aspen. At the other extreme, rays up to 40 cells wide and thousands of cells high are quite conspicuous to the unaided eye. Single rays 4 in. in height have been measured in white oak.

The pronounced ray structure in some woods is important to the woodworker. The rays, made totally of parenchyma cells, represent planes of structural weakness in the wood. In drying wood, stresses often create checks in the plane of the rays. The weakness of the rays may provide a natural cleavage plane to assist in splitting firewood, roughing out furniture parts from green wood and riving shingles. However, in attempting to machine a smooth radial surface, rays may be an impediment. Because the plane of the rays seldom coincides perfectly with the surface and because chip formation tends to follow the plane of the rays, minute tear-outs may result. As expected, such problems are related to ray size.

On the other hand, rays may add noteworthy figure to the edge-grain surface of wood. When ray appearance is distinct on the radial surface it is termed **ray fleck** *(5)*. Ray fleck is characteristic of certain species, for example, sycamore, oak and beech. In lacewood, also called silky oak (*Cardwellia sublimis*), the prized figure is produced by slicing the veneer just off true radial to produce an attractive ray fleck *(6)*. In some species, such as cherry and yellow poplar, the ray fleck appears lighter than the background wood whereas in others, such as maple and sycamore, it is darker. Together with ray size, this feature is quite helpful in identification.

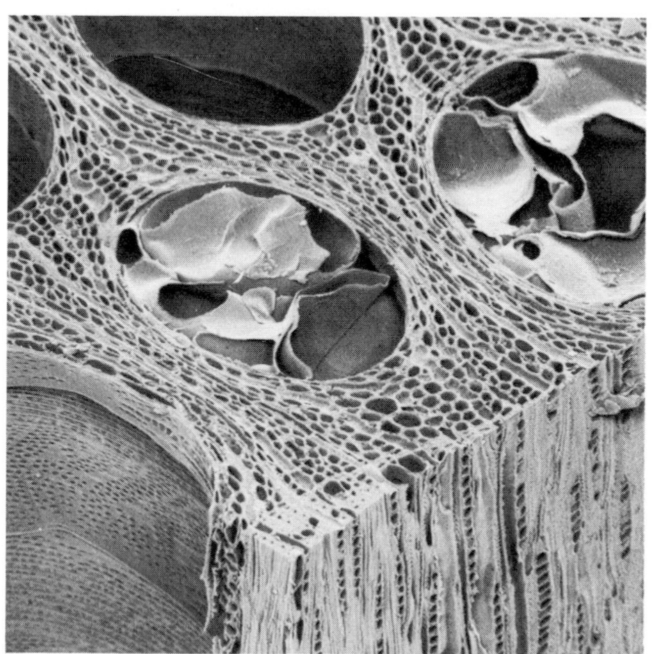

4—Tyloses in white oak.

5—Pronounced ray fleck is characteristic of the radial surface of sycamore.

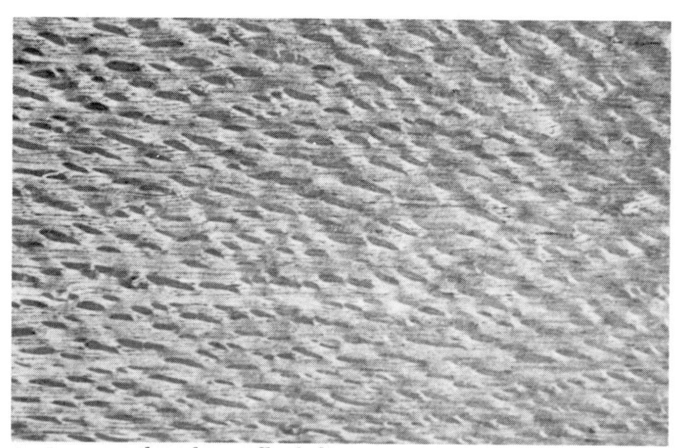

6—Lacewood with ray fleck figure.

Figure in Wood

Appreciation and understanding of the appearance of wood trace back to the anatomical structure and physiological functions of the tree, and since appearance is one of the principal values of wood to the woodworker, this subject deserves special attention.

The layman often refers to distinctive surface appearance of wood, especially that resulting from growth-ring structure, as "grain." How many times have we heard such statements as, "Oh, look at the beautiful grain in that table"? To avoid continued confusion with that already overworked word, the term **figure** is used to refer to distinctive or characteristic markings on longitudinal or side-grain surfaces of wood. In commercial parlance, the term figure is generally reserved for the more decorative woods. Figure in wood results from a combination of particular anatomical features (from normal growth structure to various abnormalities and extractives) plus the orientation of the surface that results from cutting.

Lumber is sawn to produce all surface orientations, from flatsawn to edge-grain *(2)*. Veneer can likewise be cut in any plane relative to the growth rings, from flat-slicing to quarter-slicing. Veneer can also be "peeled," that is, cut by the rotary process whereby a log is rotated against a gradually advancing knife—a continuous sheet of veneer is removed as one would unwind a roll of paper towels. Figure arises from the visual variabilities of normal wood structure, namely growth rings and rays. The more uneven the grain and the larger and more conspicuous the rays, the more distinctive the figure.

In quartersawn lumber or quarter-sliced veneer, the plane of cut is 90°, more or less, to the growth rings. Depending on the unevenness of the grain, the figure will be a parallel-line pattern *(3)*. The distinctive appearance of the rays, which is known as **ray fleck**, may vary from the intricate cross-striping of cherry to the large, showy and lustrous rays of the oaks, known as **silver grain**. To reduce the size of the ray flecks and yet produce an interesting pattern, woods such as lacewood and oak are sometimes cut just off the true radial in order to produce **rift-cut** or **rift-sawn** surfaces.

2—The method used in primary manufacture of boards and veneer determines the growth-ring orientation and the resulting surface figure.

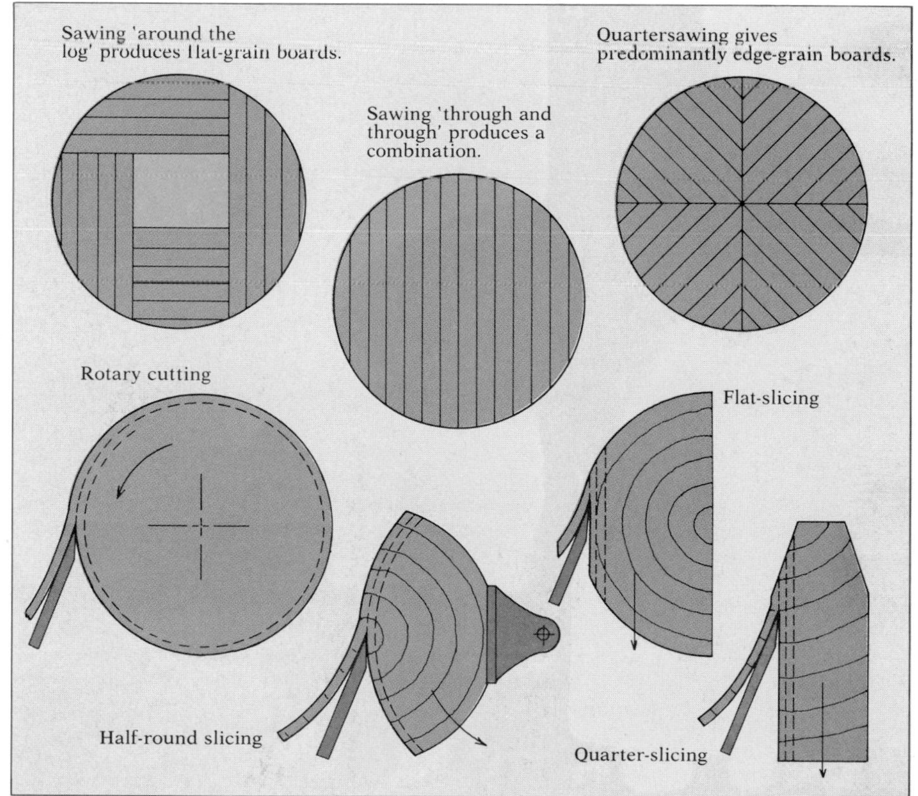

3—Quartersawn surfaces of uneven-grained conifers, such as this Douglas fir board, have parallel-line figure.

1—The feather grain of walnut crotch is highly prized.

If a log were a perfect cylinder with uniformly thick growth layers, the figure on the surfaces of boards cut in tangential planes would be parallel markings. Fortunately, every tree grows with enough irregularity to distort the otherwise static markings *(1)*. On flatsawn surfaces, then, the growth rings intercept the surface, forming ellipses, or *U*-shaped or *V*-shaped markings down through the area of closest tangent to the rings *(2)*.

Toward the edges of a flatsawn board, depending on how wide the board is and how much curvature is in the growth rings, the figure approaches edge grain. Where growth rings are fluted slightly, as in butternut, basswood and sometimes in black walnut, an irregular but interesting figure results *(3)*. Rotary veneer-cutting pro-

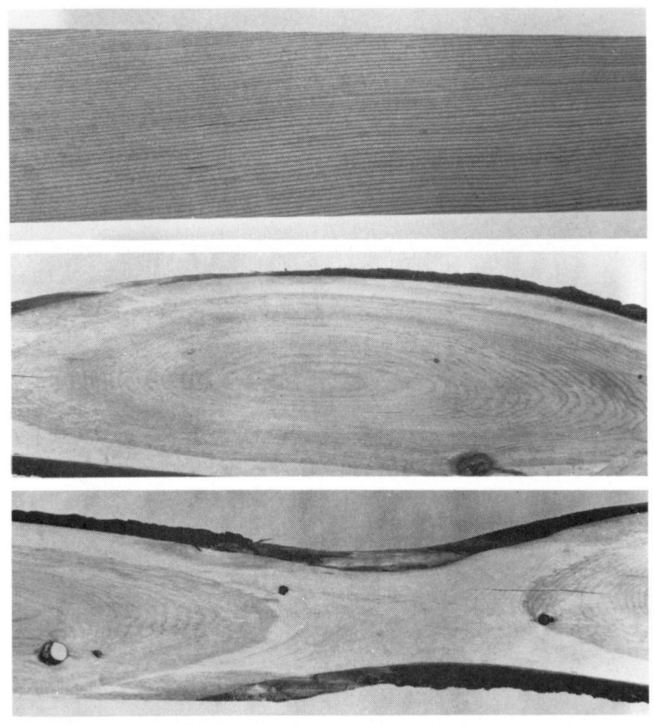

1—Quartersawn surfaces of perfectly straight logs reveal uninteresting parallel lines, top. Log with severe sweep has ellipsoid figure on flatsawn belly surface, center, and paraboloid figure on back surface, above.

2—Flatsawn figure is characterized by oval, U-shaped and V-shaped patterns. Boards (left to right) are ash, cherry, southern yellow pine, chestnut, white birch and eastern spruce.

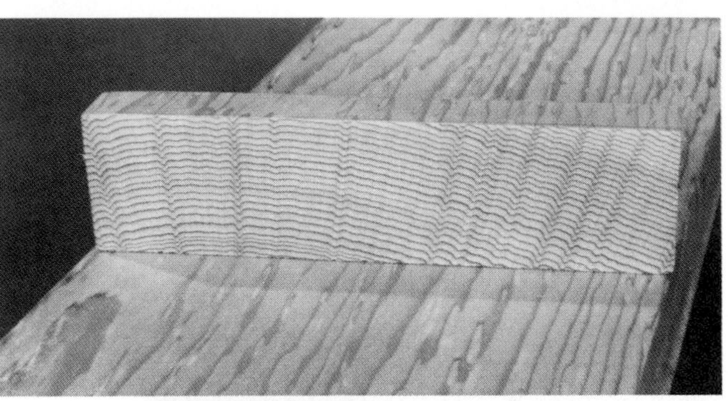

3—Indented rings in Sitka spruce, top, produce bear-scratch figure. Jagged contours of flatsawn butternut, above, come from fluted growth rings.

duces a continuous, repetitively merging series of tangential figure *(4)*. **Cone cutting**, which is a method of removing a circular bevel layer of veneer analogous to sharpening a pencil, produces interesting circular pieces bearing cone figure.

The grain direction (that is, direction of the longitudinal cells) of the average tree trunk is more or less straight, with some normal variation from strict geometric form. Wood from other parts of normal trees may exhibit interesting figure resulting from grain distortion. Crotches, for example, where a somewhat equal forking of the trunk has occurred, develop an unusually twisted and intergrown structure where the two stems merge. A cut passing through the center of the crotch produces feather crotch figure. Black walnut feather crotch (see Figure *1*, page 18) is especially prized for gunstocks. If the cut is toward the outside of the crotch, a **swirl crotch figure** results. The term **moonshine crotch** is also applied to crotch figure. The stump region of the tree also produces interesting figure.

In some species, deviation from normal structure may occur, either as a common feature or as a rare exception, which produces interesting and attractive figure in the wood. **Curly figure** results when longitudinal cell structure forms **wavy** or **curly grain**. A split radial surface looks like a washboard *(5)*. When this surface is smoothly machined to a flat plane, a cross-barred effect is produced by the variable light reflection of the cell structure intersecting the surface at various angles *(6)*. In maple this is termed **tiger maple**, or **fiddleback**, because it is

4—The continuous tangential figure of Douglas fir plywood is typical of rotary-cut veneer.

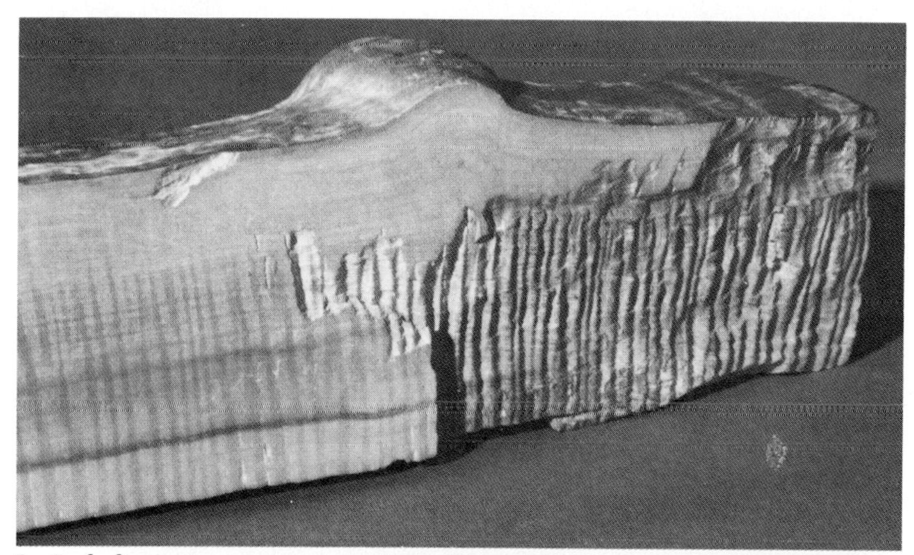

5—Curly figure is most pronounced when cut radially, as in this sugar maple.

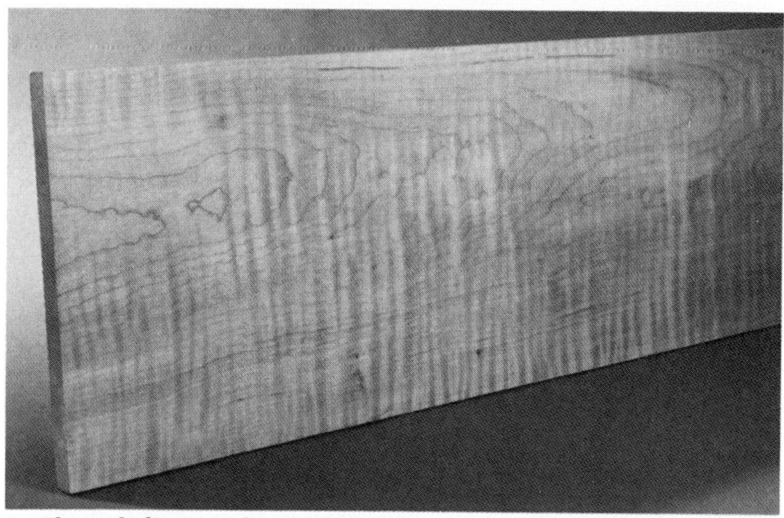

6—The curly figure in this maple board is produced largely by the changing angle of light reflection.

preferred for violin backs *(1)*. Some of the most remarkable pieces of curly figure can be seen in old Kentucky long-rifle stocks.

Ribbon or **stripe figure** is produced when wood having **interlocked** grain is cut radially. Interlocked grain is the result of repeated cycles of spiral growth, varying back and forth from left to right-hand spirals *(2)*. Except for short pieces, such wood is virtually impossible to split. When a radial surface is smoothly machined, however, these reversing spirals create a characteristic visual effect, due in part to the variation in the length of the severed vessels at the surface. The lines are long where the grain direction is parallel to the surface, but reduced nearly to pore diameter where the vessels intersect the surface at a considerable angle. The varying light reflectiveness of the fiber tissue also contributes to the overall appearance *(3)*. Figure that shows short stripes is often called **roe figure**, and the interlocked grain referred to as

2—Interlocked grain is revealed on split radial surface of American elm, far left. Top edge, where wood was split, is straight; bottom edge shows degree of interlocking. Turning down such a log would show cyclicly reversing spiral grain, left.

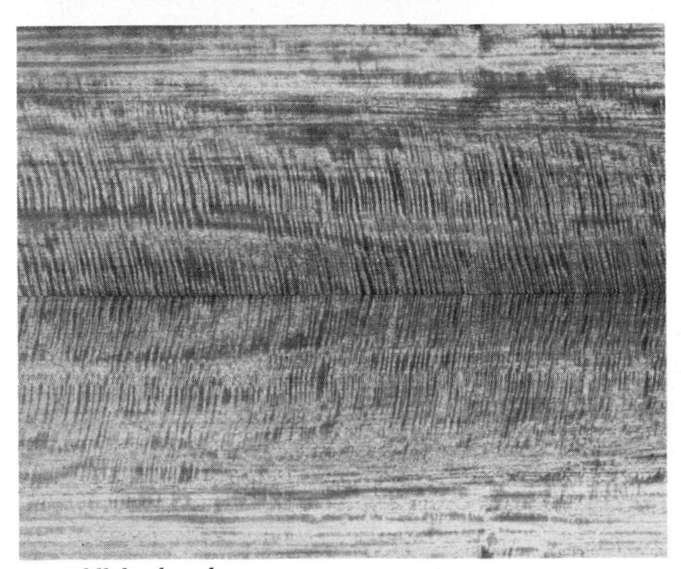

1—Fiddleback mahogany.

3—Ribbon or stripe figure in mahogany veneer. Pattern is created by varied vessel lengths and light reflection.

roey grain. Where wavy grain occurs in combination with interlocked grain so the ribbon figure is interrupted at intervals, the figure is termed **broken stripe** *(4)*; when curly figure predominates, a **mottled** figure results *(5)*.

The cambium sometimes has localized indentations, bumps or bulges that leave behind wood of characteristic figure. In ponderosa pine, lodgepole pine and Sitka spruce, **dimpling** sometimes occurs as numerous, small conical indentations of the plane of the growth ring *(6)*. In other woods, notably sugar maple, localized small swirls

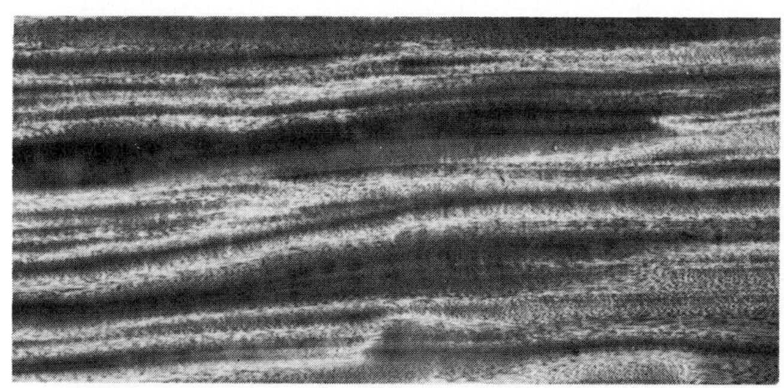

4—Broken stripe figure in this mahogany veneer is produced by a combination of interlocked and curly grain.

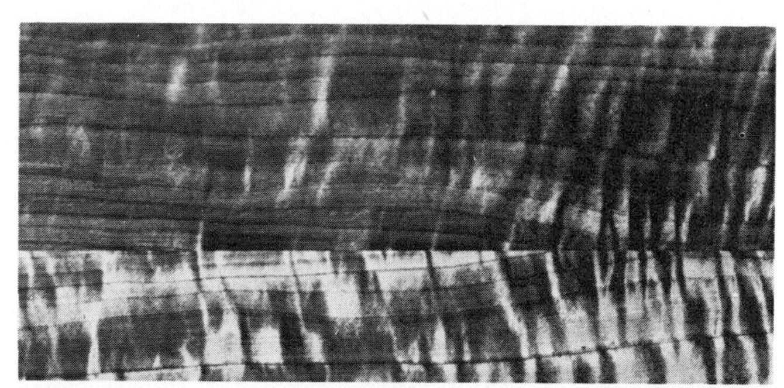

5—Mottled figure in maple veneer.

6—Depressions in the growth rings, called dimples, are obvious on the split tangential surface of this piece of lodgepole pine.

of grain direction produce **bird's-eye** figure *(1)*, so called because each swirl looks like a tiny eye.

Large, closely crowded bulges in the growth layers produce **blister figure** *(2)*, which is called **quilted figure** *(3)* if the mounds are elongated. When finished smoothly, the variation in light reflectiveness because of grain distortion creates a very unusual three-dimensional effect, so that the figure seems to roll when the piece is moved. Quilted and blister figures are usually associated with bigleaf maple but are occasionally formed in other species as well.

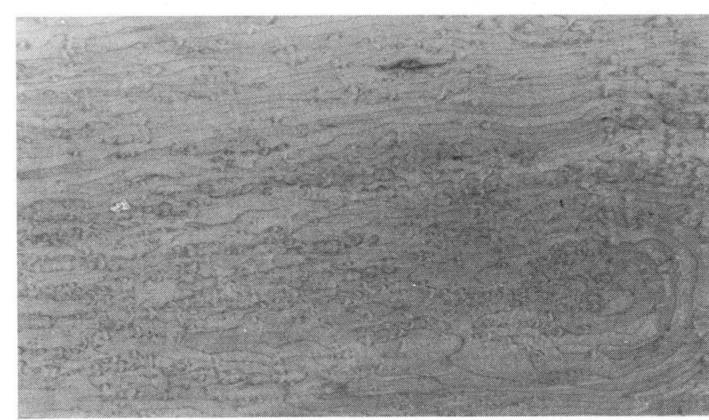

1—Bird's-eye maple.

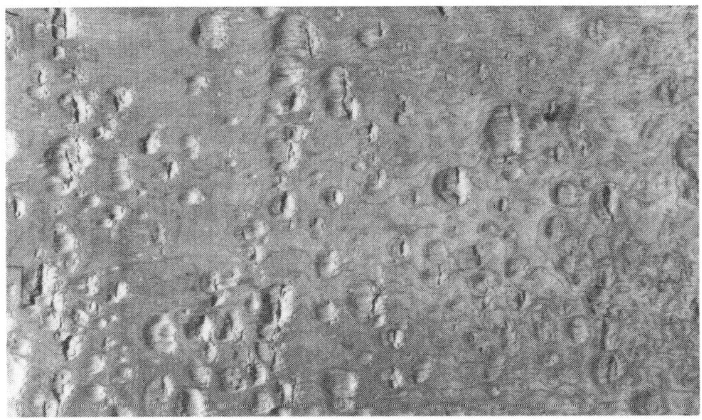

2—Grain deviation associated with blister figure.

3—Quilted figure in maple.

Burls *(4)* are large knoblike projections or bulges formed along the trunks (or sometimes limbs) of trees. The wood tissue within the burl is extremely disoriented and often contains numerous bud formations. The resulting figure is quite attractive and traditionally has been used in small articles such as bowls and turnings. Frequently, veneers cut from burls are used to display the fascinating figure *(5)*.

Figure also may be produced by irregular coloration. In hardwoods especially, the heartwood coloration is preferred and the sometimes lighter sapwood may be considered a defect and discarded. However, a striking effect in single and multiple pieces can be created by joining boards of such species as rosewood, black walnut or eastern redcedar where sapwood is prominent. Within the heartwood alone, the term **pigment figure** refers to distinctive patterns formed by uneven extractive deposits. Some examples are rosewood, zebrawood and figured red gum *(6)*. It is important to understand that the layering effect of pigmentation may be quite independent of the growth-ring layering.

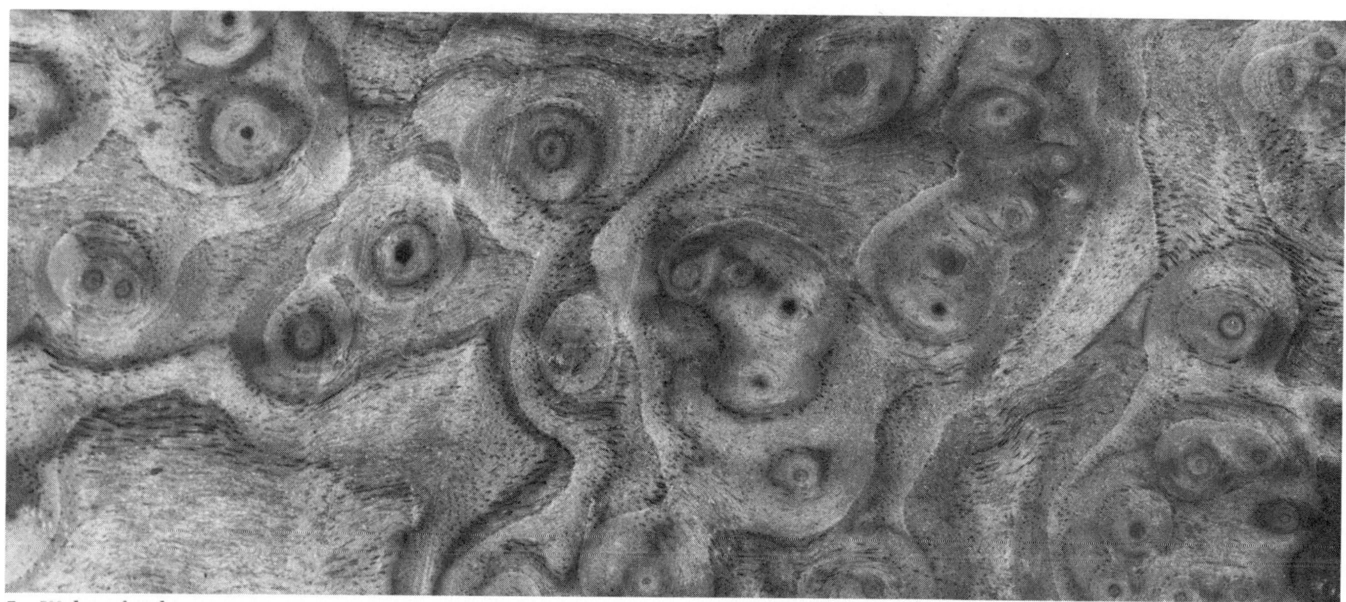

5—Walnut burl veneer.

6—Pieces of pigment-figured red gum.

4—Burl on a red oak stem.

Knots

While some irregularities in wood may increase value, as when a distinctive figure is produced, others decrease value. By tradition, any irregularities that decrease value are branded as **defects**. Although some of the features described below seem to be negative in woodworking, the woodworker is urged to reserve judgment on nature's irregularities. These were indeed defects when hand tools could not deal with them, but now many of these irregularities can be routinely machined using power tools.

Knots are a case in point. The commercial hardwood lumber-grading system assumes that every knot is a defect and bases grade on the size and number of clear areas among the knots (and other blemishes). On the other hand, many beautiful works of craftsmanship and art have been produced using, or even featuring, knots.

The woodworker should first of all understand what knots are and how their structure relates to the rest of the wood. Knots are simply the parts of limbs that are embedded in the main stem of the tree *(1)*.

As the tree grows, branching is initiated by lateral bud development from the twig *(2)*. The lateral branch thus was originally connected to the pith of the main stem. Each successive growth ring or layer forms continuously over the stem and branches, although the growth ring is thicker on the stem than on the branches and the branch diameter increases more slowly than the trunk. As the girth of the trunk increases, a cone of branch wood—the **intergrown knot**—develops within the trunk. Such knots are also termed **tight knots** because they are intergrown with surrounding wood, or **red knots**, especially in conifers where they often have a distinct reddish tinge. At some point the limb may die, perhaps as a result of over-

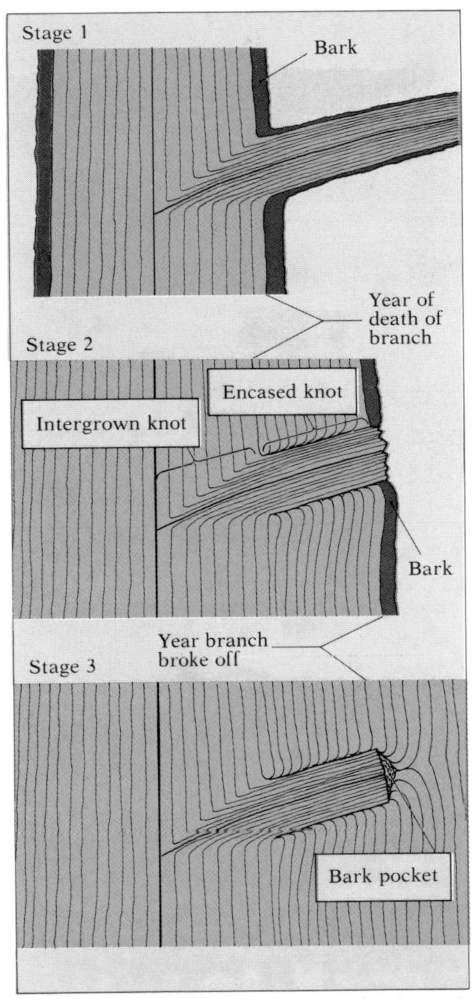

1—A knot is the basal portion of a branch whose structure becomes surrounded by the enlarging stem. Since branches begin with lateral buds, knots can always be traced back to the pith of the main stem.

2—Coniferous trees, above, are characterized by excurrent form, i.e., a dominant stem from which whorls of lateral branching occur at regular intervals, or nodes. In a softwood board, below left, knots in clusters indicate the whorls, separated by the clear wood of the internodes. The same pattern is also often present on plywood made from rotary-cut softwood veneer, below right.

shadowing by limbs higher up. The limb dies back to approximately the trunk surface, its dead cambium unable to add further girth. So subsequent growth rings added to the main stem simply surround the dead limb stub, which may begin to rot. A number of years of growth may be added to the main stem, surrounding the branch stub. The dead part of the stub becomes an **encased knot**. It is not intergrown and therefore is called a **loose knot**, often with bark entrapped. **Knotholes** result when an encased or loose knot falls out of a board. Enclosed knots also are called **black knots** because they commonly are discolored by stain and decay. In time the stub may become weakened by decay and fall or be broken off, or it may be pruned back flush with the trunk. Further growth layers will enclose the stub, and eventually the cambium will form a continuous layer. From this point on, solid layers of wood and bark will be formed beyond the overgrown knot. But as the cambium moves outward, the knot-scarred bark layers persist for an amazing number of years, providing a clue to the buried blemish.

Knots may be classified by how they are cut from the tree. If they are split by radial sawing and extend across the face of the board they are termed **spike knots**. On flatsawn boards they usually appear round or oval and are called **round knots**. Knots smaller than ¼ in. in diameter are called **pin knots**. Figure *3* shows various types of knots.

Understanding knots can be useful to the woodworker. Nothing is more devastating to a carver than to work halfway through a block of wood only to uncover an interior knot flaw. Yet the trained eye can usually predict such a blemish. If a pie-shaped section is taken from a log and the first few growth rings near the pith are removed, any branches will be seen at least as tiny knots. If none are present, there will be no knot-related defects in the piece.

3—Types of knots.

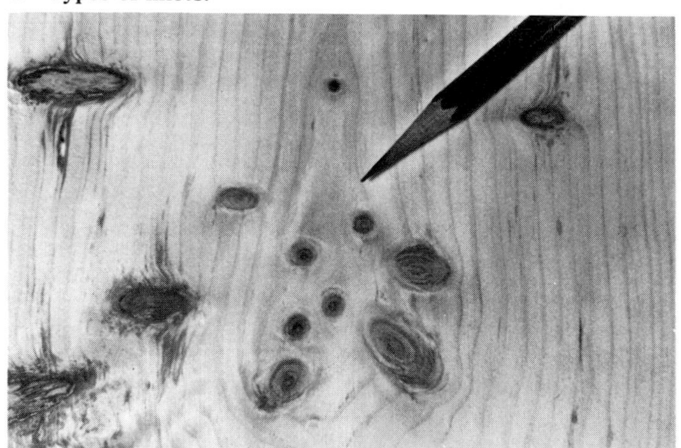

Tight round knots

Loose round knot (knothole)

Spike knots

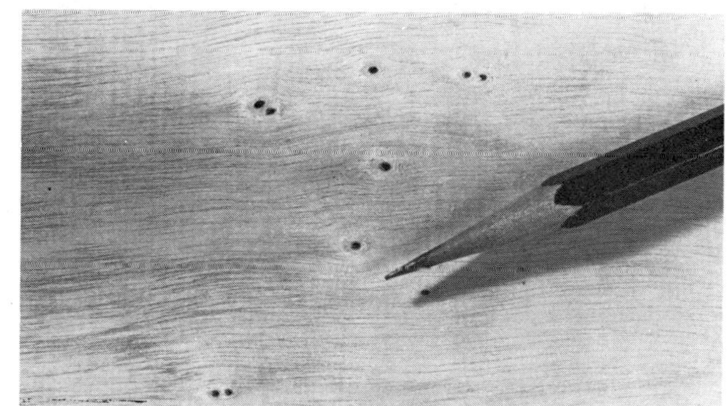

Pin knots

If any are located, the bark should be carefully examined for scars. Experience can tell much about the size and depth of such defects *(1)*.

It should be noted that since every knot originates at the pith, every knot that appears on the bark side of a flat-sawn board will also appear on the pith side of that same board. On the other hand, some knots on the pith side may have ended and grown over before reaching the bark surface. Therefore, the bark side is often the clearer, higher-quality face.

There are a variety of reasons why knots commonly are considered defects. The wood of the knot itself is different in density (usually higher), and its grain orientation is more or less perpendicular to the surrounding wood. Because shrinkage is greater across the knot than in the surrounding wood, encased knots may loosen and drop out. Although intergrown knots remain tight, they may develop radial cracks. Encased knots are usually considered worse defects because of the discoloration and the entrapped bark associated with them *(2)*. From the standpoints of strength and machining properties, the disorientation of grain direction is troublesome not only because of the knot itself but because the entire area is influenced by the knot. For example, a spike knot extending across a board may cause it to break in half under small loads.

Knots may also be an asset, and have been valuable features of figure in many ways. Knotty pine is often thought to be characteristic of Colonial decor, though in reality, knots were mostly avoided, plugged or painted over by early cabinetmakers. Knotty pine as wall boarding seems to be a 20th-century invention to use the increasing stocks of common grades of lumber. Other species that exhibit knots with some degree of regularity, such as spruce, cedar and other western softwoods, have been successfully marketed to feature their knots. Individual pieces of wood with knots increasingly are fashioned into masterpieces of cabinetry and sculpture.

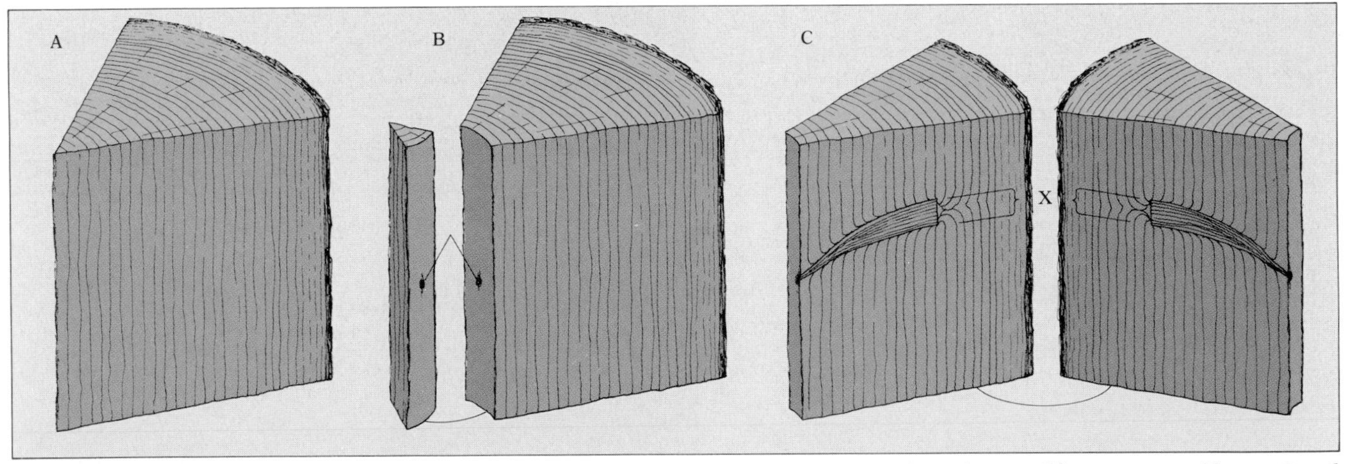

1—A carving chunk may appear flawless at first glance (A). Removing wood adjacent to the pith (B) will locate internal knots (C and photo below left). Scars in the bark (at X) may be a sign of knots large enough to ruin the block for carving.

2—Included bark adjacent to knot in red oak. This defect may also result from injury to the growing tree.

Abnormal wood

The first few growth rings added around the pith may not be typical of the mature wood formed by the tree. This core of atypical tissue is termed **juvenile wood** *(3)*. It is prevalent among conifers, especially plantation-grown trees, which grow rapidly until eventual crown closure. Then competition with other trees slows the growth to a more normal rate. Juvenile wood is characterized by wider growth rings of lower-density wood and less strength. It may also have abnormal shrinkage properties which result in greater tendency to warp, especially by twisting. Pieces of wood including (or very near) the pith should be suspect. Some trees and species show little or no juvenile-wood abnormality.

Reaction wood is a term applied to abnormal wood formed in tree stems and limbs that are other than erect, that is, parallel to the pull of gravity. The principal concern to woodworkers is the occurrence of reaction wood in leaning trunks *(4)* from which otherwise defect-free wood might be expected. Causes for leaning stems include partial uprooting by storms, severe bending under

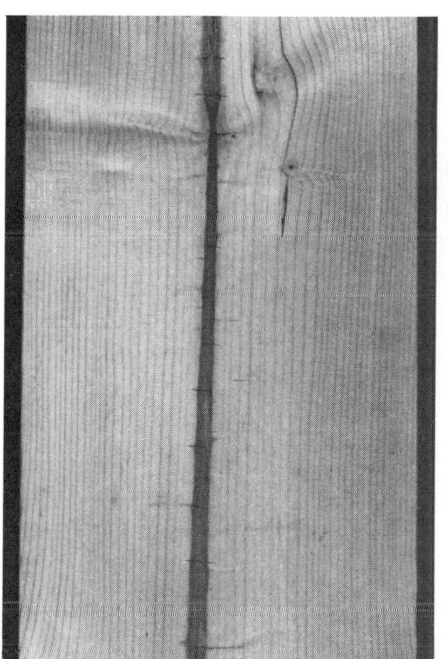

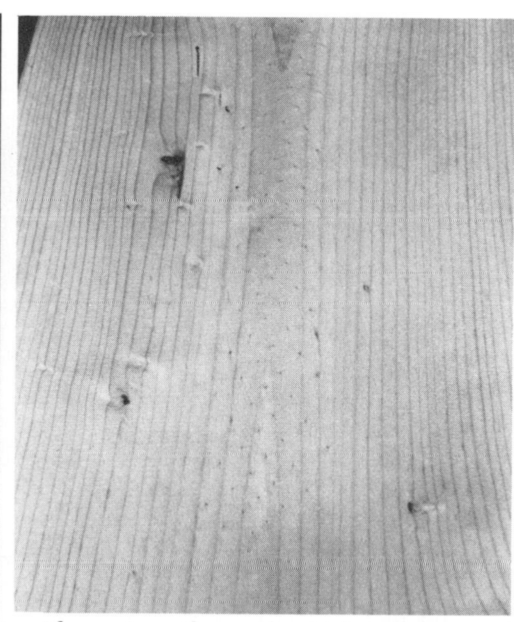

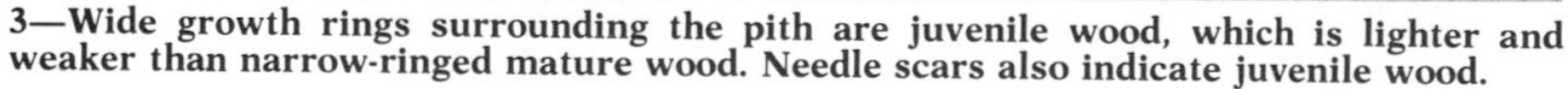

3—Wide growth rings surrounding the pith are juvenile wood, which is lighter and weaker than narrow-ringed mature wood. Needle scars also indicate juvenile wood.

4—Reaction wood forms in trees that lean. The curving sweep of the tree, above, although picturesque, means that unpredictable compression wood will be found within. Below left and right, pronounced reaction wood from leaning spruce and hemlock trees, shown in cross section.

snow or ice, and tree growth toward sunlight available from only one direction. Reaction-wood formation seems to include a mechanism for redirecting stem growth to the vertical, resulting in a bowing of the stem. Therefore boards or pieces from a log with noticeable bow should be suspected of containing reaction wood and should be examined very closely for it.

Reaction wood has different traits in softwoods and hardwoods. In softwood species, reaction wood forms principally toward the underside of the leaning stem. Because the pull of gravity presumably puts the lower side of the leaning trunk in compression, reaction wood in conifers is termed **compression wood**. Several visual features aid in its detection *(1)*.

The part of the growth ring containing reaction wood is usually wider than normal, resulting in an eccentrically shaped stem with the pith offset toward the upper side. The abnormal tracheids usually appear to form wider than normal latewood. Even-grained woods, such as eastern white pine, therefore appear uneven-grained. However, in woods that are notably uneven-grained, such as southern yellow pine, the latewood is duller and more lifeless and tends to even out the contrast.

The two main disadvantages of compression wood for the woodworker are its effects on strength and shrinkage. Since reaction-wood tracheids are thick-walled, the wood is usually denser than normal. But because they contain less cellulose than normal, and the cellulose

1—Compression wood in pine may appear as a dark streak on a flatsawn board, left, or as an abrupt change from normal light-colored sapwood to dark, above, on a quartersawn board. Right, abnormal appearance of earlywood and latewood on a flatsawn surface indicates compression wood.

chains are not as parallel to the long direction of the cells, the wood is weaker than normal. The woodcarver is especially aware of the abnormally hard but brittle qualities of compression wood. In finishing, compression wood may not stain uniformly with normal wood. The carpenter notices the difficulty in driving nails and the greater tendency to split. For structural uses where load-bearing capability is vital, as in ladder rails, unknowing use of reaction wood has resulted in fatality, because the wood breaks suddenly when bent and at lower than expected loads *(2)*.

The second major problem with compression wood is its abnormal longitudinal shrinkage. Normal wood shrinks so slightly along the grain that it is usually negligible. Compression wood shrinks up to 10 to 20 times the normal amount. What's more, because reaction-wood formation is non-uniform in a given board, the shrinkage is uneven, resulting in greater problems. Drying of reaction wood, or changes in moisture content, creates uneven shrinkage stresses in the wood, often resulting in warp *(3)*. This, along with juvenile wood, is a major cause of warp in framing lumber. Most distortions that develop in stud walls probably result from reaction wood. In woodworking, attempts to ripsaw pieces containing reaction wood may result in the wood's pinching against the saw or its splaying widely apart as the cut progresses.

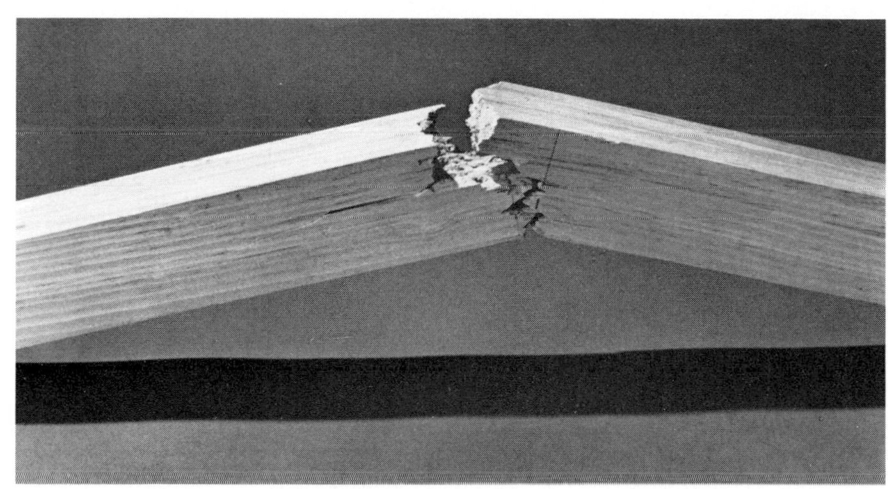

2—Brash failure in reaction wood.

3—Abnormal shrinkage in compression wood, right, is a frequent cause of warp. Below, reaction wood on edge of pine board has split and bent away.

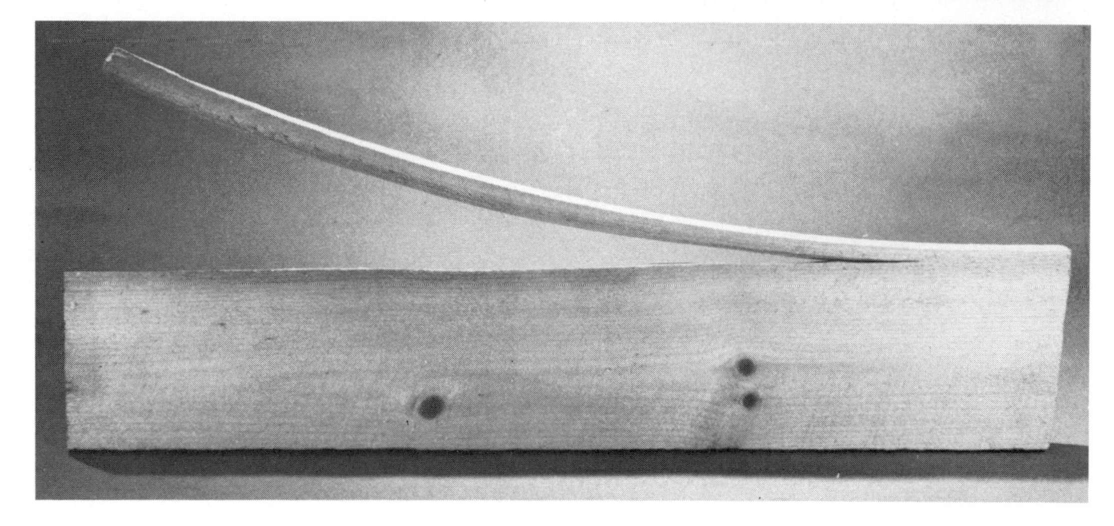

In hardwood trees, reaction wood forms predominantly toward the upper side of the leaning stem. Because gravity causes the upper side to be in tension, it is termed **tension wood**. In hardwoods, however, there is less tendency than in softwoods for the pith to be off-center in the stem, and tension wood may develop irregularly around the entire stem. Tension wood is often quite difficult to detect. Sometimes it looks silvery *(1)*, other times dull and lifeless, and in some cases there is little if any visual difference. Indications of crookedness or sweep in the log are signals of possible tension wood. The abnormal fibers of tension wood actually contain a greater than normal amount of cellulose. This wood is commonly stronger than normal. Of concern to the woodworker is the way this wood machines. Fiber structure does not sever cleanly but leaves a fuzzy or woolly surface *(2)*. Aside from the immediate problem of machining tension wood, seemingly successful efforts to smooth the wood leave a microscopic woolliness upon the surface. Upon finishing, stain is absorbed irregularly and the surface appears blotchy. As with compression wood, longitudinal shrinkage in tension wood is both irregular and greater than normal, resulting in warping and machining problems.

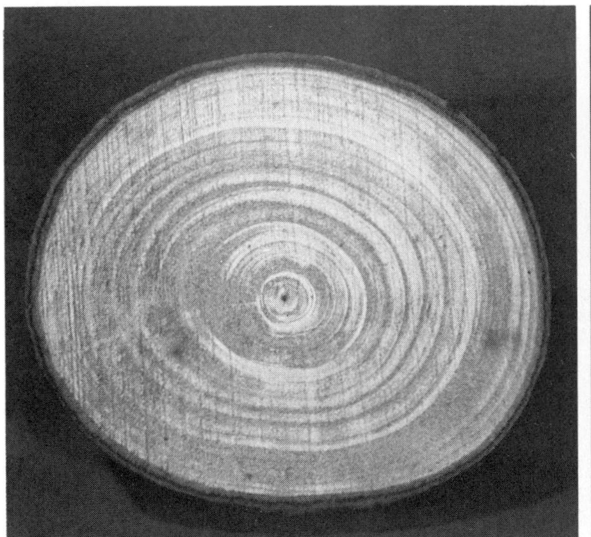

1—Tension wood looks silvery in aspen, above left. Eccentric rings in cherry, above, and red oak, below left, also indicate tension wood.

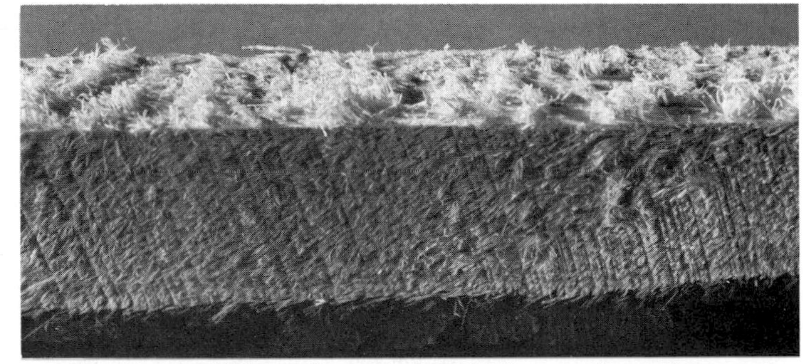

2—The abnormal fibers of tension wood left a woolly surface on this cottonwood board when it was sawn from the log.

Fungi

Wood kept under favorable conditions apparently lasts indefinitely—artifacts in excellent condition have been recovered from ancient Egyptian tombs. However, it is equally important to note that wood is biodegradable—that is, subject to deterioration by natural agents. Elimination of dead wood by decay is as necessary to the continuation of the forest as is the sprouting of the seed, for if fallen trees and limbs remained, there would eventually be no open ground for new reproduction.

Biodegradation of wood is accomplished in part by insects and marine borers *(3)*, but the greatest degree of deterioration is the work of **wood-inhabiting fungi**.

Fungi are low forms of plant life. Incapable of producing their own foods as do green plants, these parasites derive their sustenance from a host plant. Fungi that inhabit wood are classified as molds, stains or decay. Molds live mainly on the surface of the wood, while stains invade the cell structure. Both live principally off carbohydrates stored in parenchyma cells. Because their work is confined essentially to sapwood, they are termed **sapstains**. Since they commonly produce a bluish grey discoloration, the term **blue stain** is often applied. The main problem with sapstains is this discoloration *(4)*, not structural cellular damage.

Decay fungi invade the cell structure, not only consuming any available stored materials, but actually dissolving the cell-wall material by enzyme action. The early stages of infestation, called **incipient decay**, are characterized mainly by discoloration—strength may not be significantly altered. Eventually, proliferation of fungi

3—Wood damaged by marine borers.

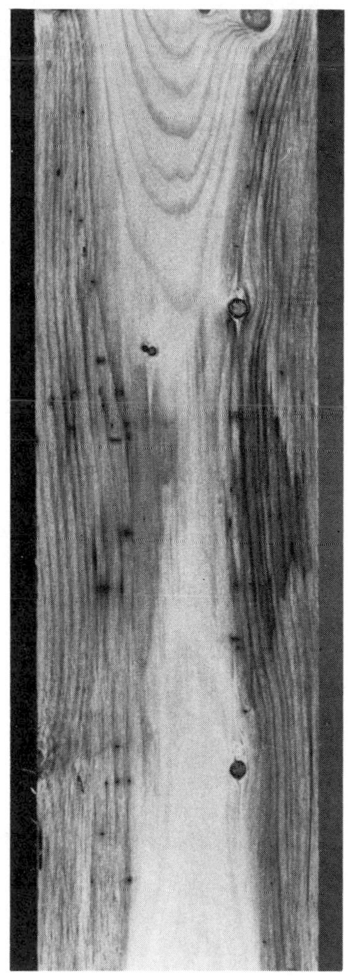

4—Severe blue-staining on this white pine board indicates the presence of sapstain fungi.

1—Infection with brown rot, above, eventually results in cubical breakup of the wood, below.

results in **advanced decay,** wherein discoloration is accompanied by obvious softening of the wood. Wood thus affected is termed **rot**, but the result varies somewhat according to the species of fungi involved. **Brown rots** attack principally cellulose, leaving a brown powdery residue, often characterized by cross-checks similar in pattern to charred wood *(1).* **White rots** consume both lignin and cellulose, leaving a whitish, sometimes stringy residue. Certain fungi concentrate in localized areas or pockets, as in **pecky cypress**, a brown pocket rot. **White speck**, which occurs in softwoods, is characterized by small pockets filled with whitish residue *(2).*

Although wood with decay is generally considered defective or worthless, decay may sometimes be a valued feature. The rot pockets in both pecky cypress and white-speck-infected lumber are often displayed decoratively, mainly in wall boarding. Such infested wood is especially attractive if the rot pockets are reasonably uniform in size and distribution.

In some wood attacked by certain white rots, the development of decay is accompanied by attractive dark brown or black staining *(3).* Thin layers of stain are apparently formed at intervals indicating successive advances of the incipient decay in the wood. When the wood is machined, the layers are exposed as thin lines on the surface and are called **zone lines**. Between the zone lines, wood in varying stages of decay may range from normally hard to punky. The term **spalted wood** is sometimes used to describe wood with zone-lined decay. Items of unusual beauty such as bowls or plaques can be made from

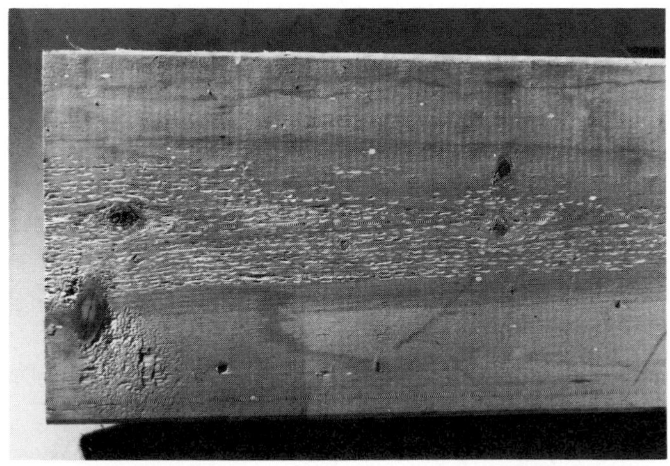

2—Spruce plank is infected with white speck, a white pocket rot that invades the heartwood of living trees, but ceases to develop once the tree is cut.

3—As certain white rots develop, dark zone lines form, as on this piece of sugar maple, above. This type of decay is called spalting. Below, turning of spalted maple, by Mark Lindquist.

spalted wood, as long as the wood is firm enough to be machined and finished.

Dealing with fungi requires understanding their life cycle and their requirements for development *(4)*. Fungi spread by tiny single-celled spores. A spore, borne by the winds, is brought into contact with a wood surface. If conditions are favorable, the spore develops a threadlike filament called a **hypha**. By branching and rebranching, a network of hyphae is formed. A mat of hyphae, which is called a **mycelium**, may appear as a cottony mass on wood surfaces.

As invasion of the hyphae continues, fruiting bodies may form. In blue-stain fungi, these are tiny structures bearing spores, which appear as dark specks to the unaided eye. In other fungi these fruiting bodies, sometimes called **conks**, appear as knoblike or shelflike projections on the decayed wood. Spores develop within crevices or holes in the undersurface of the fruiting body. The life cycle is thus perpetuated. Spores are produced in such excessive numbers that it is estimated that our every breath contains fungal spores. We must assume that the air always contains fungal spores and that any wood surface in a condition favorable to decay will be inoculated by them.

There are four basic requirements for wood-inhabiting

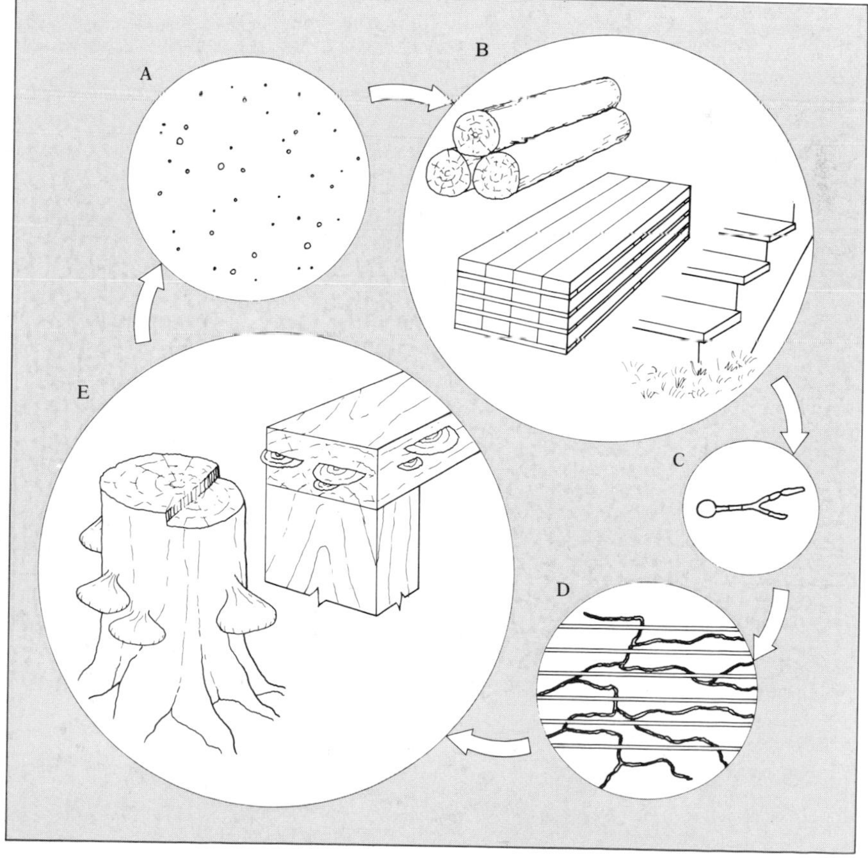

4—Life cycle of a typical wood-inhabiting fungus: Microscopic airborne spores (A) are carried by winds and air currents to potential hosts (B)—logs, lumber, wood products, etc. If conditions are suitable, the spores produce filamentlike hyphae, which elongate and branch (C), then multiply, spreading through the wood (D) or forming a cottony surface mat or mycelium. Advanced stages of the fungus produce fruiting bodies, often appearing as shelflike or bractlike conks (E) on wood surfaces. In tiny crevices on the undersides of the conks, myriads of spores are produced which, when mature, are released into the air and carried to the next potential host.

fungi to thrive. Fungi can be controlled by rendering any one of these unsuitable.

Temperature is the first condition—between 75°F and 90°F is optimum. Beyond the extremes of 40°F and 105°F, growth essentially stops. Unfortunately, humans also thrive at temperatures favorable to fungi, so regulating temperature is not a practical way to prevent attack. However, temporary storage in refrigerators or freezers is a most effective way, albeit expensive, to prevent deterioration, especially for green wood.

Fungi also need oxygen. Waterlogged wood does not decay because of the absence of oxygen. Approximately 20% air volume in the wood is needed for fungal development. If green wood of high moisture content is kept from drying out, fungi can be held in check. But again, most wood we use cannot be kept in this manner.

Moisture content is also a factor. The optimum level is at or slightly above the fiber saturation point (about 30% moisture content), in which the cell walls are saturated but the cell cavities are essentially empty. Fungi can develop at moisture contents as low as 20%. Drying wood quickly down to below 20% moisture content and keeping it dry is the principal way to prevent fungal deterioration, and one of the main reasons for drying wood is to prevent fungi from developing.

In dealing with moisture requirements, as with oxygen and temperature, making the conditions unsuitable merely causes a fungus to go into dormancy, in which state it can survive for many years. If favorable conditions are restored, development may resume. This is perhaps the explanation for the fallacy about **dry rot**—there is really no such thing. A few species of fungi capable of infecting wood of low moisture content are called **dry-rot fungi**. Although they are capable of transporting moisture to the areas of infection, these fungi cannot develop in wood with total lack of moisture. Wood must be moist to decay. However, sometimes wood that undergoes intermittent wetting and decay is inspected during a dry period. The obvious powdery dry rotted wood is interpreted as dry rot.

Food is the fourth requirement. The sapwood of most species is suitable, both because it lacks extractives and because it contains carbohydrates stored in parenchyma cells. The heartwood may be naturally decay-resistant if extractives are toxic or repellent to fungi. Woods vary considerably in decay resistance or durability *(Table 4).* Where it is impossible or impractical to keep wood below 20% moisture content, the next best approach is to choose a durable wood, or wood that has been impregnated with a chemical preservative. The subject of fungal control will be further discussed in connection with drying wood and with finishing and treating it.

Table 4—Comparative resistance of heartwood to decay.

Resistant or very resistant	Moderately resistant	Slightly or nonresistant
Baldcypress (old growth)	Baldcypress (young growth)	Alder
Catalpa	Douglas fir	Ashes
Cedars	Honeylocust**	Aspens
Cherry, black	Larch, western	Basswood
Chestnut	Oak, swamp chestnut	Beech
Cypress, Arizona	Pine, eastern white	Birches
Junipers	Pine, longleaf	Buckeye**
Locust, black*	Pine, slash	Butternut
Mesquite	Tamarack	Cottonwood
Mulberry, red*		Elms
Oak, bur		Hackberry
Oak, chestnut		Hemlocks
Oak, Gambel		Hickories
Oak, Oregon white		Magnolia
Oak, post		Maples
Oak, white		Oak (red and black species)**
Osage-orange*		Pines (most other species)**
Redwood		Poplar
Sassafras		Spruces
Walnut, black		Sweetgum**
Yew, Pacific		Sycamore
		Willows
		Yellow poplar

* These woods have exceptionally high decay resistance.
** These species, or certain species within the groups shown, have higher decay resistance than most of the other woods in their respective categories.

Insect damage

Many different insects cause damage to wood; some in living trees as they grow, some in harvested logs and in sawn lumber, others in finished products. Here are a few of the familiar types of damage that are encountered by woodworkers.

Pith flecks, which occur in living trees, are caused by larvae of tiny flies (*Diptera*) belonging to the genus *Agromyza*. These flies burrow downward in the tree stem along the cambial layer during the growing season. The tunnels damage the cambium and become occluded with scar tissue. When the wood eventually is machined, the tunnels are revealed on exposed surfaces as dark brown blemishes. They appear as an oval or oblong spot in cross section, but as a somewhat irregular or intermittent streak along longitudinal surfaces *(1)*.

At first glance, these defects may resemble the pith of some woods—hence, the name **pith flecks** or **medullary spots**. Pith flecks commonly occur in a number of species, notably in paper, grey and river birches and in red and silver maples. They also occur sporadically in other maples and birches and in many other hardwoods, including basswood, willow, cherry and aspen. Because pith flecks are small in size and well distributed, their structural damage is insignificant—whether they should be considered visual defects or simply visual characteristics is arbitrary.

Bore holes in various sizes are mainly the work of beetles. While many attack logs after harvest or sawn lumber during air drying (and may inadvertently be included in processed material), some will attack the dry wood in finished products. In the latter case, **powder-post beetles** are one of the most common types. These small, cylindrical beetles lay their eggs in minute seasoning checks or in open pores, especially in coarse-textured hardwoods such as ash, oak and chestnut. The larvae burrow through the wood in quest of the carbohydrates stored in parenchyma cells, hence their damage is concentrated in sapwood. Their presence usually goes undetected until the adults emerge from the surface and small amounts of the powdery frass sift out of the exit holes. If the adults reinfest the same piece of wood repeatedly, the interior of the wood eventually becomes a maze of tunnels packed with powdered wood.

Infected wood should not be used, for once established, the beetles are extremely difficult to exterminate. Control measures should therefore focus on prevention. For most products, a continuous film of finish or paint is highly effective in sealing the pores or checks in which the eggs can be laid. Attacks are most common in unfinished hardwoods used in structural members, farm implements and tool handles.

Two additional insects, **termites** and **carpenter ants,** deserve mention in view of the devastating damage they cause in stored lumber and wood structures. Both insects are social, and their colonies have a preference for moist wood. In fact, the most common species of termites (**subterranean termites**) must maintain contact with the soil. Termites live in complete seclusion and remain unseen except when some winged adults emerge and swarm in flight, usually in the spring, to establish an outpost colony. Termites actually consume wood for sustenance and where conditions are favorable, damage will continue to spread. Carpenter ants, on the other hand, do not eat wood, but excavate galleries to provide shelter. As an ant colony becomes large, the galleries can become extensive enough to cause serious structural damage, usually in a concentrated location. The large black adults can often be seen scavenging for food in the vicinity of an active colony.

For both termites and carpenter ants, the key to prevention is keeping wood dry and isolated from the ground. Wooden parts of buildings should be well above soil surfaces and construction should be properly designed and maintained to avoid leakage of rainwater.

1—Pith flecks on tangential surface of this plank of paper birch were formed by the larvae of tiny flies that burrowed down the tree stem along the cambial layer.

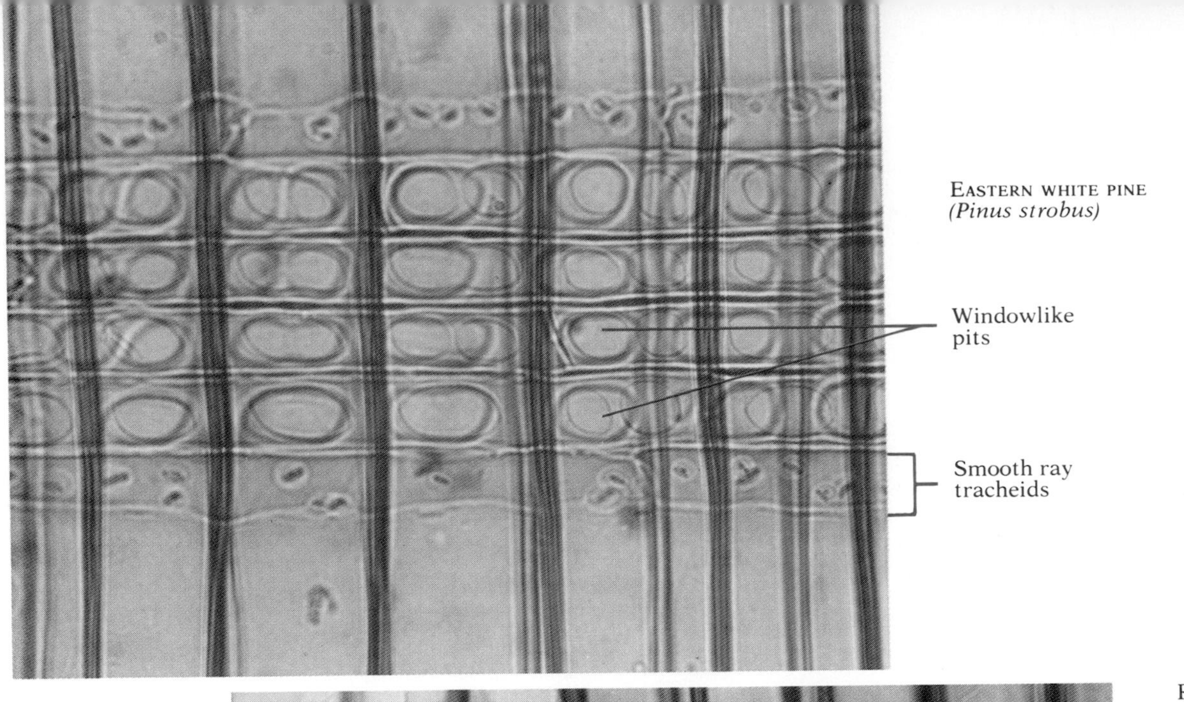

Eastern white pine
(Pinus strobus)

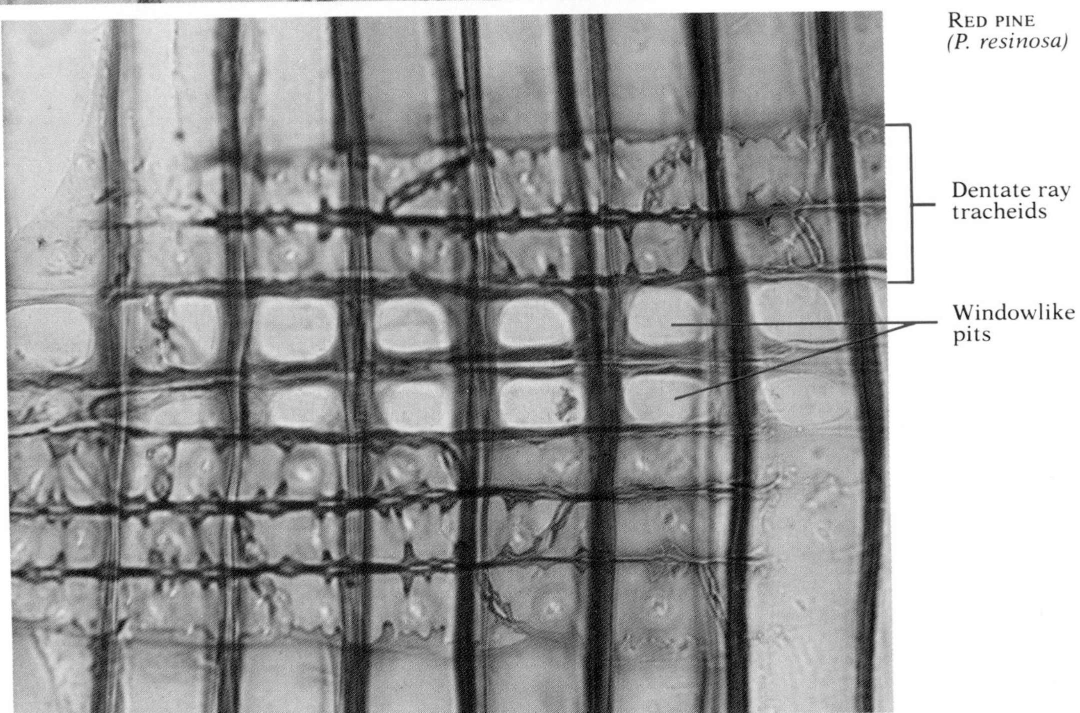

Red pine
(P. resinosa)

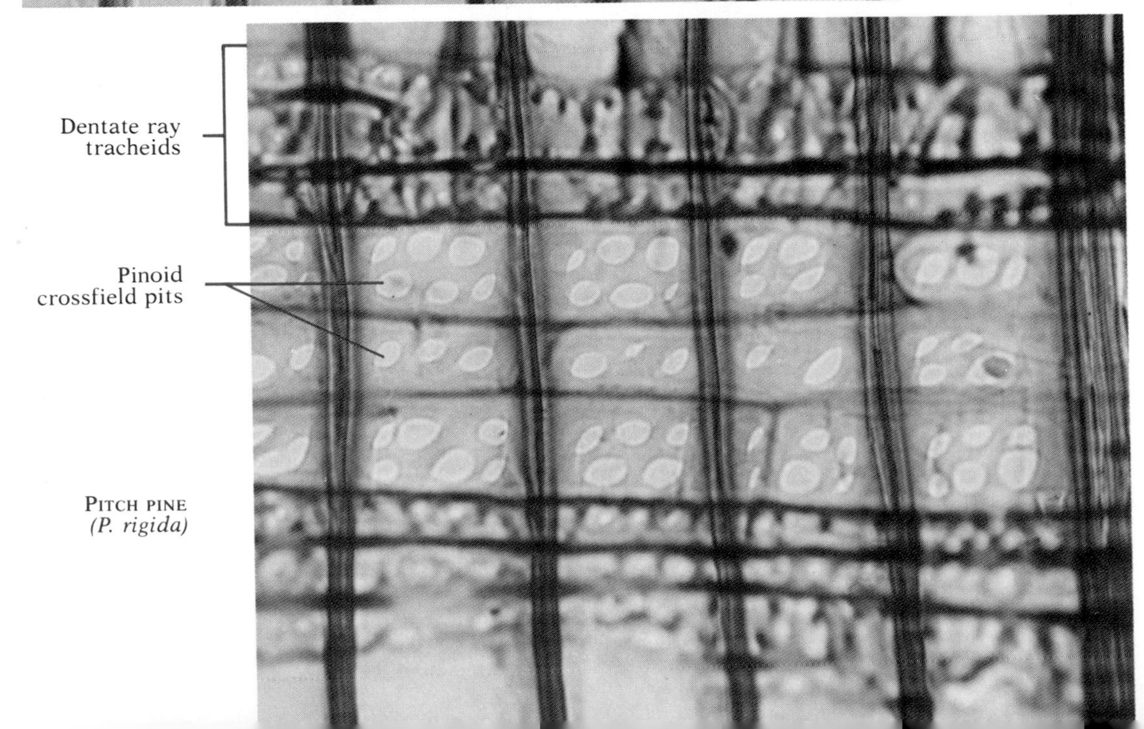

Pitch pine
(P. rigida)

Wood Identification

Because of the wide range of woods, success in identification is quite unpredictable. The beginner in this fascinating game should be forewarned that although some species can be identified accurately at a glance, others defy final separation even by experts using the most sophisticated equipment. However, the vast majority occupy a middle ground where systematic examination and comparison with known wood features are usually successful, and a satisfying level of skill in identification can be attained. By diligent examination of wood features, particularly those found on end-grain surfaces, any woodworker can learn the distinguishing characteristics of commonly encountered woods.

It is tempting to take the "looks-like" approach, where you simply look at a piece of wood and decide what it resembles. I have known old lumbermen who could plod through a drying yard and identify nearly every board at a glance, but they weren't always right, and when they were wrong they were often off by a mile. The other problem with the "looks-like" method is that you usually don't know when you're looking at a wood you've never seen before. And every wood is apt to look like another wood *(3)*. A far better approach is to study anatomical features. Identification by gross features (visible to the unaided eye) is usually possible only after considerable experience and knowledge gained through study under magnification.

The most reliable approach is based on minute or microscopic features, which means observing thin sections of wood under a microscope. Microscopic examination is most important with softwoods, where visual features are characterized more by similarity among woods than differences; in many instances the only basis for separation is minute features *(1, 2)*. Considerable

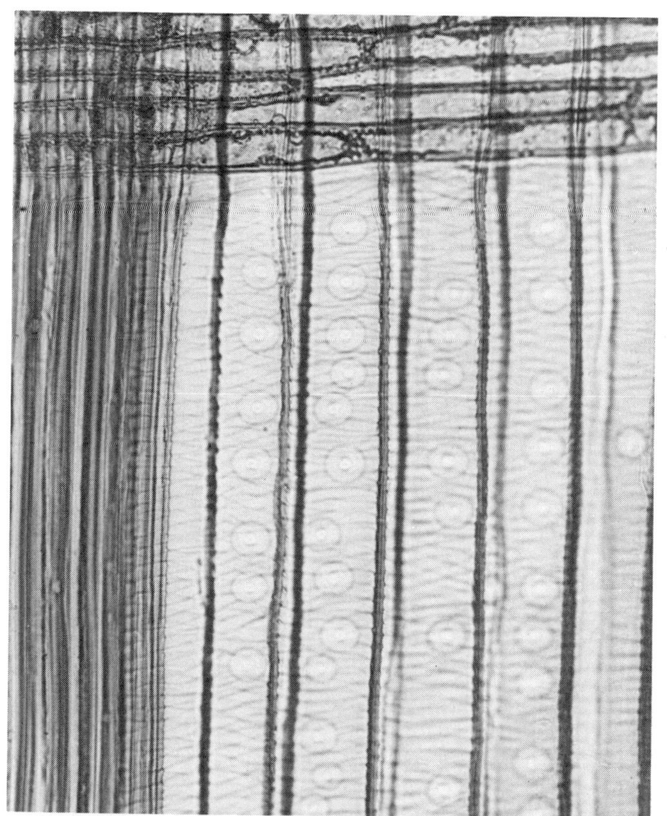

2—Spiral thickenings on the inner walls of earlywood tracheids, the most definitive identification feature of Douglas fir, appear under the microscope as fine coil springs. Even a tiny sliver can be positively identified by this constant feature.

1—Left, three species of pine common in New England antiques are easily differentiated by microscopic examination of the radial sections (400×). The definitive features are the appearance of the ray tracheids and the nature of the pits in the ray parenchyma cells.

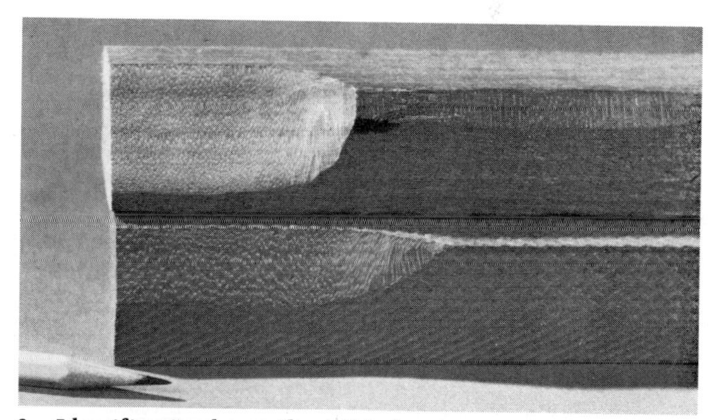

3—Identification by easily visible features can be quite unreliable. For example (photo above) the similar appearance of a light-colored ray fleck against a darker background can lead to confusion between yellow poplar (above), and black cherry (below), especially when the original heartwood color has been changed by staining and aging (as in antiques). When viewed in cross section (photo below), however, the light line of terminal parenchyma will easily distinguish yellow poplar, right.

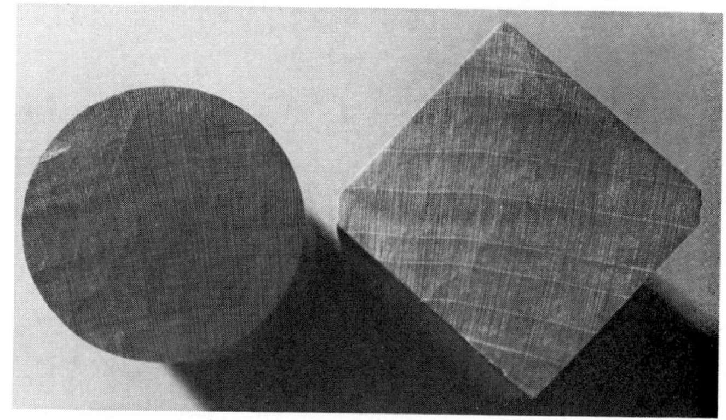

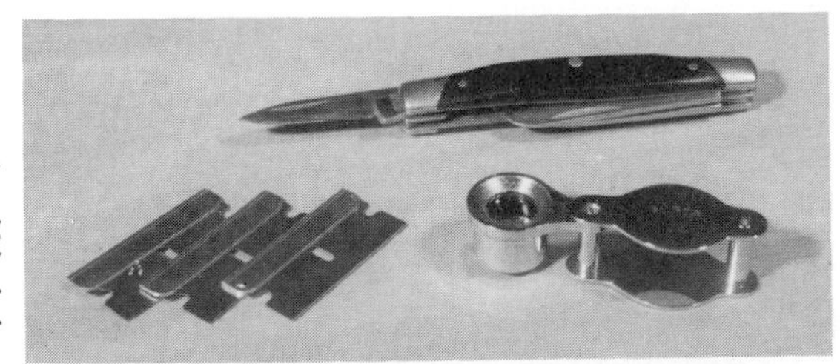

1—Tools for macroscopic identification are few and simple: a 10× hand lens and a means of cutting wood surfaces cleanly. A very sharp knife or an industrial single-edged razor blade is quite satisfactory.

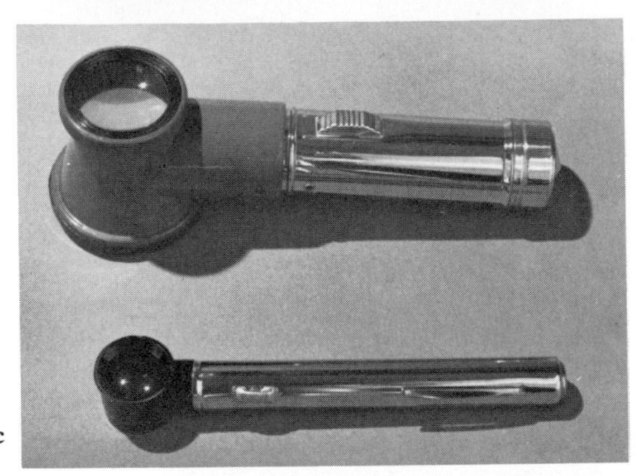

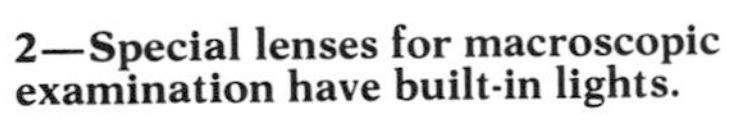

2—Special lenses for macroscopic examination have built-in lights.

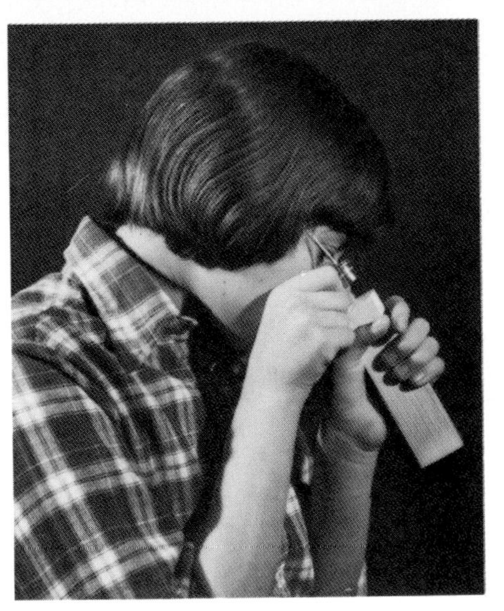

3—To examine the surface of a wood sample, hold the lens close to the eye, then bring the sample toward the lens until the surface comes into focus. Focus is maintained by butting the hands together and bracing them against the cheek.

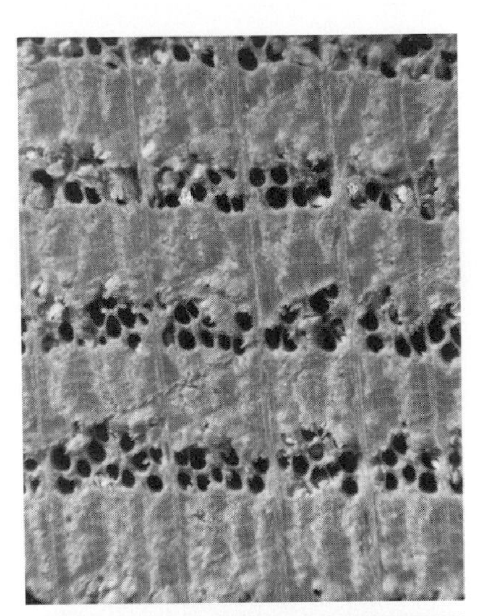

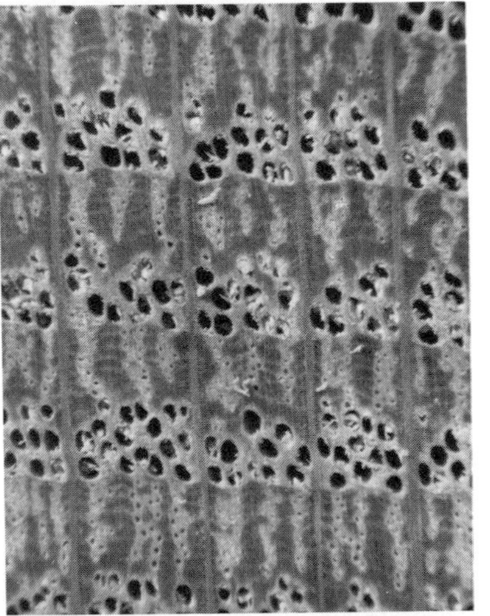

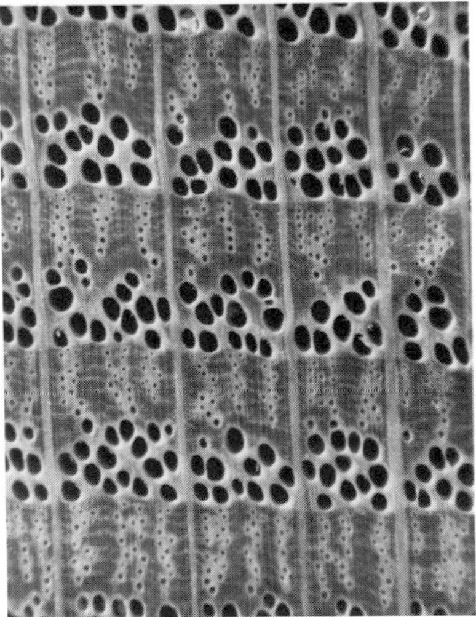

4—The cleaner the cut, the more detail revealed. Compare these cross sections of red oak which were cut with saw (top), dull knife (center) and razor blade (above).

study is required to gain skill in using the microscope and sufficient acquaintance with all the necessary features, as well as with the extent of variation to be expected among individual pieces of the same species. Many excellent references on the subject are listed in the bibliography.

As a compromise, however, identification based on **macroscopic features** (those discernible under slight magnification) is perhaps best suited to the needs of the average woodworker. It has the advantage of requiring only simple equipment *(1)* and yet the number of woods that can be identified far surpasses what can be mastered with the naked eye. The consistent and unique combinations of anatomical features in hardwoods make macroscopic identification quite effective. In addition to the end-grain "cellular fingerprint," other obvious features such as color, luster, density and hardness are also considered.

The standard magnifier used for macroscopic identification is the 10-power (10×) lens. The most common type is shown in Figure *1*; some special lenses have built-in lights *(2)*. Hold the lens close to the eye in good light, then move the piece of wood toward the lens until it comes into focus. By butting hand to hand and hand to cheek, you'll be able to hold the eye, the lens and the wood sample in constant position with maximum visibility of the cell structure *(3)*.

The most common pitfall in identification work is not producing a cleanly cut surface. In order to see cell structure clearly, the end grain must be severed by a flat, clean slice as free as possible of cellular disturbance. Figure *4* shows the importance of smooth cutting. In principle, any sharp tool that can make a slicing cut will suffice. An ordinary pocketknife, if superbly sharpened, will do nicely. Few of us, however, do that good a job of sharpening. My favorite routine is to whittle down the wood so the area to be cut forms a small plateau of perhaps 3/16 in. square. I then make a pass or two with a single-edge razor blade for the final surface. In most woods, the area to be cleaned up need be only a small area within a single growth ring. In other cases, as when looking for resin canals, a larger area of the sample may have to be surfaced.

The best razor blades to use are those sold for industrial purposes like slicing leather or scraping; the edge on blades sold for shaving is too fragile for any but the softest woods. By holding the wood with fingers well below the surface to be cut and by butting the knuckles of the third, fourth and fifth fingers against the piece, the single-edge blade can be safely slid across the wood surface *(5)*. On uneven-grained woods, it is important to orient the blade edge perpendicular to the growth rings so that at any moment of cutting, the blade edge engages equal amounts of earlywood and latewood. If the blade is held parallel to the rings, it will advance unevenly as the blade alternately encounters earlywood and latewood, and a washboard surface will result. The secret is sliding the edge. Most woods can be cut as they are, although it helps to moisten some woods or to soak the most difficult ones in hot water. Transverse surfaces are the most difficult to cut cleanly; radial and tangential surfaces are much easier.

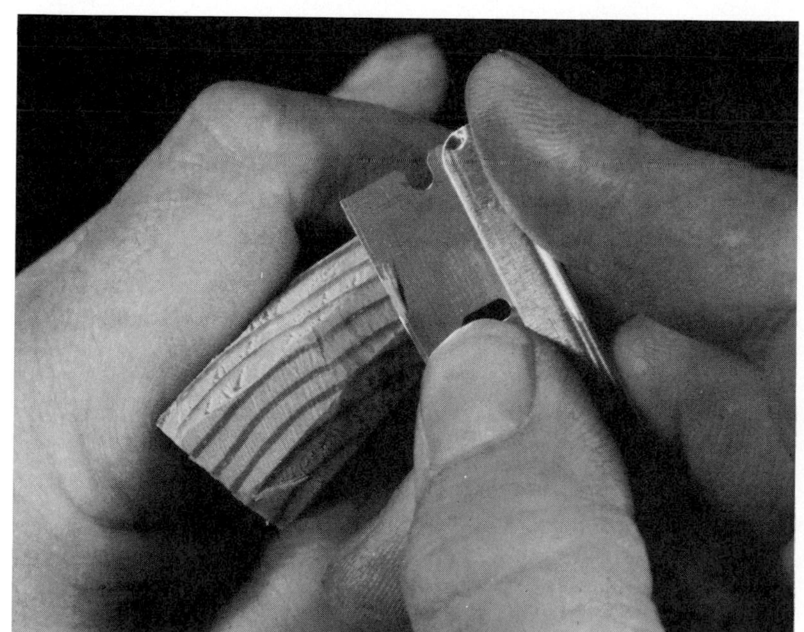

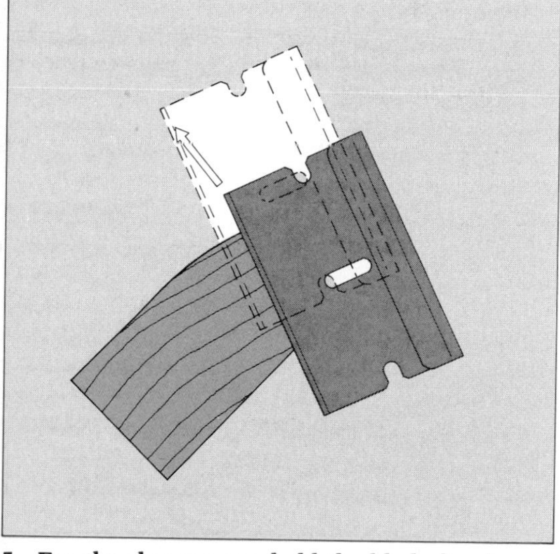

5—For the cleanest cut, hold the blade firmly and move it in the direction of the arrow to produce a sliding, slicing cut. Make a thin cut on a very small area. Moistening the surface may help.

What to look for

In learning to identify woods, first gain an acquaintance with the pertinent structural features, both gross and macroscopic. These features are discussed below, with summary review of those I've already described. The features described are pointed out in photos of representative species, pages 47 to 65.

Sapwood/heartwood—If a piece has both heartwood and sapwood, the combination is usually discernible. If only one is present, a distinct or dark color may indicate heartwood; a light, neutral color may be sapwood. Heartwood color is important in identification.

Growth rings—The width of growth rings is sometimes a salient characteristic; for example, rings are typically narrow in ponderosa pine, usually wide in willow. In some species wide variation may exist, as in redwood, and wood samples used for identification should avoid extremes where possible, because their anatomical features may be atypical.

The degree of evenness of grain is usually characteristic for a species. Where reasonable unevenness exists, the nature of the transition from earlywood to latewood may also be significant. An **abrupt transition** indicates a sharp delineation between the larger, thin-walled cells of the earlywood to the smaller, thick-walled cells of the latewood; the term **gradual transition** indicates no clear delineation between earlywood and latewood.

Special features of growth rings, such as the fluted contours typical of butternut and blue beech *(1)* and basswood may be important.

Texture—Texture refers to the relative tracheid diameters in softwoods or the relative pore size in hardwoods. As a macrofeature for classification, fine, medium and coarse texture are the customary terms.

Resin canals—Resin canals are present in four genera of softwoods—pines, spruces, larches and Douglas fir. The number, distribution and size of resin canals are characteristics used for identification *(2)*.

Pores—In hardwoods, pore size, distribution, arrangement and abundance are important identification aids. Distribution of pores within the growth rings—i.e., ring-porous, semi-ring-porous or diffuse-porous—is the customary initial consideration. As seen in cross section, localized groupings can also have characteristic arrangements. **Solitary pores** appear singly, bounded entirely by other cell types. **Wavy bands of pores**, as in elm or hackberry, form undulating concentric lines in the

1—Fluted or wavy growth rings are characteristic of butternut, left, and blue beech, right.

latewood and are most convenient identification features. **Pore multiples**, common in hickories and birches, refer to two to several pores in close radial arrangement. **Pore chains** are long radial rows in actual contact or apparent close arrangement, as in holly. Less specific radial arrangements appearing as flame or fan-shaped groupings or irregular lines (oaks, chestnut, chinkapin) may likewise be characteristic.

Parenchyma—Because of their small size, parenchyma cells are individually indiscernible even with a hand lens. In cross section, numbers of cells arranged as lines or characteristically shaped patches may show up as lighter-colored tissue in contrast to the darker fiber masses. (Tracheids, which are of about the same size and cell-wall thickness, may also be intermixed with parenchyma in lighter-colored tissue in oaks, ashes and chestnut.) **Terminal parenchyma** designates formation of this tissue at the end of the growth ring. Yellow poplar, for example, is the epitome of diffuse-porous structure. Growth rings might not be recognizable, were it not for the distinct line of terminal parenchyma. This feature is also present in true mahogany *(Swietenia* spp.). Lines of parenchyma within the growth ring are termed **banded parenchyma**. These are clearly visible in all species of hickory. Shorter or less distinct lines of banded parenchyma are features of some species, such as the oaks. Light-colored tissue associated with pores (paratracheal parenchyma) is said to be **paratracheal vasicentric** if it encircles a single pore. In some cases, the parenchyma that encircle adjacent pores merge, forming **confluent** parenchyma, as in the latewood of white ash. The distinction between latewood pores joined by confluent parenchyma and the wavy bands of pores (as in elm) should be noted as an identification aid.

Rays—In softwoods, rays are not distinctive macroscopic features. In hardwoods their appearance is important in all three planes. In cross section their size (especially in relationship to pore size), apparent number and spacing, and visual distinctiveness (they range from distinct to the naked eye to invisible even with a lens) are important. In some species the apparent straightness of the rays versus a tendency to weave through the pores may be characteristic. Ray height and the relative distribution of rays can best be judged on a tangential surface *(3)*. On radial surfaces **ray fleck** may be distinctive.

Tyloses—The presence or absence of tyloses as well as the relative abundance of tylosis formation is usually a consistent feature of hardwood pores.

2—Numerous, uniformly distributed resin canals can be easily seen in this piece of eastern white pine.

3—Distinctive ray patterns identify rulers as maple (above), beech (center) and sycamore (below).

Physical properties

Several physical properties that can be perceived with the naked eye or by unsophisticated methods are sometimes unique, either alone or in combination with other features, so as to aid greatly in identification.

Color—In terms of both **hue** and **shade**, color should routinely be noted in examining wood. It is sometimes unique for a particular species. In eastern redcedar, for example, the purple hue and deep shade of the heartwood make identification quite easy. Similarly, the heartwood color of redwood, black walnut, amaranth (purpleheart) and padauk (vermilion) are renowned. In some species wide variation may be normal. Yellow poplar has a fairly broad layer of creamy white sapwood. The heartwood is commonly medium to deep green, although variations ranging from pinkish or purplish to almost inky black are sometimes encountered. In some species, the color, although consistent, is common to a host of other species and therefore not distinctive. It should also be noted that surface color is greatly affected by exposure to daylight and air and may change drastically with time. For example, the deep purple of redcedar ages rapidly upon exposure to daylight to a nondistinct medium brown.

Luster—Light reflectiveness or sheen is occasionally distinctive for a particular species. The side-grain surfaces of spruce are considered lustrous in relation to those of redwood, which by comparison are dull.

Odor—Although distinctive for only a few woods, odor may be a most useful feature. Species for which it is an especially noteworthy trait include eastern redcedar, Douglas fir, sassafras, teak and incense cedar. The ability to detect odor varies among persons, but some find the more subtle odors of basswood, bald cypress and catalpa distinctive enough to be helpful in identification. Odor is most pronounced in freshly cut material and may be strengthened by moistening the wood. Unfortunately, it can also be transferred to adjacent wood by close contact. Many a wood identification student is dismayed to learn that an entire collection of wood samples packed in a tightly closed box all smells like the one eastern redcedar sample.

Density and hardness—These physical properties relate to the relative weight of wood and its surface compressive strength properties, respectively. Specific numerical data on these properties are available *(Table 3,* page 8), but they are also helpful in a grossly general sense. A close correlation exists between density and hardness, and if a numerical rating of density is known for a species, hardness can often be approximated. Running one's thumbnail perpendicular to the grain direction across a side-grain surface gives a surprisingly meaningful measure of hardness. The difference in hardness between butternut and black walnut, for example, can usually be detected with a little practice.

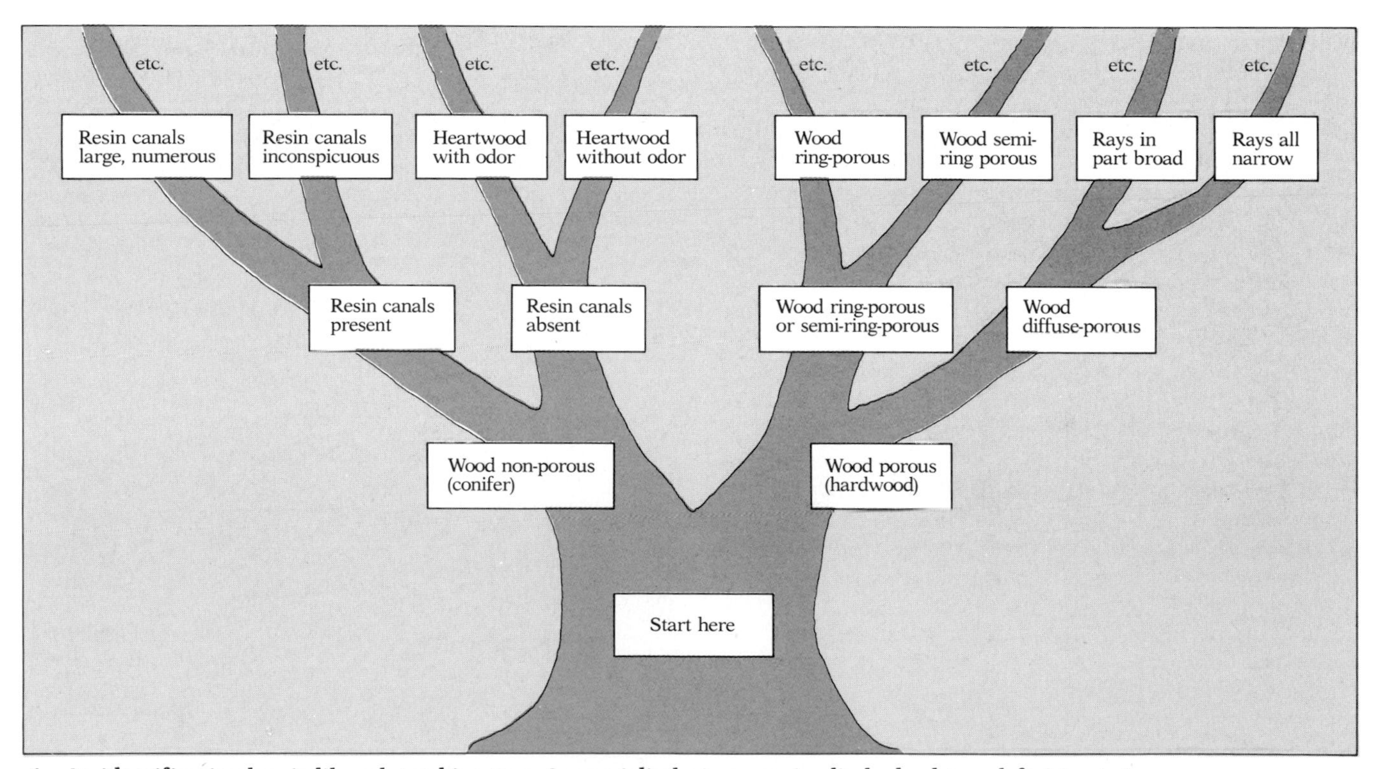

1—An identification key is like a branching tree. As one 'climbs,' successive limbs lead to a definitive twig.

Identification techniques

There is no best or single technique for identifying wood. What one does depends on the degree of difficulty of the particular wood and the experience and skill of the individual. The available aids include known samples of wood, manuals and photographs of wood characteristics, and identification keys. Each has advantages and disadvantages.

Collections of known samples are extremely valuable, for they allow direct viewing of features that may be difficult or impossible to describe accurately in words. They can be used for direct visual comparison by holding them side-by-side with the unknown being considered. They offer a palpable sense of the wood's physical properties. But collections of single samples are risky, because the observer has no way of measuring how typical each sample is. It is a good idea to save small pieces of stock, drill holes in them and string them together. Accumulating such multiples will aid immeasurably in sensing the variation within a species.

Good photographs of wood structure taken at an appropriate magnification are nearly as helpful as actual samples. An advantage is that the woods photographed were probably chosen to typify the species.

Verbal descriptions of the anatomical features of particular species appear in various texts and identification books. They may include general information, physical properties and source, as well as gross, macroscopic and microscopic features, and commonly are accompanied by macro and microphotographs of wood surfaces or sections. The information is systematically arranged for convenient comparison with the sample in question. Such material is most useful in confirming a tentative identification or in comparing two possible choices. The difficulty is in narrowing down the choices enough to locate the definitive description.

The use of **keys** is another approach to identification. An identification key can be thought of as an information tree *(1)*. The tree always branches into forks and the identity of a particular species is found at the tip of one of the twigs. To use the key you start at the bottom and proceed upward. At the first fork, you decide which limb to take by considering the two choices given, each a statement of wood characteristics. You take the limb that correctly describes the wood in question and ascend to the next fork. And so on. Eventually you reach a twig where the identity of your wood is given.

Keys usually appear as paired, numbered statements *(2)*. Based on the features of the wood being examined, the user of the key chooses the appropriate statement in each pair. This choice in turn leads to another choice and so on until a species is reached. The portion of a typical key reproduced at right indicates their general form. (Several excellent keys are mentioned in the bibliography.)

If you follow a key through either forward or back-

IDENTIFICATION KEY

1. START: Wood without pores; rays not distinct without using lens. **2**
1. START: Wood with pores; rays sometimes visible to the naked eye. **14**

2. Resin canals visible as light or dark colored dots on cross section or as tiny interrupted streaks on radial or tangential surface. **3**
2. Resin canals not easily visible to the naked eye or with a hand lens. **7**

3. Wood rather hard to cut across grain, latewood usually prominent on cross section. **4**
3. Wood softer, easier to cut, latewood not as prominent. **5**

4. Numerous resin canals, wood shades of yellow brown. . . . . **Southern pines**
4. Resin canals not numerous, wood with an orange-red cast; when moistened, wood has a distinct odor. . . . . **Douglas fir**

5. Latewood distinct, sharp boundary with earlywood in the same growth ring, latewood clearly contrasted on tangential surface. Wood frequently dimpled (appear as indentations). Resin canals appear as pale dots, not holes, on cross section. . . . . . **Ponderosa pine**
5. Latewood indistinct, gradual boundary with earlywood in the same growth ring, not contrasty on tangential surface. Resin canals appear as open holes on cross section. **6**

6. Resin canals large and numerous, appearing as brown streaks on radial or tangential surfaces. . . . . **Sugar pine**
6. Resin canals smaller and less numerous, not as prominent on radial or tangential surfaces. . . . . **Western or eastern white pine**

7. Heartwood a dark shade of brown, red brown, or purplish brown. **8**
7. Heartwood a light shade of tan or cream. **11**

8. Purplish to rose brown, frequently with small spots of cream-colored wood enclosed. Cells small, wood smooth-cutting. Distinctive cedar-chest odor. . . . . **Eastern redcedar**
8. Dull brown to reddish brown, cells larger; or if orange brown, more resistant to cutting. **9**

2—Part of a typical identification key. Enter the key at start and make a choice between the two options. That choice will lead you to another choice, and so on.

ward, you can compile a list of wood features for a particular species. Some keys are based solely on macroscopic features, some solely on microscopic features, and others on a combination of both.

There are two serious drawbacks to keys, however. The first is that if a wrong choice is made, you can proceed on an incorrect path to an incorrect solution without realizing it. The novice often goes astray because many of the necessary choices are based on subjective considerations and judgment developed through previous acquaintance with wood anatomy. The second pitfall is trying to identify a wood that isn't even on the key being used. This can happen, especially with a totally unknown wood.

Choose keys carefully. Try to find fairly detailed ones that include the woods of a given geographic area. Beware of keys that are abbreviated or cover a small number of species. A key for the "ten most common woods of the United States" would surely invite trouble, since you would be quite likely to encounter a wood that isn't included.

After considerable practice in identification and after becoming familiar with routine wood features, you can usually tell when a wood "isn't keying out right" because it isn't in the key. But this further suggests that keys are better for the advanced than for beginners. As you get more and more practice with common woods, you quickly recognize an exotic wood when you encounter it.

Identification is important in the woodshop. A woodworker may want to identify a piece of scrap stock or need to know the identity of a broken chair part for a correct replacement. Perhaps the species of wood in an antique must be determined. We must assume the woodworker is acquainted with gross and macroscopic anatomical features and with the woods he has used before. When the piece is examined, certain features should be noted immediately—hardwood or softwood, evenness of grain, tyloses in the pores, parenchyma arrangements, size of rays, and so on. Certain species are automatically eliminated as the list of possibilities is mentally narrowed down. The exercise usually concludes at one of these three points:

1. The wood appears to be species *A*, based on features familiar to the observer from previous experience and which are clearly definitive of the species. Comparison with a known sample and review of a written description help confirm the choice—or reject it.
2. The wood is probably species *A* or *B* or *C*, or might even be species *D*. Again, samples of the possible choices should be examined and descriptions reviewed, being especially alert for distinguishing features. In both of these cases, the person's experience has in effect mentally traversed most of the keys, arriving at or near the end. Plowing through a key would have been a tedious waste of time.
3. The features observed do not suggest any answer. The wood is definitely unfamiliar. At this point the keys are the thing to use. When an answer is finally arrived at, the choice should be carefully compared with check samples, if available, or with written descriptions or photographs. Extreme care must be taken in deciding whether the identification is indeed correct or whether the wood might in fact be a species not even included in the key.

The importance of circumstantial evidence cannot be overemphasized. For example, in the identification of timber excavated from the site of an early 19th-century sawmill in western Massachusetts, the half-decayed fragments could be identified only as hemlock *(Tsuga* sp.). Because of where the wood was found, it was easy to conclude that it was eastern hemlock *(T. canadensis)* rather than western hemlock *(T. heterophylla)*. Likewise, under similar circumstances, the sections shown in Figure *1* on page 38 could be narrowed to eastern white pine *(Pinus strobus)* rather than western white pine *(P. monticola)*, red pine *(P. resinosa)* rather than Scotch pine *(P. sylvestris)* and pitch pine *(P. rigida)* rather than other hard pines *(Pinus* spp.).

Macrophotographs

Presented in the next 18 pages are macrophotographs of 54 common species. Although inclusion of a complete wood identification manual and keys is beyond the scope of this book, this section should give the reader a survey of basic macroscopic anatomical features, which are critical to identification and understanding wood in general. The reader is urged to use the photos, along with known or unknown samples viewed with a 10× hand lens, as a practice exercise. Photos approximate the resolution possible with an ordinary 10-power hand lens. The accompanying descriptions cover both gross and macroscopic features and properties.

DOMESTIC SOFTWOODS

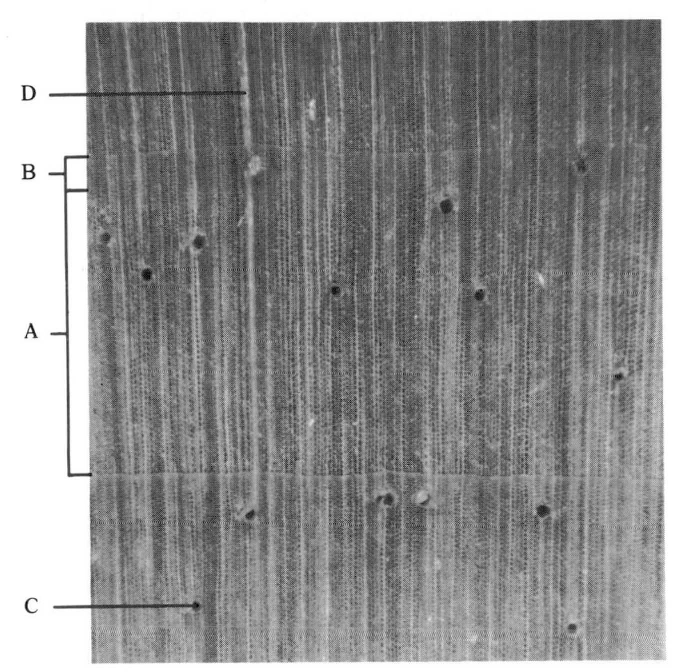

Eastern white pine *(Pinus strobus)*

Moderately soft and light (average specific gravity 0.35). Heartwood creamy white to light brown, often with a reddish tinge, aging to a much darker color. Pleasant resinous odor. Medium texture. Fairly even grain, growth rings indistinct to distinct. Earlywood (A) occupies majority of growth ring, with gradual transition to narrow band of latewood (B). Resin canals (C) medium-sized, visible to naked eye, numerous and uniformly distributed, but confined mainly to outer portion of growth rings. Rays of two sizes: narrow rays visible with hand lens; rays containing horizontal resin canals (D) sometimes visible without lens.

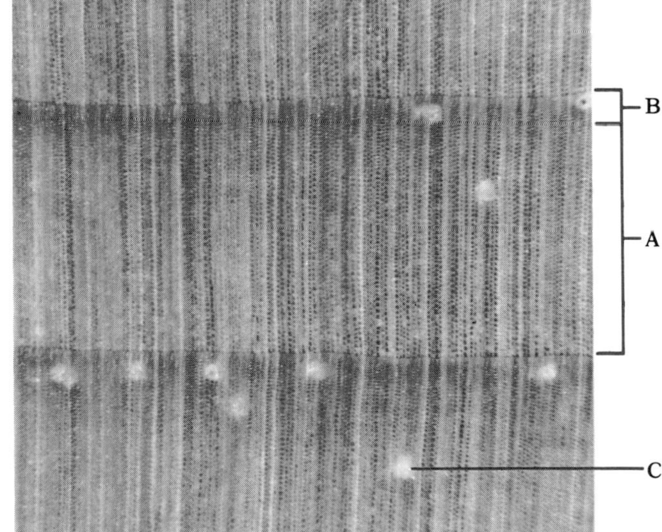

Western white pine *(Pinus monticola)*

Moderately soft and light (average specific gravity 0.38). Heartwood cream-colored to light or reddish brown, darkening with age. Faint resinous odor. Medium to medium-coarse texture. Fairly even grain, growth rings distinct with gradual transition from earlywood (A) to denser but narrow latewood (B). Resin canals (C) medium large, visible to unaided eye, numerous, uniformly distributed in outer portions of growth rings. Rays of two sizes: narrow rays visible only with hand lens; large rays containing resin canals visible with unaided eye.

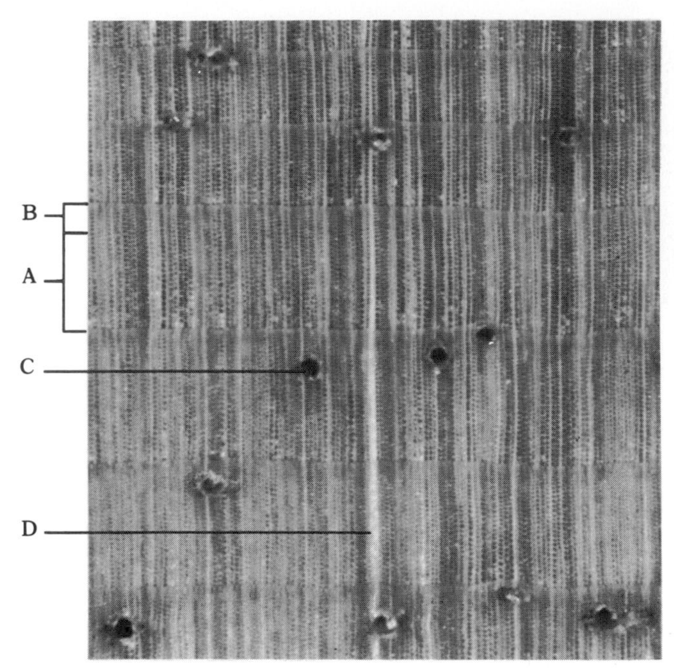

Sugar pine *(Pinus lambertiana)*

Moderately soft and light (average specific gravity 0.36). Heartwood buff or light brown, sometimes with reddish tinge. Faint resinous odor. Texture medium-coarse to coarse. Fairly even grain with gradual transition from earlywood (A) to narrow latewood (B); growth rings distinct. Resin canals (C) numerous, uniformly distributed, largest among coniferous species, clearly visible to naked eye; contents migrating onto wood surfaces or into adjacent cell structure often make resin canals even more conspicuous. Rays of two widths: larger rays containing transverse resin canals (D) visible with unaided eye. Transverse resin canals usually appear as distinct dark specks on tangential surfaces.

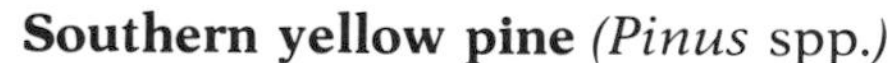

Southern yellow pine *(Pinus spp.)*

Moderately hard and heavy (average specific gravity 0.48 to 0.59). Heartwood ranging from shades of light or yellow brown to orange or reddish brown. Wood resinous, and with distinct resinous odor. Texture medium. Extremely uneven grain; annual rings distinct and prominent on all surfaces. Abrupt transition from earlywood (A) to very dense latewood (B) which commonly occupies one-third to one-half the annual ring. Resin canals (C) numerous, medium-large, visible to naked eye, uniformly distributed through latewood portions of growth rings. Rays of two widths: larger rays containing transverse resin canals visible to unaided eye.

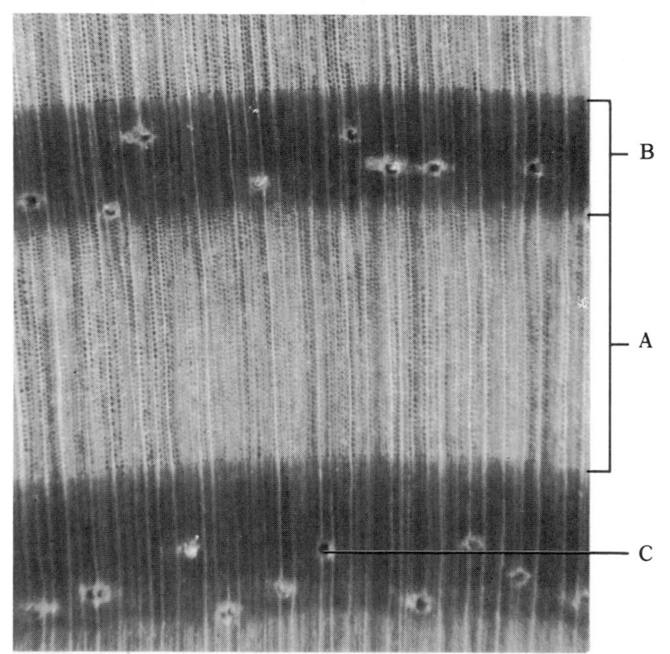

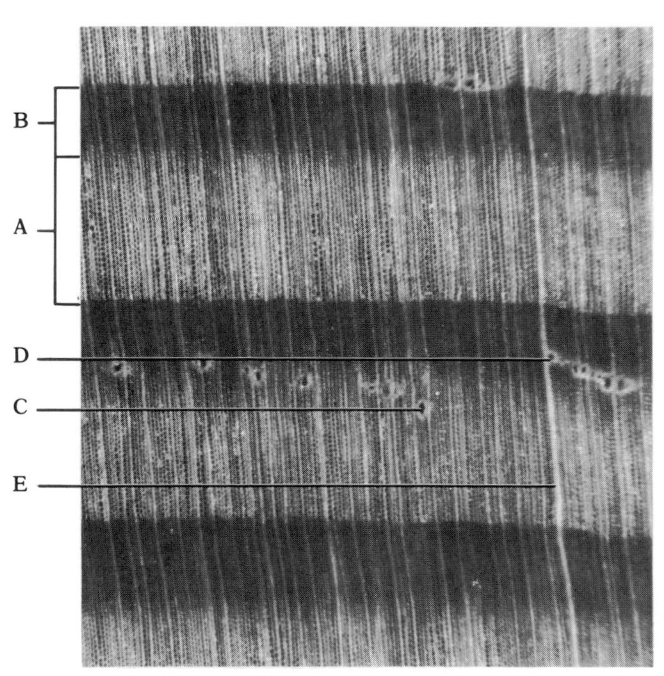

Douglas fir *(Pseudotsuga menziesii)*

Moderately hard and heavy (average specific gravity 0.48). Heartwood orange-brown to deep reddish brown or sometimes yellowish brown. Characteristic resinous odor. Texture medium to medium-coarse. Uneven grain, earlywood (A) usually wider than latewood (B), with abrupt transition. Resin canals (C) indistinct to visible to unaided eye, but distinct with hand lens, sporadically distributed, frequently in tangential groups (D) in the latewood. Rays of two widths: those with transverse resin canals (E) barely visible to the naked eye.

Tamarack *(Larix laricina)*

Moderately hard and heavy (average specific gravity 0.53). Heartwood yellowish brown to reddish or russet brown. Wood has a somewhat waxy or greasy feel. Texture medium. Uneven grain, abrupt transition from earlywood (A) to latewood (B). Resin canals small, inconspicuous or invisible to naked eye, sparse or sporadically distributed, mainly in latewood, solitary (C) or in tangential groups. Rays of two sizes visible with hand lens.

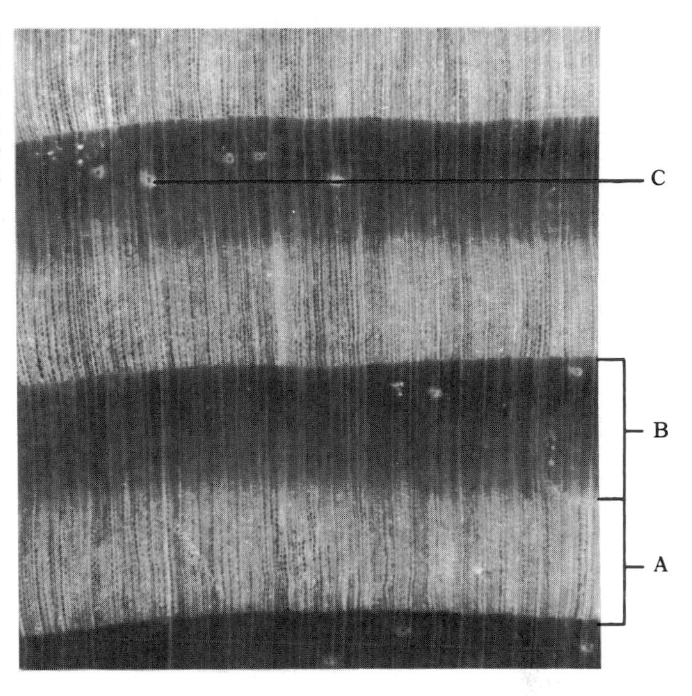

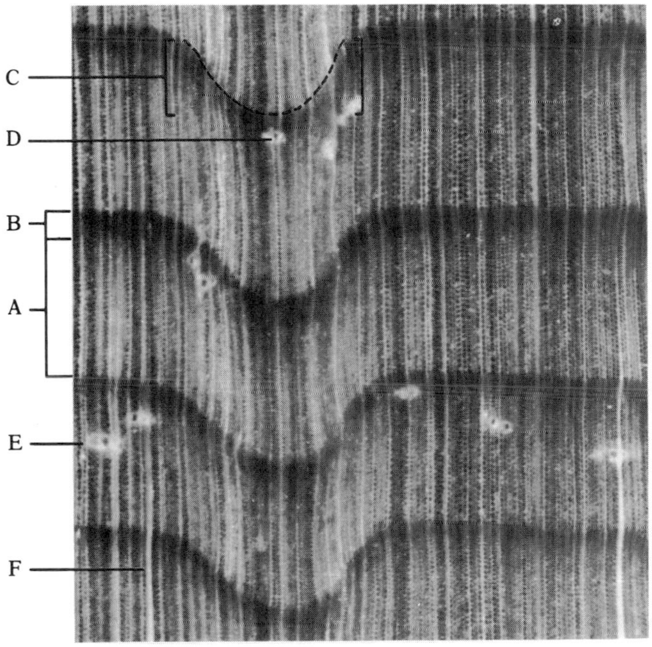

Sitka spruce *(Picea sitchensis)*

Moderately soft and moderately light (average specific gravity 0.40). Heartwood yellowish to pale brown, commonly with a pinkish or purplish tinge. Texture medium. Moderately even grain, earlywood (A) occupying one-half to two-thirds of the growth ring, transition to latewood (B) gradual to semigradual. Ring indentations (C) sometimes present. Resin canals (D) present, medium-sized but sporadically distributed, often in multiples (E). Rays of two sizes: larger rays (F) with transverse resin canals visible to the unaided eye.

Eastern spruce *(Picea* spp.*)*

Moderately soft and moderately light (average specific gravity 0.40). Wood creamy white or pale yellowish brown (sapwood not distinct from heartwood), lustrous. Texture medium-fine. Moderately even grain, growth rings fairly distinct, transition gradual from earlywood (A) to narrow latewood (B). Resin canals very small, usually not visible to unaided eye, appearing as pale yellow or whitish dots with hand lens, unevenly distributed, solitary (C) or in small tangential multiples (D). Rays of two widths: narrow rays visible only with hand lens, larger rays with resin canals barely visible to the unaided eye.

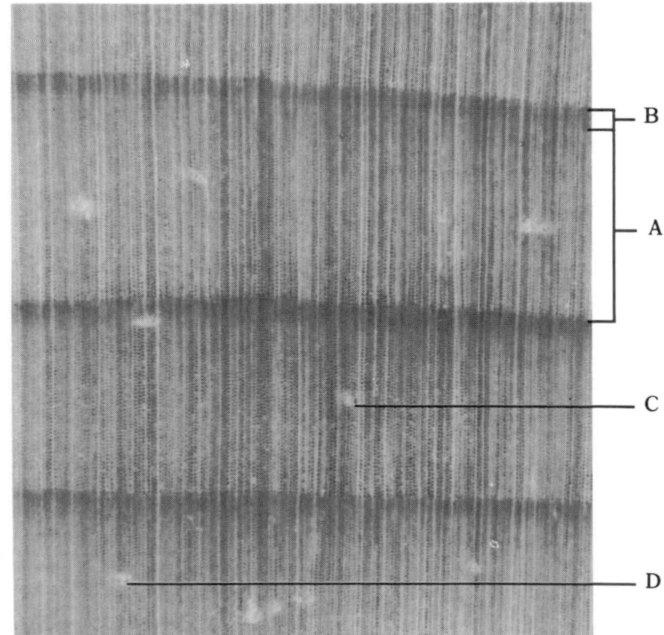

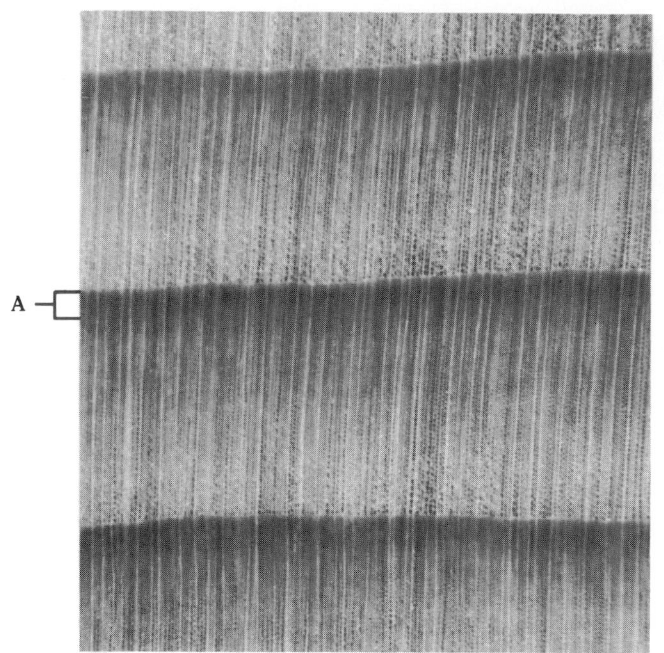

Eastern hemlock *(Tsuga canadensis)*

Moderately hard but moderately light (average specific gravity 0.40). Wood buff or pale brown, sometimes with a faint reddish or purplish tinge. (Sapwood not always distinct from heartwood.) Texture medium. Fairly uneven grain. Latewood (A) distinctly dense, occupying one-third or less of the ring; transition variable, semi-abrupt to abrupt. Normal resin canals absent. Rays very fine, not visible to unaided eye.

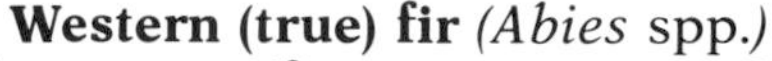

Western (true) fir *(Abies spp.)*

Moderately soft and light (average specific gravity 0.35). Wood pale buff to light brown, the darker latewood portion of the growth ring having a purplish cast (sapwood not distinct from heartwood). Texture medium-coarse, moderately even to moderately uneven grain, growth rings distinct, transition generally gradual to latewood (A). Normal resin canals absent. Rays very fine, indistinct to the unaided eye.

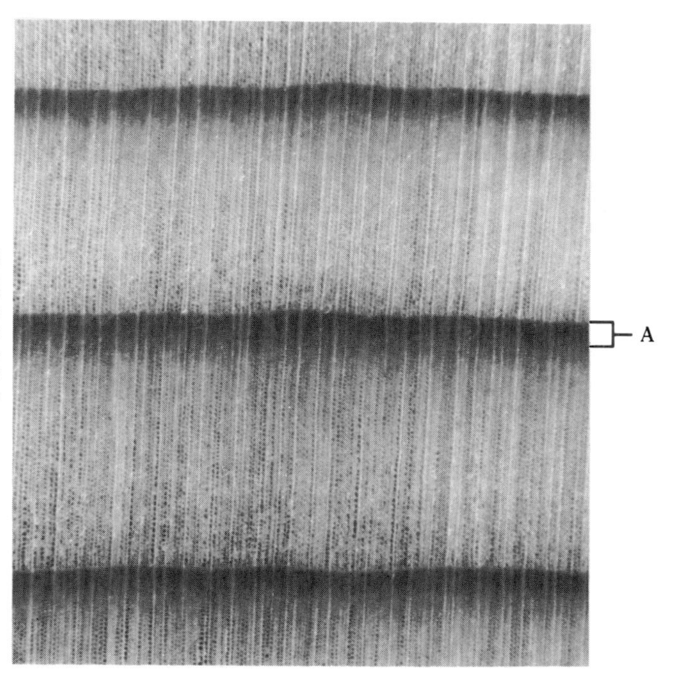

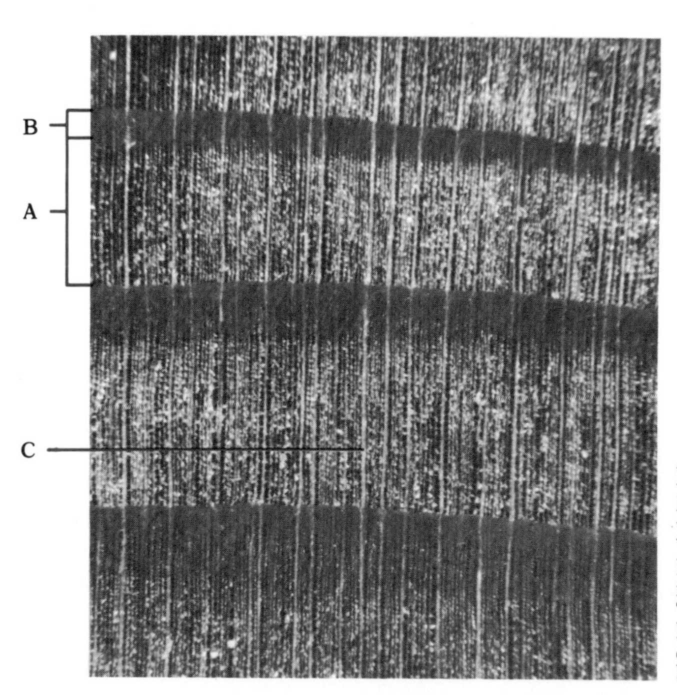

Redwood *(Sequoia sempervirens)*

Moderately soft and light (average specific gravity 0.37). Heartwood variable in color from light cherry to deep reddish brown. Very coarse texture; individual tracheids clearly visible with hand lens on cleanly cut cross-sectional surfaces. Ring width variable from wide to very narrow; grain fairly uneven to moderately even. Transition from earlywood (A) to latewood (B) abrupt. With hand lens, parenchyma visible on longitudinal surface as strands of dark reddish specks. Rays (C) distinct as light lines against darker background on cross section.

Western redcedar *(Thuja plicata)*

Relatively soft and light (average specific gravity 0.32). Heartwood medium to dark coffee-brown. Characteristic cedar odor. Medium texture. Moderately uneven grain, growth rings distinct. Abrupt transition from earlywood (A) to narrow latewood (B). Parenchyma not visible or barely visible with hand lens, sometimes appearing as an indistinct tangential line in latewood. Rays (C) fine, inconspicuous without hand lens.

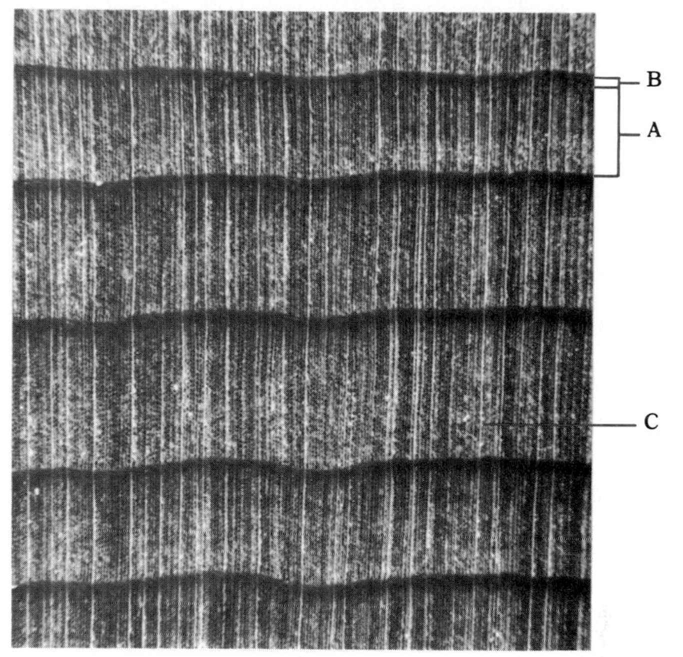

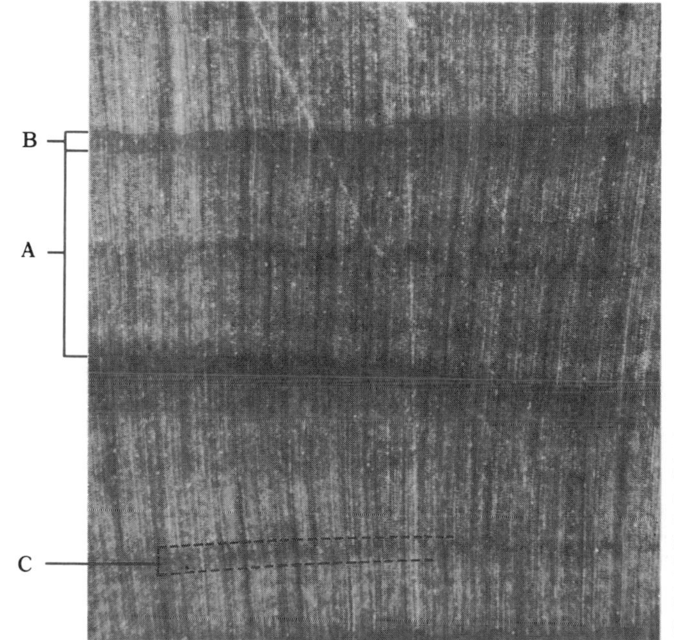

Eastern redcedar *(Juniperus virginiana)*

Moderately hard and moderately heavy (average specific gravity 0.47). Heartwood vivid purplish red when freshly cut, fading to drab reddish brown with exposure; heartwood often with streaks or pockets of included creamy light sapwood. Characteristic fragrant odor. Texture very fine. Even grain, growth rings evident due to slightly denser latewood. Transition from earlywood (A) to latewood (B) gradual to semi-abrupt. Parenchyma with dark contents numerous, forming one or more dark zones (C) approaching false rings within the growth ring, sometimes visible to the unaided eye or distinct with hand lens. Rays very fine.

DOMESTIC HARDWOODS

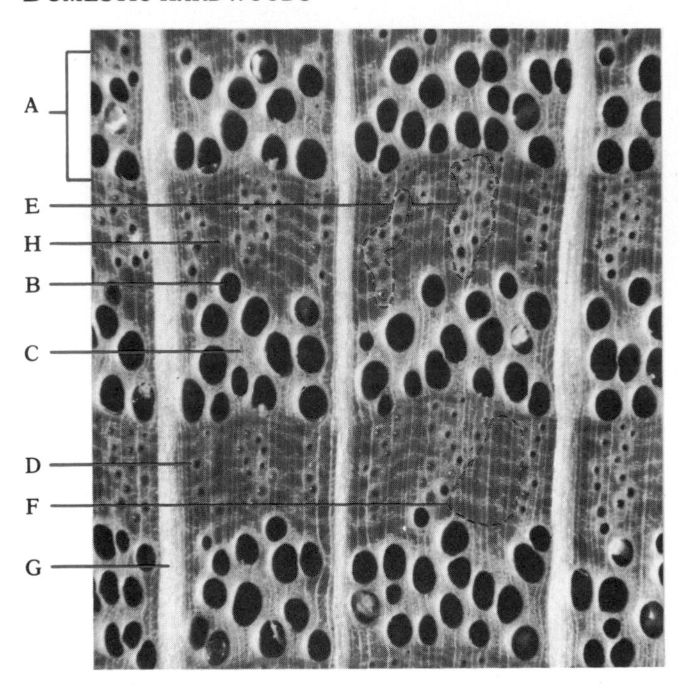

Red oak *(Quercus spp.)*

Hard and heavy to very heavy (average specific gravity 0.63). Heartwood light reddish brown, often with a pinkish or flesh-colored tinge. Ring-porous, with distinct earlywood (A), up to four pores in width, transition from earlywood to latewood abrupt. Earlywood pores distinct to the naked eye, usually lacking tyloses (B), and surrounded by lighter tissue (C). Latewood pores (D) distinct with hand lens, "few enough to be countable," and arranged in single or branching radial rows, surrounded by lighter tissue (E). Tangential lines of parenchyma often visible against darker fiber mass in latewood (F). Extremely large rays (G) visible easily to the naked eye; small rays (H) just discernible with hand lens. (On tangential surfaces, large rays average ⅜ in. to ¾ in. in height, tallest only occasionally more than 1 in.)

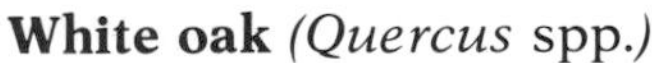

White oak *(Quercus spp.)*

Hard and heavy to very heavy (average specific gravity 0.67). Heartwood light to dark brown, often with a greyish cast. Ring-porous with distinct earlywood (A) up to three pores wide, transition to latewood usually distinct. Large earlywood pores (B), usually with abundant tyloses in heartwood, and surrounded by lighter-colored tissue (C). Latewood pores (D) from distinct to indistinct, "too numerous to count," arranged in radial groups intermingled with lighter tissue (E). Tangential lines of parenchyma (F) usually visible against darker fiber mass in latewood. Extremely large rays (G) visible to naked eye; narrow rays (H) barely discernible with hand lens. (On tangential surfaces, rays commonly exceed 1 in. in height, tallest may be several inches.)

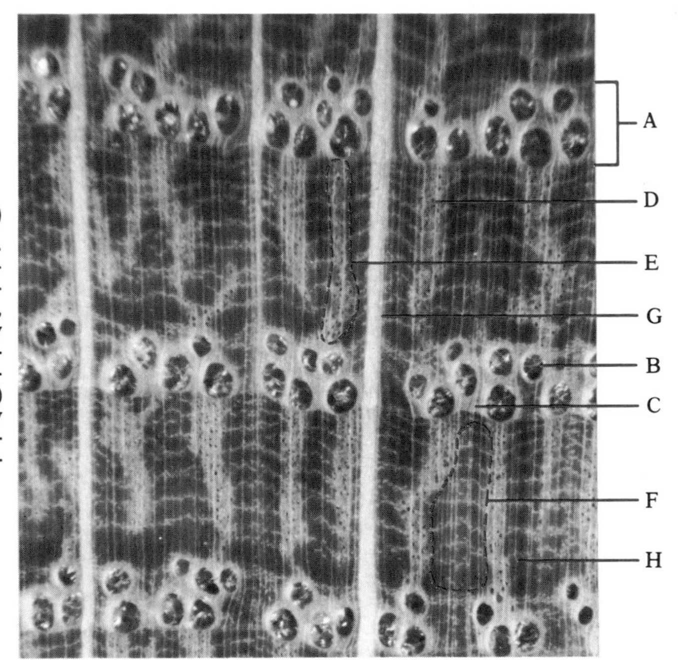

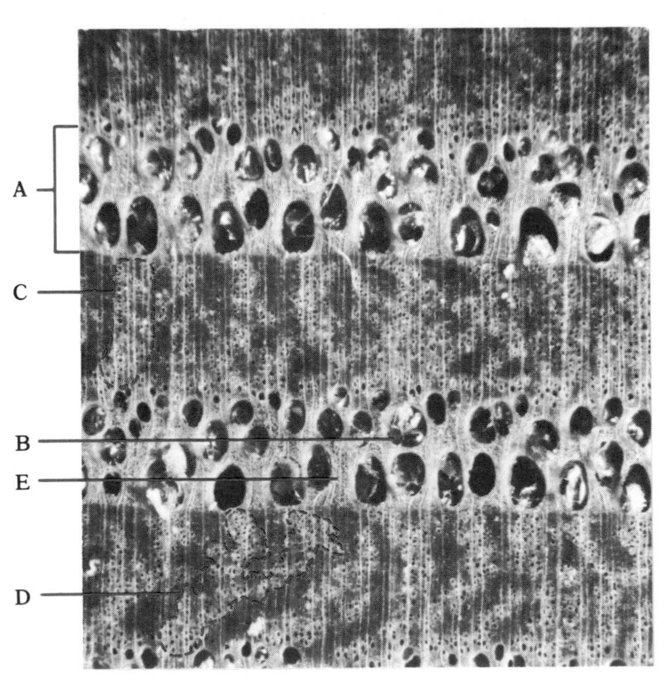

American chestnut *(Castanea dentata)*

Moderately soft and moderately light (average specific gravity 0.43). Heartwood light to medium or greyish brown. Wood ring-porous. Earlywood zone several pores wide (A), pores distinct to unaided eye, typically oval in the radial direction, with abundant tyloses (B). Latewood pores numerous, small, barely discernible with hand lens, and arranged in more or less radial (C) or wandering (D) patches. Parenchyma massed to form lighter background around earlywood (E) and latewood pores (C,D). Rays all fine, barely discernible with hand lens.

Golden chinkapin *(Castanopsis chrysophylla)*

Moderately hard and moderately heavy (average specific gravity 0.46). Heartwood light brown streaked with ginger or pinkish shades. Wood ring-porous, with earlywood (A) suggested by largest vessels in a single, but interrupted row (B). Transition to latewood distinct, or sometimes indistinct due to grading pore size (C). Latewood pores small, numerous, arranged in patches, spreading or branching radially (D). Rays uniformly fine, just visible with hand lens (E).

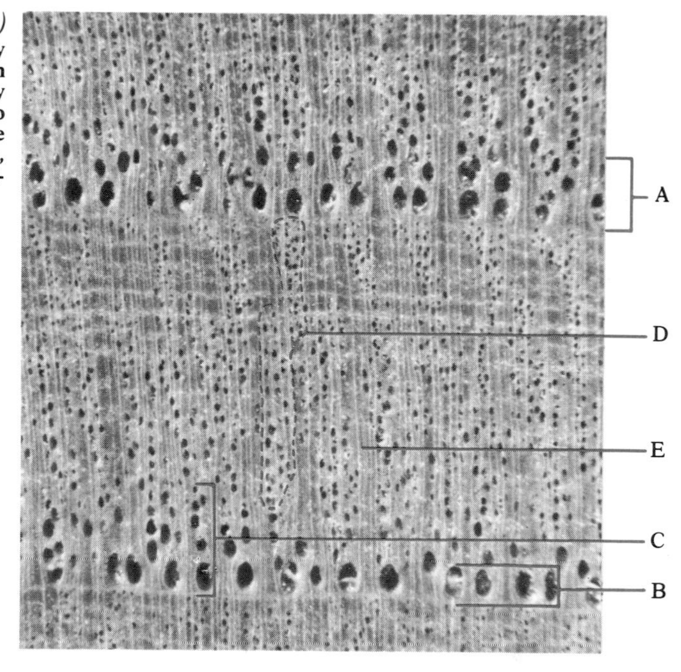

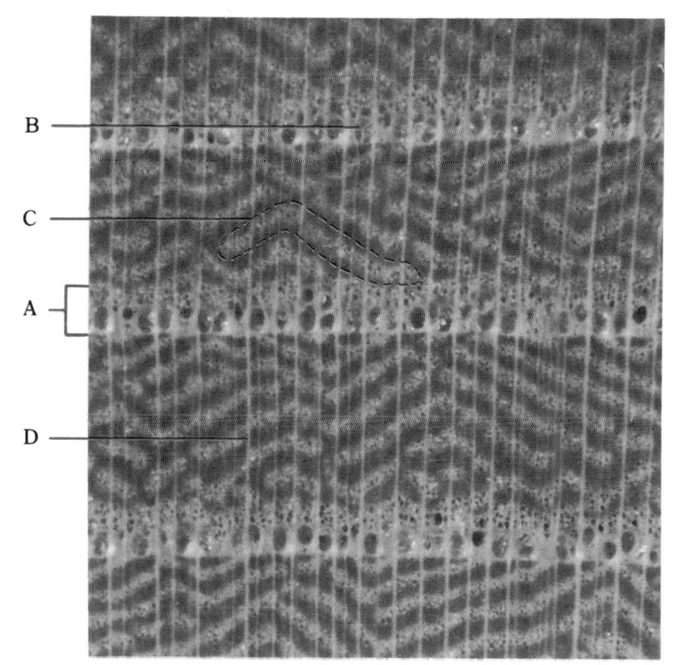

American elm *(Ulmus americana)*

Moderately hard and heavy (average specific gravity 0.50). Heartwood brown or reddish brown. Interlocked grain common. Ring-porous, earlywood comprises the largest pores usually in a single row (A); pores usually have abundant tyloses (B). Transition to latewood abrupt. Latewood pores small and numerous, arranged in wavy bands (C). Rays not distinct to the unaided eye but visible as uniformly narrow light lines with hand lens (D).

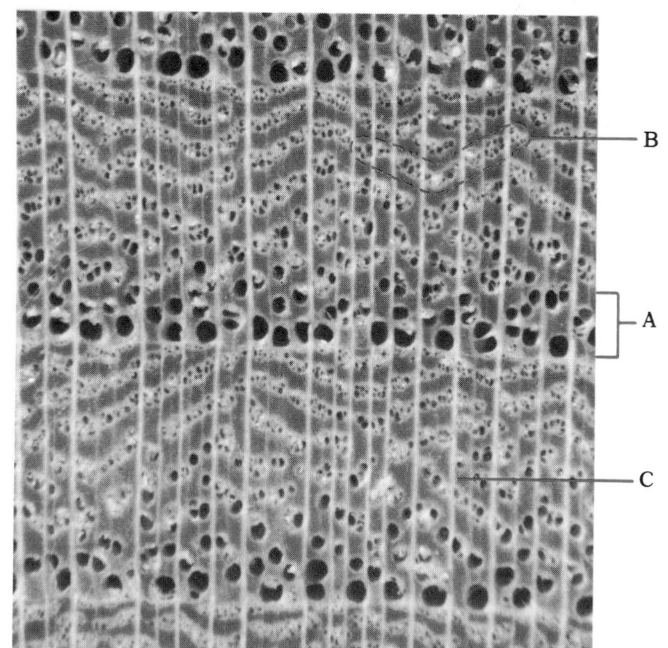

Hackberry *(Celtis occidentalis)*

Moderately hard and heavy (average specific gravity 0.53). Heartwood light greyish brown to yellowish brown, sapwood usually wide, light straw yellow, often with greyish or greenish cast and commonly blue-stained. Grain sometimes interlocked. Ring-porous, earlywood zone two to five pores wide (A), with fairly abrupt transition to latewood. Latewood pores numerous and arranged in wavy bands (B). Rays (C) distinct to the unaided eye.

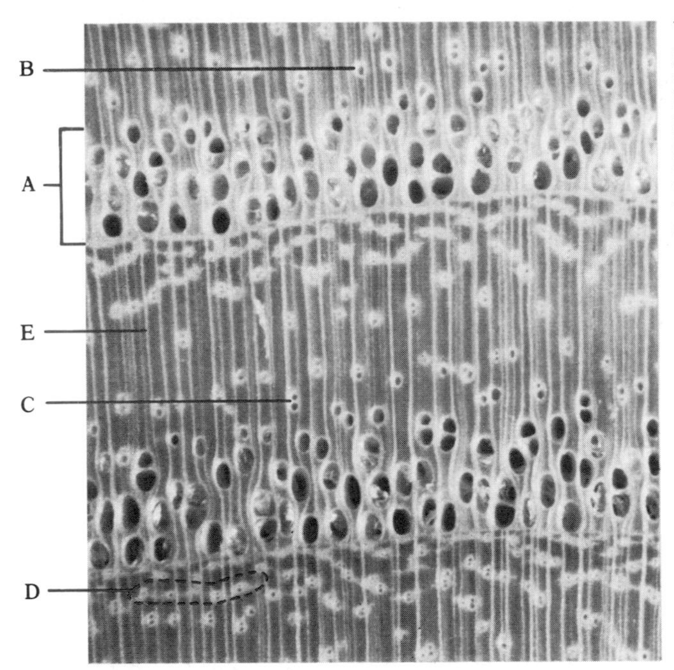

White ash *(Fraxinus americana)*

Moderately hard and heavy (average specific gravity 0.60). Sapwood creamy white, usually wide. Heartwood medium to greyish brown, often streaked or blotchy. Wood ring-porous, pores distinct. Earlywood zone two to four pores wide, pores surrounded by lighter-colored tissue (A). Abrupt transition from earlywood to latewood. Latewood pores solitary (B) or in multiples of two or three (C). Lighter parenchyma encircles latewood pores and unites them into lateral groups, especially in the outer latewood (D). Rays appear as fine light lines with hand lens (E).

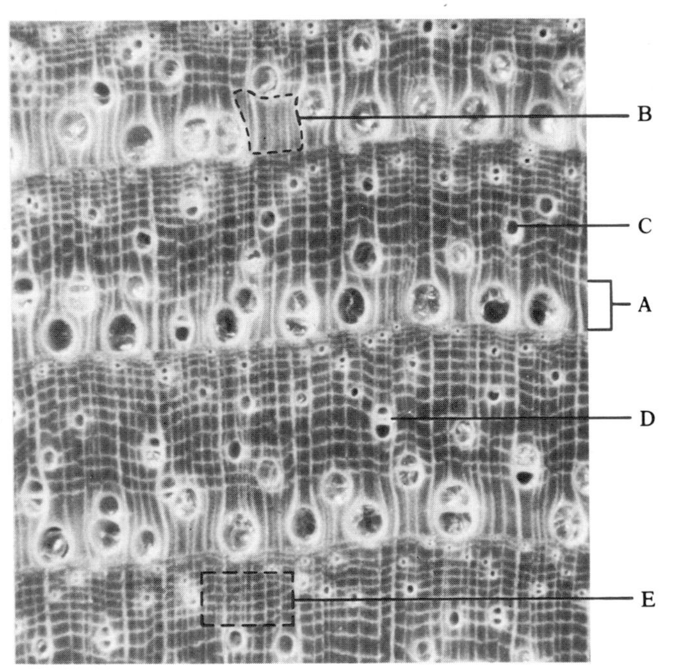

Shagbark hickory *(Carya ovata)*

Very hard and heavy (average specific gravity 0.74). Sapwood creamy white to tan, heartwood pale to medium brown, sometimes blotchy or streaky with darker color. Wood ring-porous, earlywood pores visible to naked eye, largest pores in a single row (A) interrupted by areas of fiber tissue (B). Transition from earlywood to latewood abrupt. Latewood pores smaller, solitary (C) or in multiples of two to four (D). Rays not visible to the naked eye. With the hand lens, rays appear as distinct radial lines, forming a mesh-like pattern with the equally distinct tangential lines of banded parenchyma (E).

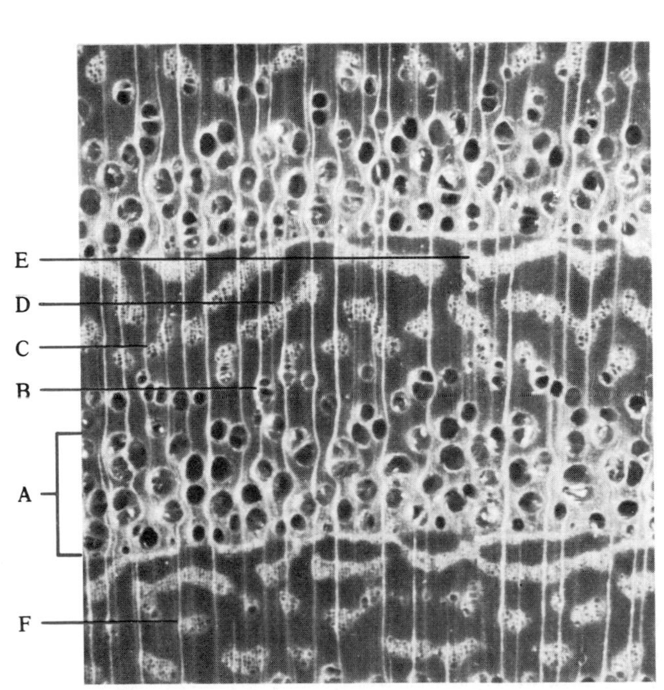

Northern catalpa *(Catalpa speciosa)*

Moderately soft and moderately light (average specific gravity 0.41). Heartwood greyish brown to medium brown, with a distinctive odor, somewhat musty or spicy. Usually ring-porous, earlywood zone usually wide, up to several pores in width (A), with abrupt transition to latewood. Sometimes semi-ring-porous, with gradual earlywood to latewood transition. Latewood pores small, grouped in small radial multiples (B) and clusters (C) near earlywood, tending to form longer clusters (D) and merging into bands (E) toward outer margin of ring. Rays (F) indistinct to unaided eye, but visible as fine bright lines with hand lens in the latewood.

Sassafras *(Sassafras albidum)*

Moderately hard and moderately heavy (average specific gravity 0.46). Heartwood medium to dark greyish brown, with distinctive spicy sassafras odor on freshly cut surfaces. Ring-porous, earlywood three to several pores wide (A), transition to latewood abrupt or somewhat gradual. Tyloses present, although variable in occurrence (B). Latewood vessels small, numerous, commonly in multiples of two or three (C), or solitary (D). Parenchyma appears as light-colored borders around latewood solitary pores and multiples (E), and especially toward outer margin of ring, extending laterally and joining adjacent pores (F). Rays uniformly narrow and barely visible to the unaided eye.

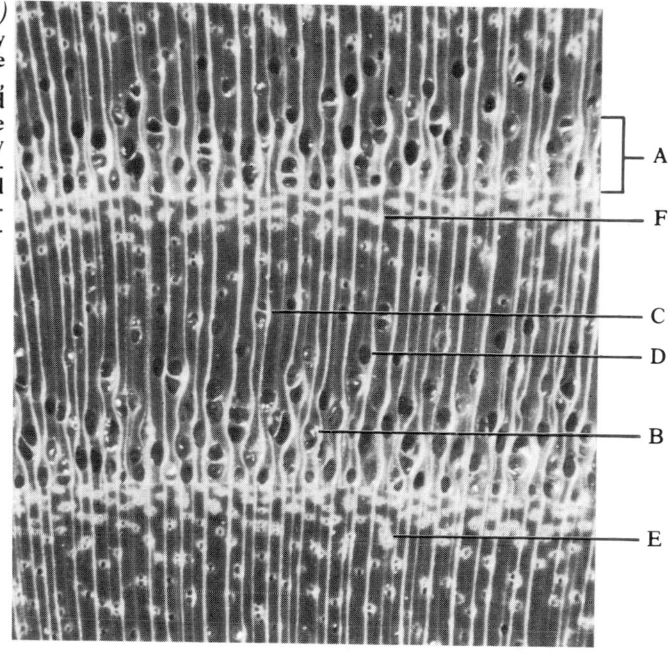

Red mulberry *(Morus rubra)*

Hard and heavy (average specific gravity 0.61). Heartwood medium yellowish brown when freshly cut, aging to deep golden brown or dark brown. Ring-porous, the earlywood zone two to many pores in width (A), abundant tyloses (B), abrupt transition to latewood. Latewood pores small, numerous, either solitary (C) or more commonly arranged in multiples and small clusters (D) and frequently with clusters extending laterally into poorly defined bands (E). Rays (F) plainly visible to naked eye as light yellow lines against the darker fiber mass.

Black locust *(Robinia pseudoacacia)*

Very hard and heavy (average specific gravity 0.69). Heartwood yellowish or golden brown, often with a greenish cast. Ring-porous, the earlywood zone two or three pores wide (A). Latewood pores small, arranged in nested clusters (B), joined laterally into loosely defined bands toward the outer margin of the growth ring (C). All pores densely packed with tyloses, giving an overall flat yellowish appearance to the pores and making adjacent pores indistinct from one another. Rays (D) distinct as sharp yellow lines against the darker brown fiber mass of the latewood.

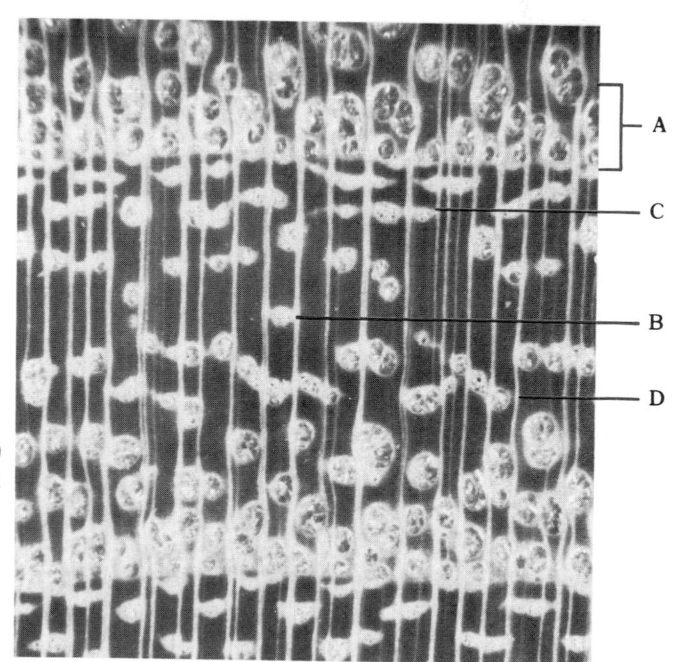

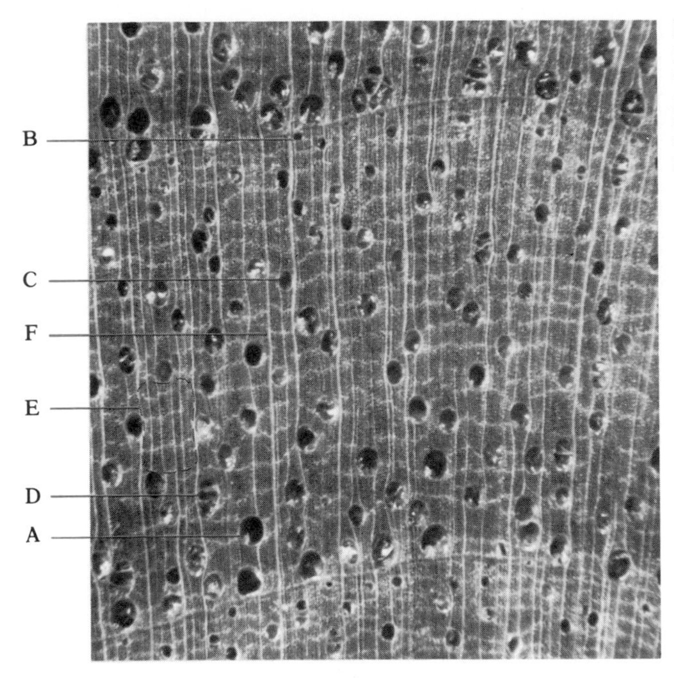

Butternut *(Juglans cinerea)*

Moderately soft and light (average specific gravity 0.38). Heartwood light to medium chestnut or ginger brown. Semi-ring-porous; pores grading in size from large and distinct (A) to small and indistinct (B), solitary (C) or in radial multiples (D). Tyloses variable in abundance. Parenchyma visible with hand lens as faint tangential lines (E). Rays (F) uniformly fine, visible with hand lens.

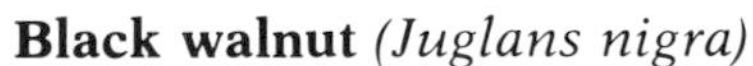

Black walnut *(Juglans nigra)*

Moderately hard and heavy (average specific gravity 0.55). Heartwood light to deep chocolate brown, occasionally with a purple tinge; heartwood color may be variable in shade. Semi-ring-porous, pores grading in size from large and distinct to unaided eye in the earlywood (A) to small and visible only with hand lens in the outer latewood (B). Pores solitary (C) or in radial multiples (D). Tyloses present, but variable in abundance (E). Parenchyma barely visible with hand lens as faint, fine, somewhat irregular tangential lines (F). Rays (G) indistinct to unaided eye, visible as fine lines with hand lens.

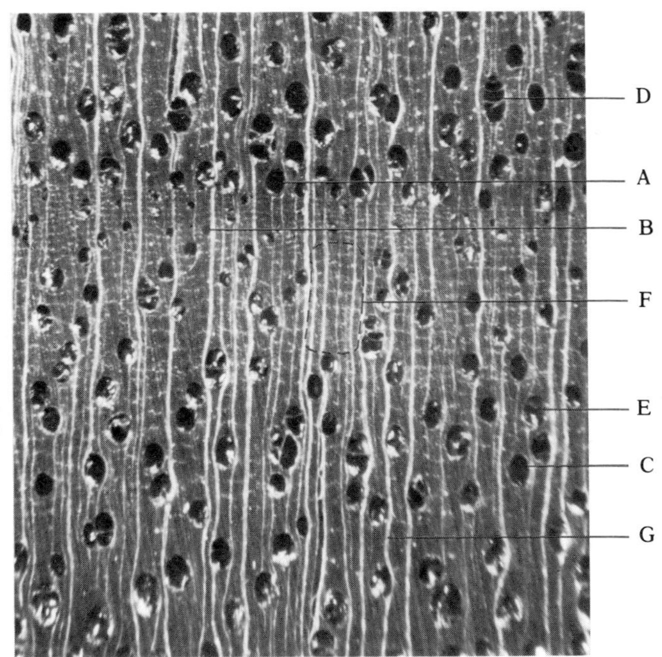

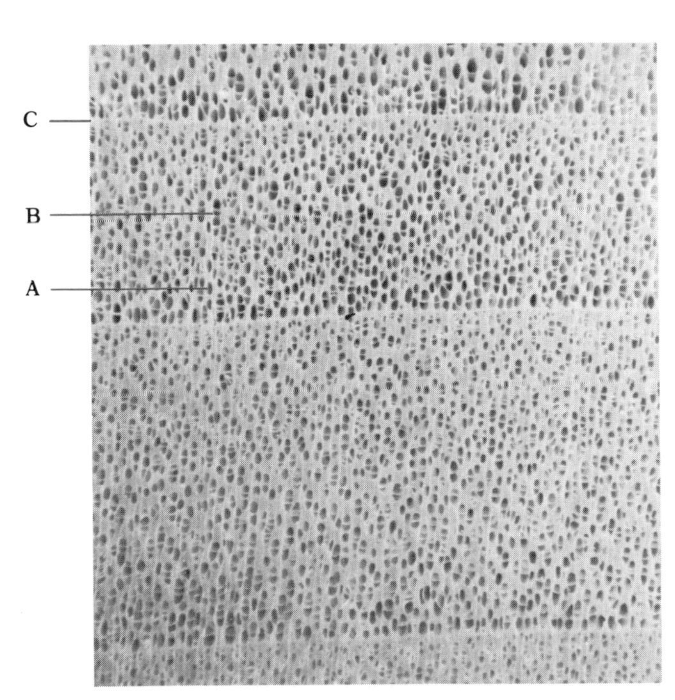

Cottonwood *(Populus* spp.*)*

Relatively soft and light to moderately light (average specific gravity *P. balsamifera* 0.34, *P. deltoides* 0.40). Heartwood light brown to light greyish brown. Green wood often has a sour, unpleasant odor. Wood generally diffuse-porous, sometimes semi-diffuse-porous. Pores numerous, densely but uniformly distributed, solitary (A) or in radial multiples (B). Largest pores barely visible to unaided eye. Terminal parenchyma form a fine light line along the growth-ring boundary (C). Rays very fine, indistinct even with hand lens.

Red alder *(Alnus rubra)*

Moderately soft and moderately light (average specific gravity 0.41). Wood light brown with reddish or peach hue. (No sapwood/heartwood distinction.) Diffuse-porous, with growth rings distinctly delineated by a fine line of contrasting (usually lighter) tissue (A). Pores numerous, small, solitary (B) or in multiples (C). Rays of two distinct sizes: numerous fine rays visible only with hand lens; large rays (D) few in number and irregularly distributed, easily visible to unaided eye.

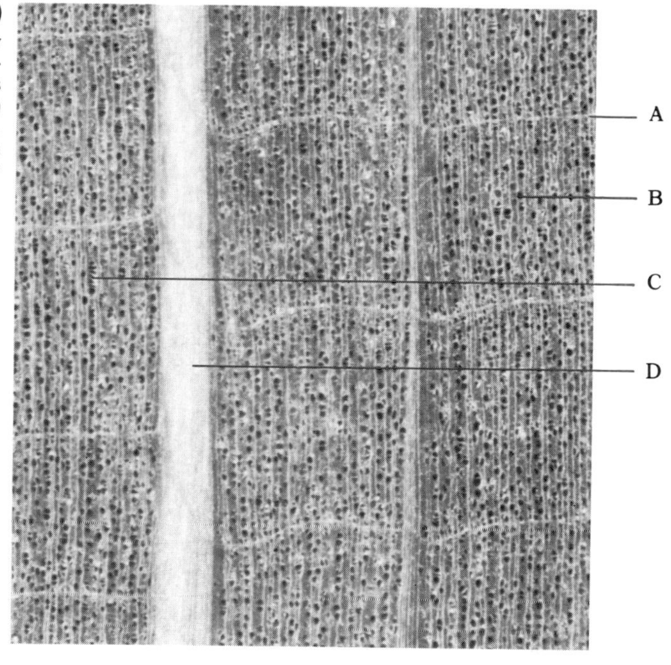

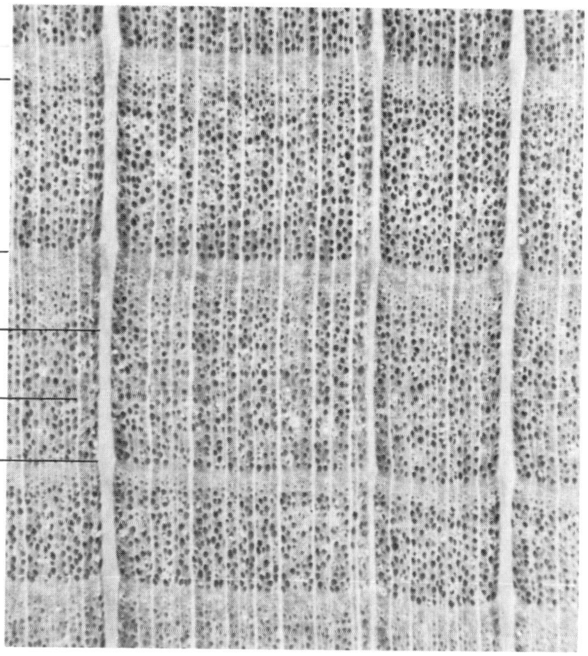

B
A
C
D
E

American beech *(Fagus grandifolia)*

Hard and heavy (average specific gravity 0.64). Heartwood light to medium brown or reddish brown. Diffuse-porous, pores small and distinct only with hand lens, evenly distributed through first-formed portion of growth ring (A); at outer margin of ring a zone of denser cell structure (B) with very small pores, indistinct even with hand lens. Rays variable in size, the largest (C) easily visible to the eye, the smallest (D) visible only with hand lens. Large rays are noded (swollen) at growth-ring boundary (E). (Rays appear on tangential surfaces as characteristic, variable-sized lines.)

Sycamore *(Platanus occidentalis)*

Moderately hard and moderately heavy (average specific gravity 0.47). Heartwood light to medium brown, often with orange or reddish tinge. Diffuse-porous; pores small and uniformly distributed through first-formed portion of growth ring (A), very small, scattered and indistinct in outer zone of growth ring (B), which appears lighter in color. Rays (C) visible to the unaided eye, uniformly large and evenly spaced, noded (swollen) at growth-ring boundary (D). (On tangential surfaces, rays appear as uniform-sized lines somewhat closely but uniformly spaced.)

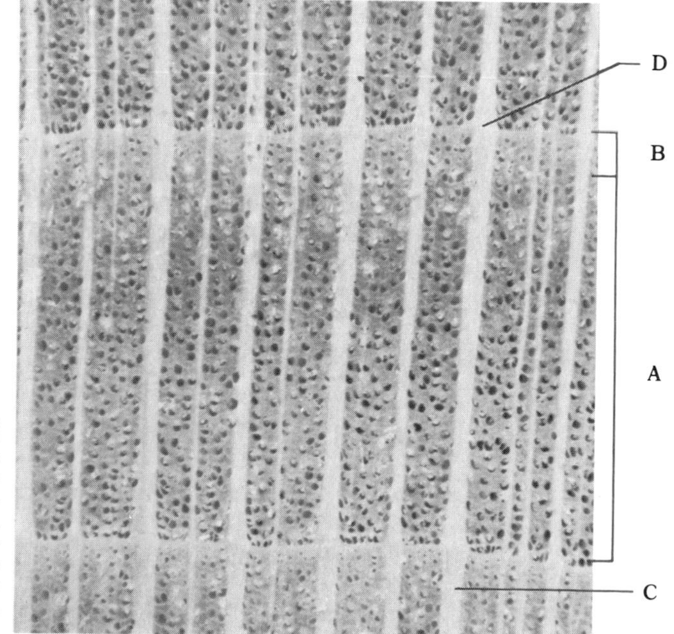

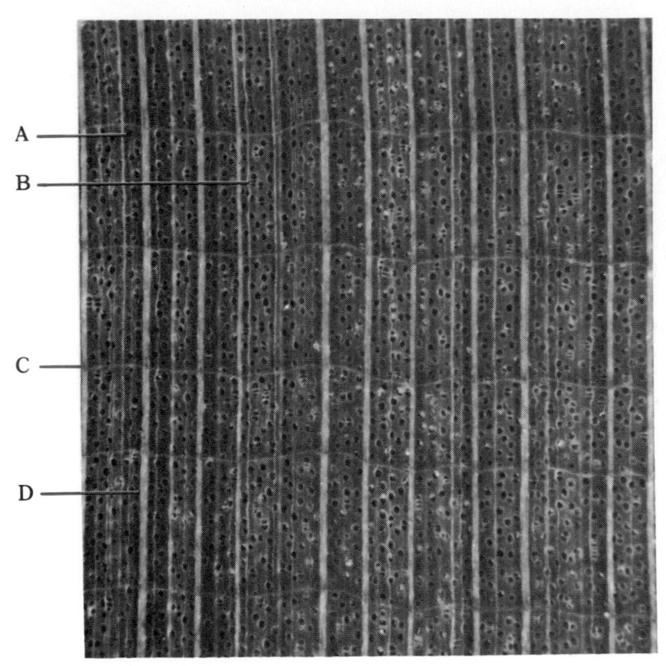

Sugar maple *(Acer saccharum)*

Hard and heavy (average specific gravity 0.63). Heartwood light to medium reddish brown. Sapwood usually wide, near white or very light, with pale brown or reddish tinge. Diffuse-porous, pores solitary (A) or in radial multiples (B), indistinct without hand lens. Growth rings delineated by darker tissue (C) at growth-ring boundary. Rays of two sizes: small rays barely perceptible even with hand lens; large ones (D) visible to naked eye, as wide as or wider than largest pores. (On tangential surfaces, rays usually appear as very fine but distinct lines, uniformly spaced, but clearly smaller than in beech or sycamore.)

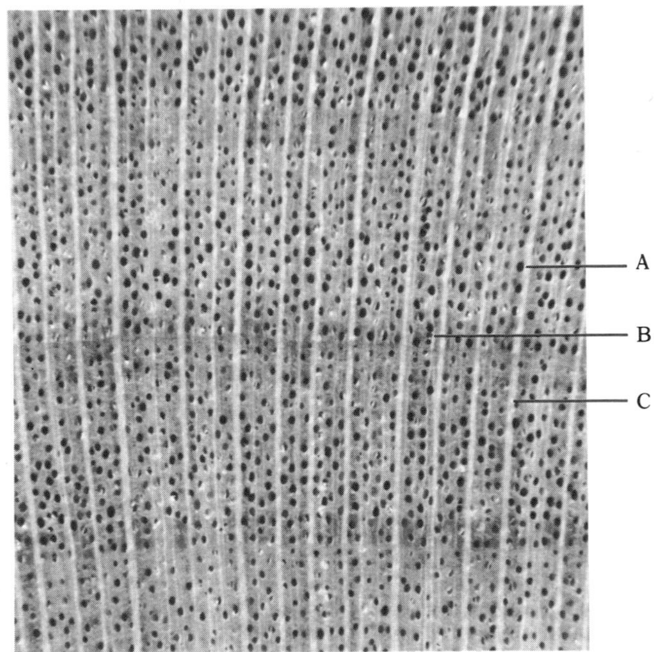

Flowering dogwood *(Cornus florida)*

Hard and heavy (average specific gravity 0.73). Sapwood very wide, pale brown, usually with pink tinge. Heartwood somewhat limited, dark brown in color. Diffuse-porous, growth rings discernible but not conspicuous. Pores solitary (A) and in radial multiples (B). Rays of two general sizes: large rays (C) as wide as largest pores, but not sharply distinct from background; narrow rays nearly invisible, even with hand lens.

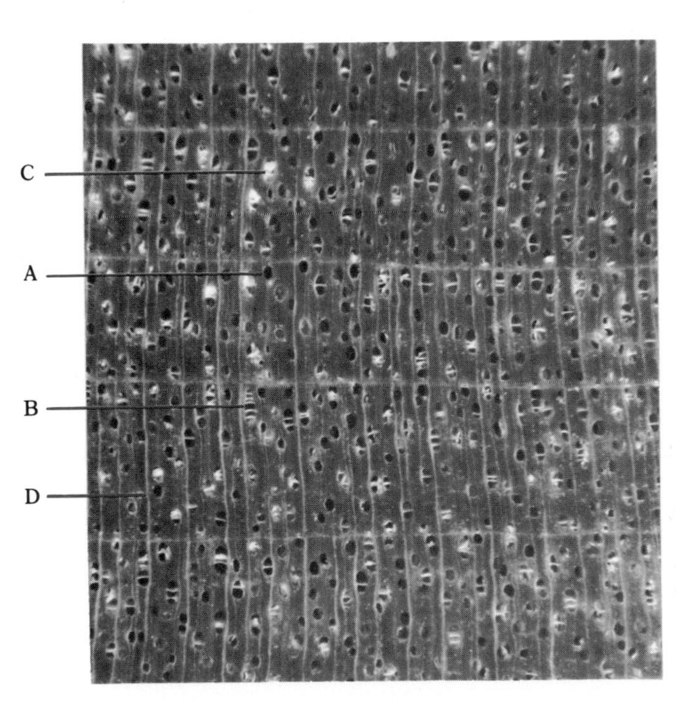

Yellow birch *(Betula alleghaniensis)*

Hard and heavy (average specific gravity 0.62). Heartwood light to medium brown or reddish brown. Sapwood often quite wide. Diffuse-porous, growth rings often indistinct. Pores barely visible to the unaided eye (vessel lines visible on longitudinal surfaces), solitary (A) or in radial multiples (B), many having whitish contents (C). Large pores clearly larger than widest rays. Rays fine (D), generally not visible without hand lens.

California laurel, Oregon myrtle
(Umbellularia californica)

Moderately hard and heavy (average specific gravity 0.56). Heartwood medium greyish brown, commonly streaked with darker brown. Characteristic spicy aroma. Diffuse-porous, pores barely visible to naked eye but distinct with hand lens, solitary (A) and in multiples (B). Growth rings apparent due to slightly darker zone of cells in outer latewood (C). Rays (D) distinct with hand lens as fine, uniformly spaced lines.

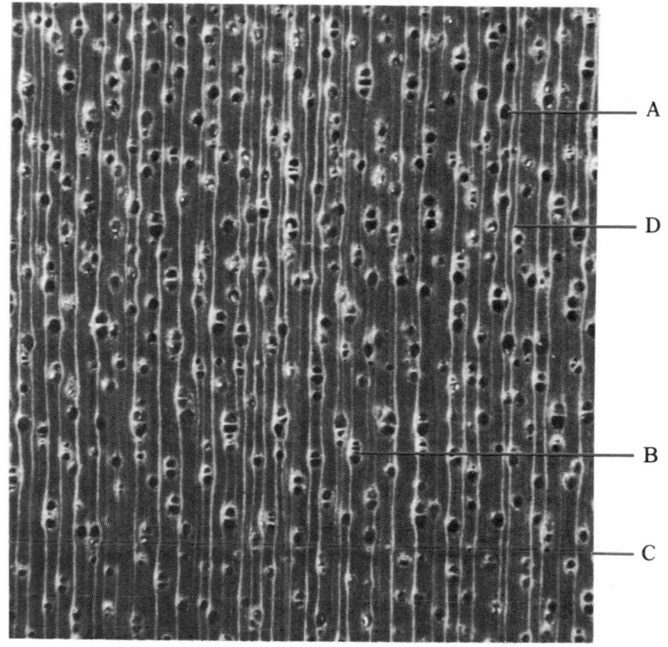

Black cherry *(Prunus serotina)*

Moderately hard and heavy (average specific gravity 0.50). Heartwood light cinnamon brown to dark reddish brown, sometimes variable. Diffuse-porous, pores small, distinct with hand lens, uniformly distributed, solitary (A) and in multiples and small clusters (B). A row of pores in the earlywood (C) delineates the growth rings. Rays (D) are distinct to naked eye, appearing as sharp light lines against the darker cell mass.

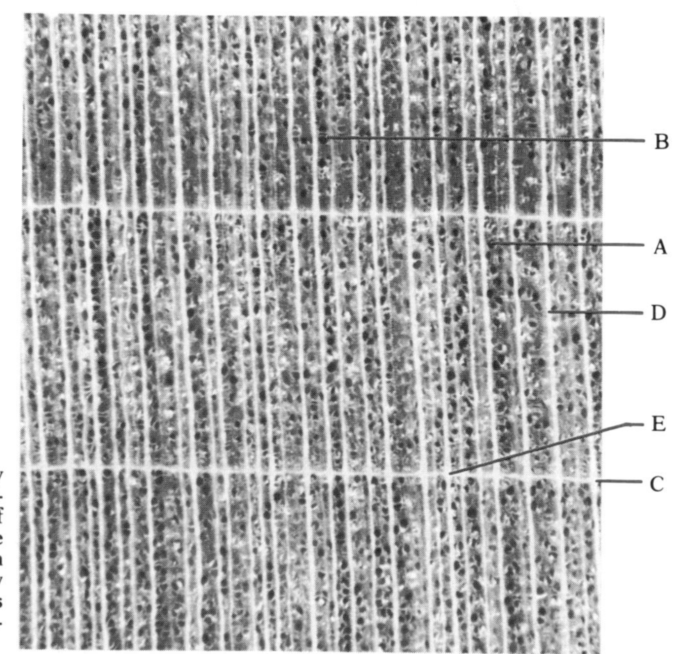

Yellow poplar *(Liriodendron tulipifera)*

Moderately soft and moderately light (average specific gravity 0.42). Heartwood commonly green or greenish brown, occasionally shaded with purple, blue, black or yellow, or with streaks of various colors. Sapwood flat creamy or greyish white ("whitewood"). Diffuse-porous, pores small, solitary (A) and in multiples (B). Growth ring distinct due to whitish or pale-yellow line of terminal parenchyma (C), clearly visible to naked eye. Rays (D) also visible to naked eye (about as distinct as terminal parenchyma), often swollen (noded) at growth-ring boundary (E).

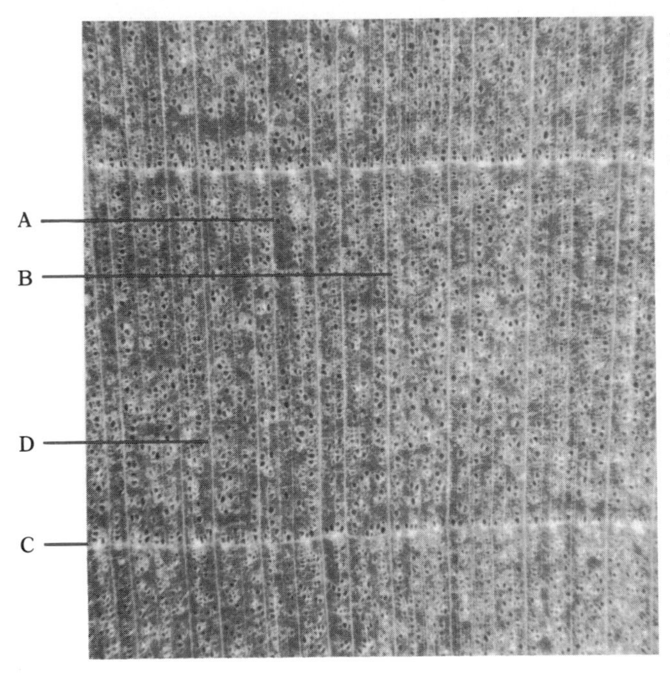

American basswood *(Tilia americana)*

Moderately soft and light (average specific gravity 0.37). Heartwood pale brown or creamy light brown. Freshly cut or moist wood has a characteristic musty odor. Diffuse-porous, pores fairly small, uniformly distributed, solitary (A) or in tangential or radial multiples (B). Growth rings delineated by terminal parenchyma (C). Rays (D) distinct with hand lens.

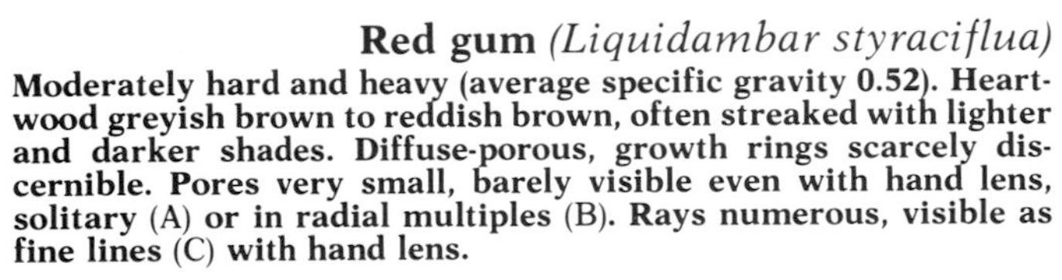

Red gum *(Liquidambar styraciflua)*

Moderately hard and heavy (average specific gravity 0.52). Heartwood greyish brown to reddish brown, often streaked with lighter and darker shades. Diffuse-porous, growth rings scarcely discernible. Pores very small, barely visible even with hand lens, solitary (A) or in radial multiples (B). Rays numerous, visible as fine lines (C) with hand lens.

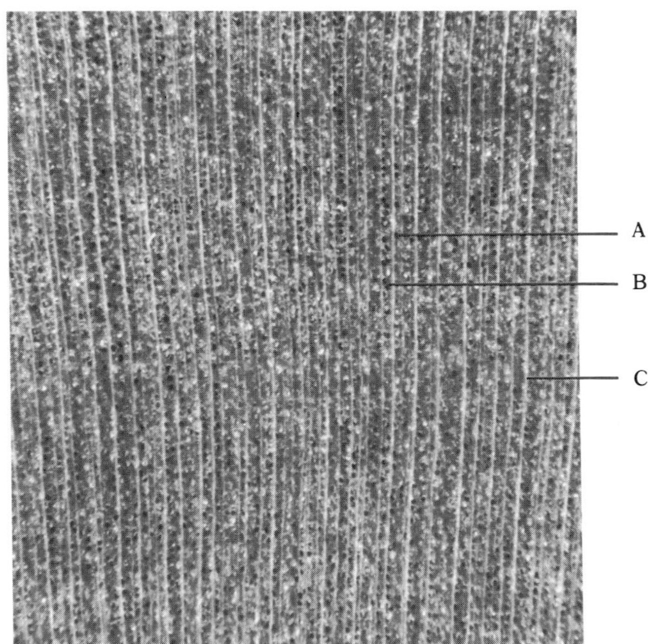

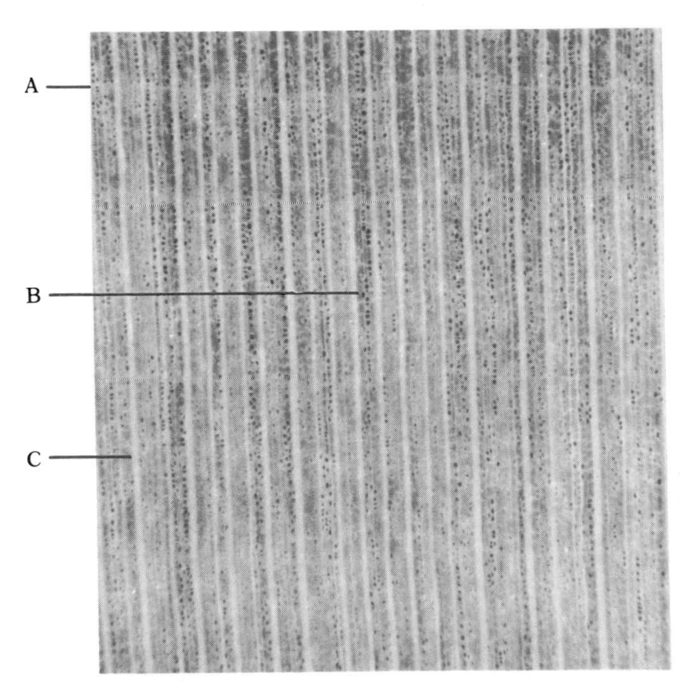

American holly *(Ilex opaca)*

Moderately hard and heavy (average specific gravity 0.57). Heartwood pale greyish white or bluish white, often streaked. Diffuse-porous. Growth rings not distinct; growth-ring boundaries faintly visible due to denser fiber masses in outer margin of ring (A). Pores very small, difficult to see even with hand lens, but characteristically arranged in long chains, or series of multiples (B). Rays of two sizes: large rays (C) the most conspicuous visual feature on the cross section, small rays indistinct and barely visible even with a hand lens.

Tupelo, black gum *(Nyssa spp.)*

Moderately hard and heavy (average specific gravity 0.50). Heartwood greyish white, or with greenish or brownish cast. Normally with interlocked grain. Diffuse-porous, growth rings only faintly delineated. Pores very small, numerous and uniformly distributed, solitary (A) or in radial multiples (B). Rays (C) visible with hand lens as fine, closely spaced lines, in some cases only one pore-width apart (D).

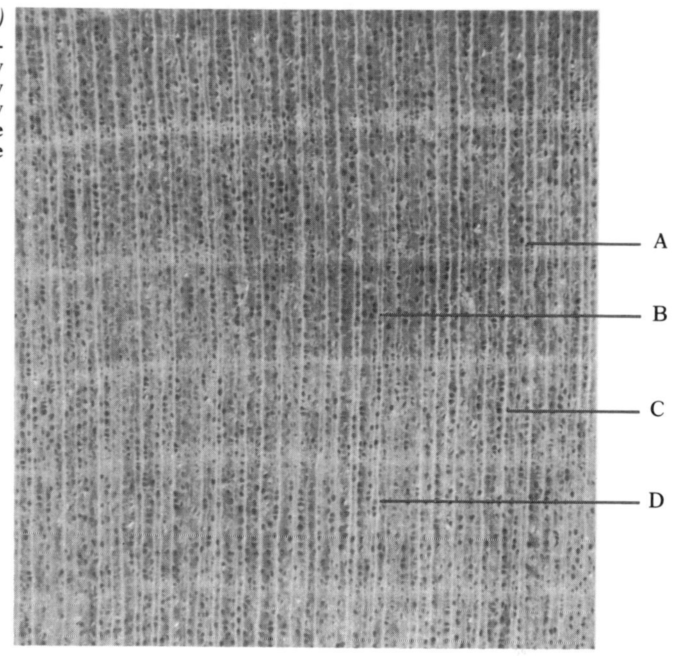

TROPICAL HARDWOODS

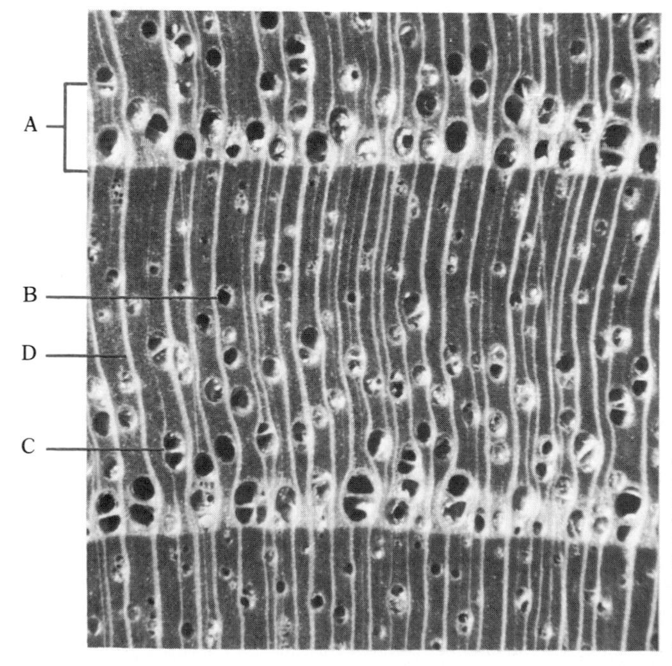

Teak *(Tectona grandis)*—genuine teak, Burma teak

Hard and heavy (average specific gravity 0.66). Golden brown, darkening to brown or almost black with age and exposure. Characteristic spicy odor and waxy feel. Grain straight, sometimes wavy. Ring-porous, growth rings distinct. Texture coarse. Earlywood (A) one to three vessels wide, latewood pores solitary (B) and in radial multiples of two or three (C). Parenchyma terminal and surrounding pores. Rays (D) distinct without lens on cross section.

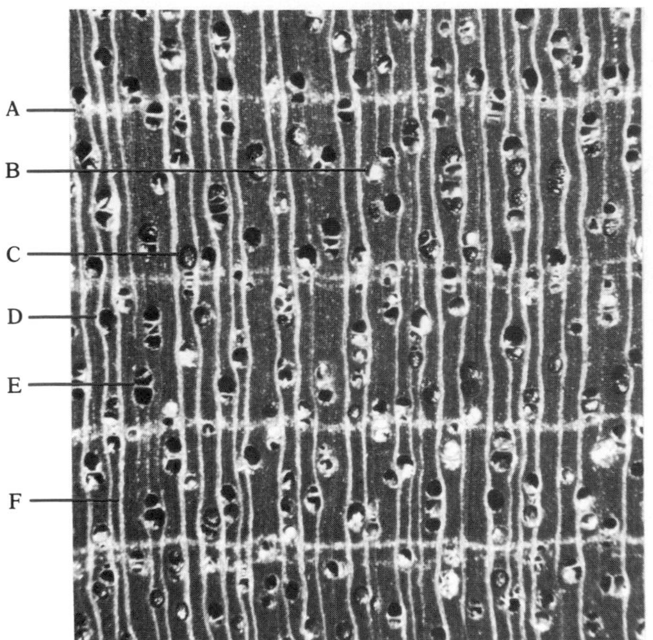

Central American mahogany *(Swietenia spp.)*—genuine mahogany, Honduras mahogany, baywood, Cuban mahogany, West Indian mahogany

Moderately light to heavy (specific gravity 0.40 to 0.83). Color variable from pale or yellow brown to dark red or dark reddish brown, often with pinkish cast. Darkens on exposure. Grain straight, interlocked or irregular. Growth rings distinct due to concentric lines of terminal parenchyma (A). Texture medium. Vessels distinct to the naked eye, numerous, evenly distributed, may have white (B) or gum (C) deposits. Pores solitary (D) or in radial multiples of two to ten (E). Rays (F) barely visible to naked eye on cross section; distinct on radial surface, darker than background.

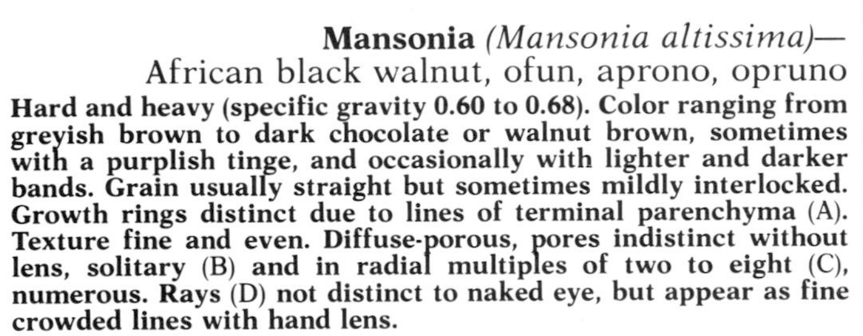

Mansonia *(Mansonia altissima)*—African black walnut, ofun, aprono, opruno

Hard and heavy (specific gravity 0.60 to 0.68). Color ranging from greyish brown to dark chocolate or walnut brown, sometimes with a purplish tinge, and occasionally with lighter and darker bands. Grain usually straight but sometimes mildly interlocked. Growth rings distinct due to lines of terminal parenchyma (A). Texture fine and even. Diffuse-porous, pores indistinct without lens, solitary (B) and in radial multiples of two to eight (C), numerous. Rays (D) not distinct to naked eye, but appear as fine crowded lines with hand lens.

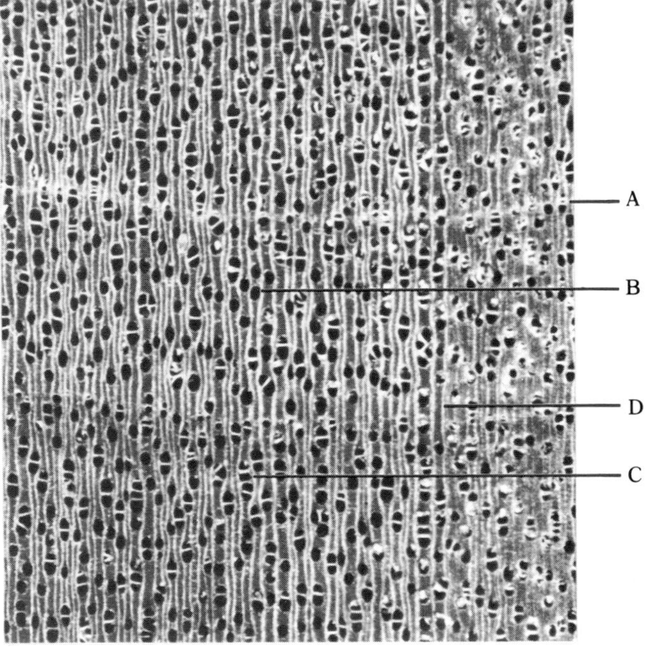

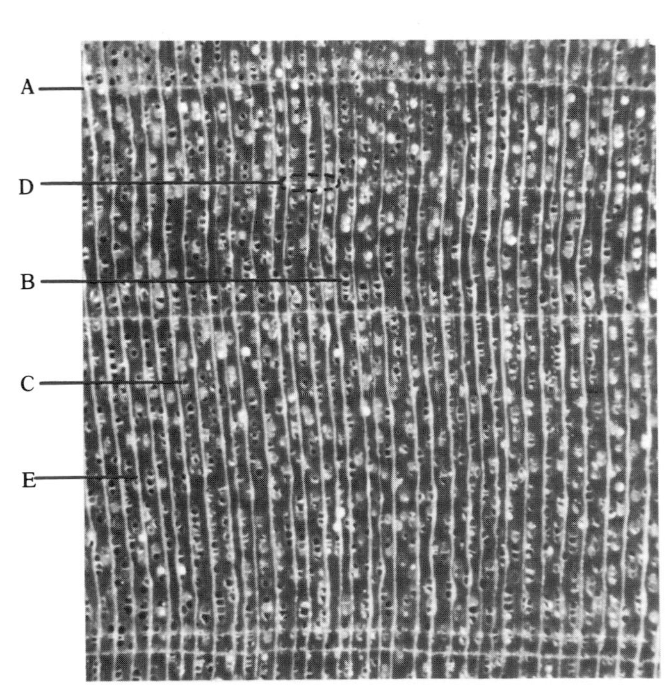

Satinwood *(Chloroxylon swietenia)*—East Indian satinwood, Ceylon satinwood

Hard and heavy (average specific gravity 0.84). Color light yellow or golden yellow and lustrous, sometimes with darker streaks or dark gum veins. Grain narrowly interlocked producing a narrow ribbon figure, often mottled. Growth rings distinct to the naked eye, owing to thin lines of terminal parenchyma (A). Texture fine and even. Diffuse-porous. Vessels small but distinct with hand lens, pores mostly in radial groups of two to six (B) or solitary (C). Parenchyma terminal and in short tangential lines (D). Rays (E) fine, distinct with lens.

Ramin *(Gonystylus* spp.*)*—melawis

Hard and heavy (average specific gravity 0.67). Color pale yellow to buff. Grain usually straight, sometimes slightly interlocked. Growth rings not apparent or delineated by lighter-colored tissue (tracheids). Texture moderately fine, wood evenly diffuse-porous. Vessels (pores) distinct with hand lens, solitary (A) or in radial pairs (B). Parenchyma aliform (C) and confluent (D) (connecting pores), and in tangential lines independent of pores (E). Rays (F) fine, distinct with lens.

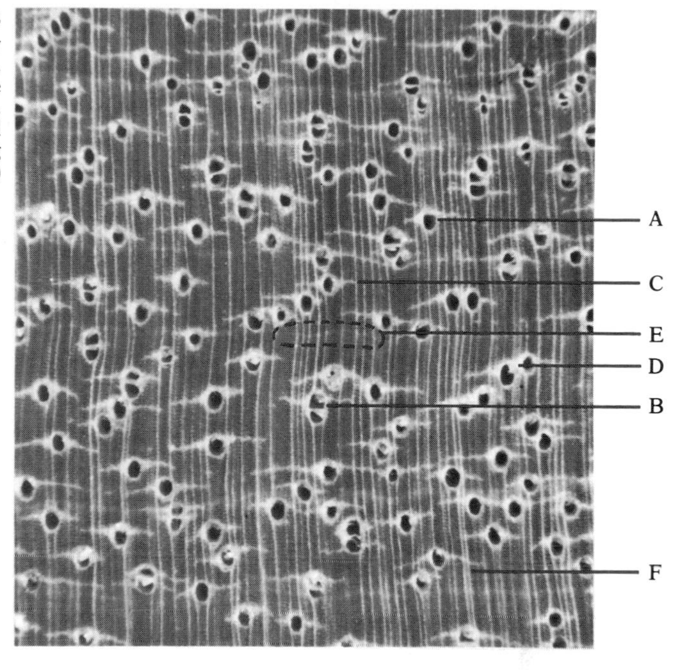

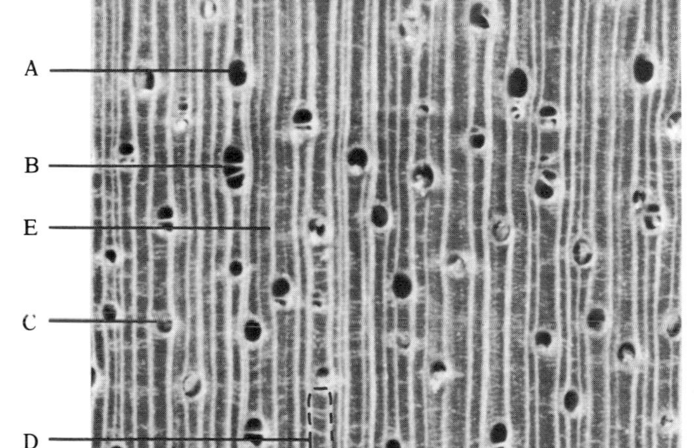

Obeche *(Triplochiton scleroxylon)*—wawa, ayous, samba, African whitewood, obechi

Moderately soft and light (average specific gravity 0.38). Color uniformly pale yellow to creamy white or pale buff. Grain characteristically interlocked, producing ribbon figure on radial surfaces. Growth rings indistinct or distinct due to change in fiber density. Texture medium to moderately coarse. Vessels distinct to eye, pores solitary (A) or in small radial multiples (B), sometimes irregularly distributed, commonly with tyloses (C). Parenchyma visible only with lens as short, fine tangential lines (D). Largest rays (E) visible on cross section; on radial surface, rays are distinct, appearing lighter than background.

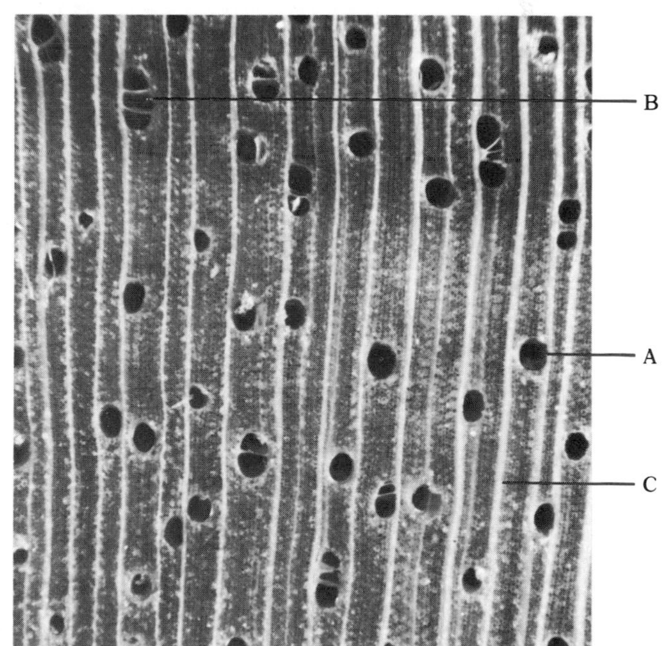

Balsa *(Ochroma* spp.*)*

Extremely light (average specific gravity 0.10 to 0.20). Color white, cream to pale brown, occasionally having a pink tinge. Straight-grained. Growth rings not apparent. Texture coarse. Diffuse-porous. Vessels large, distinct to naked eye; pores solitary (A) and in radial multiples of two or three (B). Parenchyma not distinct. Rays (C) distinct to naked eye. Extreme lightness (in weight) and softness of wood are valuable identification features.

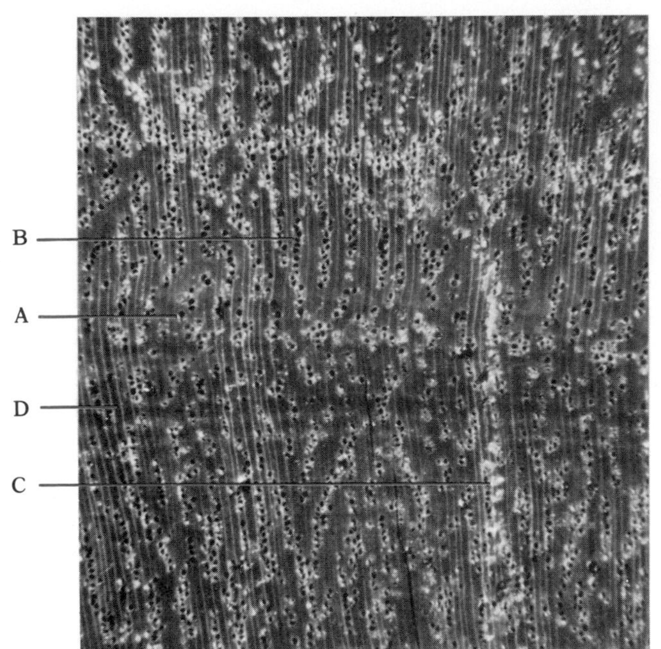

Lignum vitae *(Guaiacum officinale, G. sanctum)*

Very hard and very heavy (average specific gravity 1.15 to 1.30). Color variable from light olive green to dark greenish brown, usually with dark streaks. Grain strongly interlocked, producing closely spaced ribbon stripe pattern on radial surfaces. Growth rings indistinct. Texture very fine. Vessels indistinct without lens, solitary (A) or in radial rows (B); gum and resin deposits (C) abundant. Parenchyma not visible. Rays invisible to the naked eye; with lens rays appear as very fine lines (D) on cross section.

African blackwood *(Dalbergia melanoxylon)*—Senegal ebony, grenadillo

Very hard and very heavy (average specific gravity 1.22). Color plum or dark purple to purplish black and commonly streaked with jet black. Grain usually straight. Growth rings indistinct. Texture fine. Vessels distinct only with lens, few and evenly distributed. Diffuse-porous, pores solitary (A) and in radial multiples of two to four (B). Parenchyma faintly visible with lens as tangential lines. Rays indistinct to naked eye; visible as fine lines with lens.

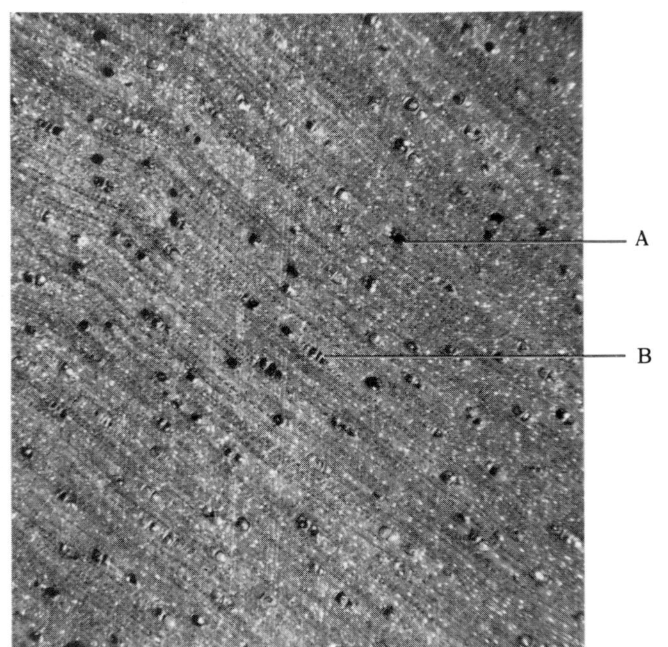

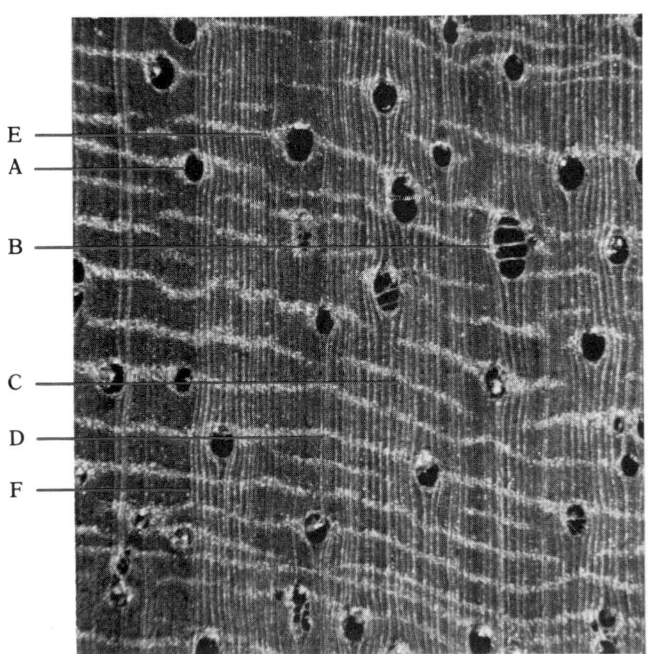

Padauk *(Pterocarpus* spp.*)*—vermilion, red narra, maidou

Hard and heavy (average specific gravity 0.77). Color variable from yellowish brown or bright orange to reddish brown or blood red, often with darker streaks. Grain usually interlocked but sometimes straight. Growth rings indistinct, or distinct due to terminal parenchyma. Texture medium to moderately coarse. Vessels distinct without lens. Mostly diffuse-porous, pores solitary (A) and in radial multiples of two to ten (B). Parenchyma visible without lens, as tangential lines independent of pores (C), connecting pores (D), or extending laterally (E) from pores. Rays (F) fine, not visible without lens.

Zebrawood *(Brachystegia* spp.*)*

Moderately hard and heavy (average specific gravity 0.60 to 0.88). Color light straw to golden or light brown with parallel darker brown stripes. Grain characteristically interlocked, producing a uniform stripe figure, but dominated by the light/dark pigment figure. Growth rings distinct due to lines of terminal parenchyma (A). **Texture moderately coarse. Diffuse-porous. Vessels distinct to naked eye, solitary** (B) **and in radial groups of two to six** (C). **Parenchyma also distinct to naked eye, vasicentric (surrounding pores)** (D), **aliform with short wings** (E), **or confluent** (F). **Rays** (G) **fine and closely spaced, visible only with lens.**

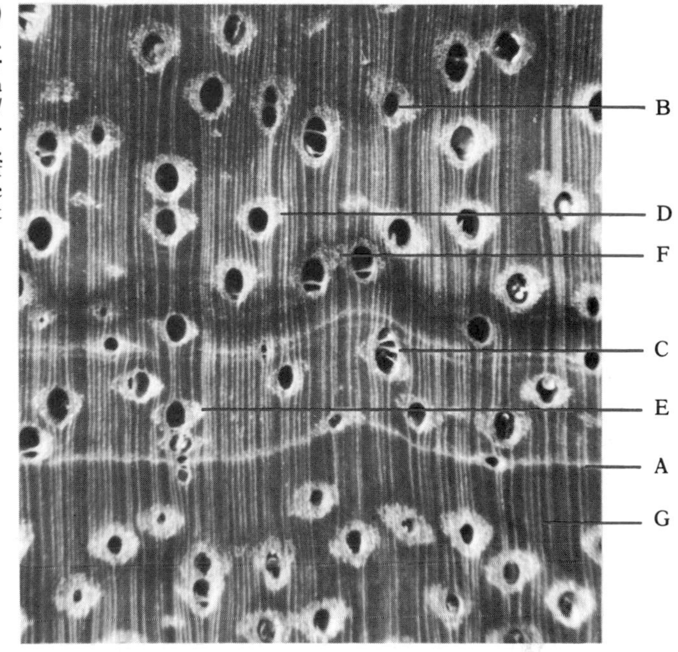

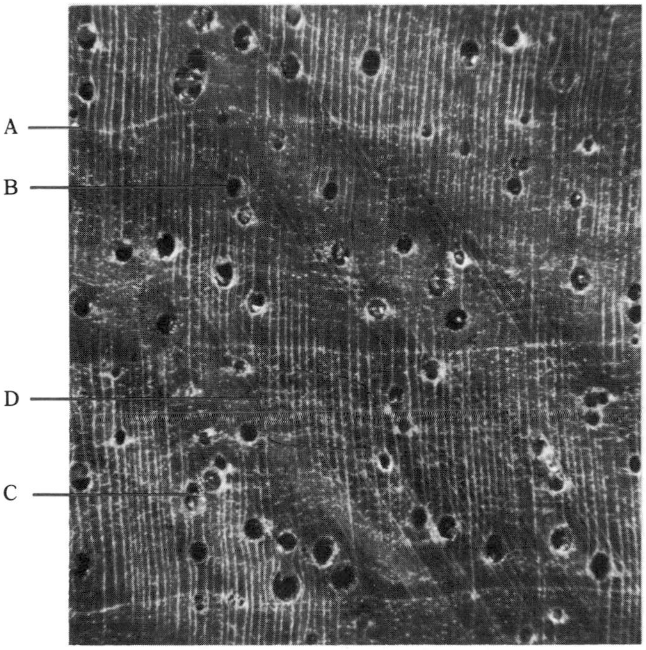

Honduras rosewood *(Dalbergia stevensonii)*

Hard and heavy (average specific gravity 0.92 to 1.08). Pinkish or purplish brown with darker purple to black streaks or markings. Grain straight or slightly interlocked. Growth rings distinct, delineated by lines of terminal parenchyma (A). **Texture medium. Diffuse-porous, vessels variable in size. Pores solitary** (B) **and in multiples of two to three** (C). **Parenchyma distinct with lens as terminal or as concentric lines** (D). **Rays fine, indistinct without lens.**

Angelim *(Hymenolobium excelsum)*

Hard and heavy (average specific gravity 0.74). Color orange brown with lighter striping. Grain interlocked. Growth rings not distinct. Texture medium to moderately coarse. Diffuse-porous, vessels distinct to the naked eye, pores solitary (A) **or in short radial groups** (B). **Parenchyma conspicuous as wide, lighter-colored tangential bands, typically aliform** (C) **and confluent** (D). **Rays** (E) **visible to naked eye, conspicuous with lens as light lines against dark fiber mass.**

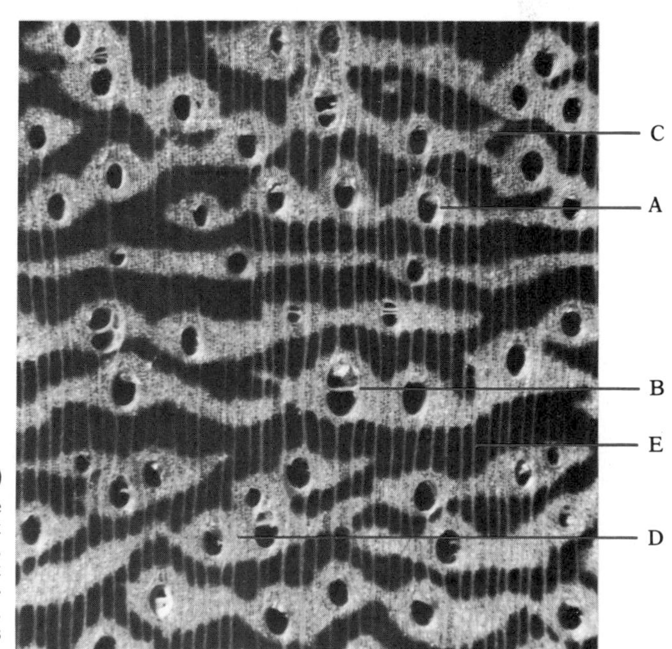

625
ATHOL

625
ATHOL

Water and Wood

Someone once quipped that more than 90% of all problems with wood involve moisture. For those who ignore basic wood-moisture relationships, that is probably a conservative estimate.

Everyone has been introduced to the interaction of water and wood, for everyone has seen the problems that result when wood shrinks and swells. The bureau drawer that slides freely in January but sticks tightly in August is an all too familiar example of the dimensional response of wood to changes in atmospheric humidity. Warp and surface checks in lumber, loose tool handles and out-of-round turnings are also common symptoms. Although other consequences of moisture—such as fungal discoloration or gluing failures—can plague the woodworker, dimensional problems are by far the most common and troublesome.

Wood in trees is wet. Very wet. The cell structure contains excessive water (sap) and is fully swollen. But under conditions where wood is commonly used, much of this water will dry out and the wood will partially shrink. Eventually a fluctuating moisture balance between the dryness of the wood and the humidity of its environment will be reached. The obvious goal is twofold: first, to dry wood (and thereby preshrink it) to a moisture content consistent with its eventual environment, and second, to control any subsequent gain or loss of moisture in order to minimize dimensional change. To overcome problems, the woodworker must understand the initial drying of sap from freshly cut wood, as well as the continuing exchange of moisture between the wood and the surrounding atmosphere.

In a nutshell, the atmospheric humidity determines the moisture content of the wood, and the moisture content, in turn, determines the dimension of the wood. Although the interrelationship of humidity, moisture content and dimensional change is somewhat complex, a working knowledge of these wood-moisture relationships can be gained step by step. Let's start with the **moisture content of wood** and the **relative humidity of the atmosphere.**

Moisture content—The moisture content (MC) of wood is measured as the ratio of the weight of water in a given piece of wood to the weight of the wood when it is completely dry. The water-free weight of wood is usually referred to as the oven-dry weight, because drying in an oven is a common method of obtaining it. This ratio is traditionally expressed as **percent moisture content**.

Suppose a piece of wood weighs 30 lb. If after drying in an oven it weighs only 25 lb., 5 lb. of water have been driven off. The moisture content would be 5/25 = 0.20, usually expressed as 20% moisture content. If the same piece had originally weighed 60 lb., the 35 lb. of water would have represented a moisture content of 140% (35/25 = 1.40 = 140%).

Relative humidity—Humidity is a general term referring to water or moisture in vapor form in the atmosphere. **Absolute humidity** refers to the actual quantity of moisture present in air. This is usually expressed in grains per cubic foot (1 grain = $\frac{1}{7000}$ lb. avdp.) or in grams per cubic meter. The amount of water the air can hold varies with temperature (*2*). At 70°F, for example, the air can hold a maximum of 8 grains of moisture per cubic foot.

Relative humidity (RH) is the ratio of the amount of moisture in the air at a certain temperature to the amount it would be able to hold at that temperature. If the air at 70°F, for example, held 4 grains of water per cu. ft., the relative humidity (RH) would be 50%, because the air is capable of holding 8 grains at that temperature. If the absolute humidity were 6 grains per cu. ft., the RH would be 75%. The **dew point** is the temperature at which water vapor condenses from the air. Air at 70°F and 50% RH (with 4 grains/cu. ft.) has a dew point of 49.3°F. Air with that much moisture, when cooled to 49.3°F, can hold no more moisture, and it is

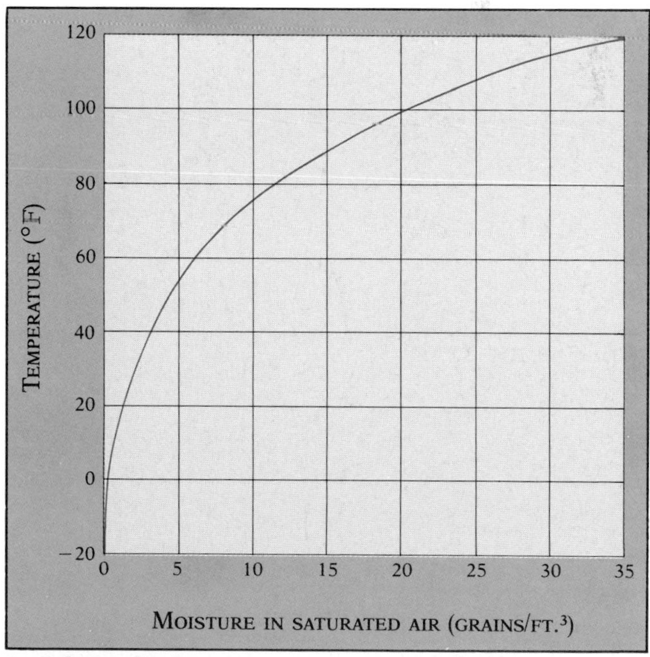

2—Relationship between temperature and moisture content of saturated air. The maximum moisture that the air can hold depends on how warm it is; as the temperature rises, so does the saturation point.

1—The moisture content of red pine sapwood may be more than 200%. Much of the free water can be squeezed out of the wood in a vise.

therefore at **100% relative humidity**. If the same air were cooled to 41°F, it could hold only 3 grains of moisture per cu. ft., so 1 grain per cu. ft. would condense out as precipitation.

Nature determines our atmospheric humidity. Weather systems bring air masses having a certain absolute humidity. Additional and significant local influences upon humidity result from moist vegetation or from surface evaporation of bodies of water. Because excess moisture in the air will condense and precipitate when air is cooled to the dew point, temperature itself may establish an upper limit to absolute humidity, because air can never contain more moisture than associated with its dew point. Consequently, we associate bitter winter weather with low absolute humidity and summer weather with high absolute humidity.

In buildings we routinely manipulate nature's air—mainly by heating it up when it is too cold, to a lesser extent by cooling it, and least of all by adding or subtracting moisture from it. It is important to realize the effect of our heating or cooling air without accompanying humidification or dehumidification. Heating air increases its ability to hold moisture. If we increase the temperature of air while the **absolute** humidity is unchanged, the **relative** humidity will be lowered. In subzero winter weather, outdoor air has a low absolute humidity as it seeps into our homes. When we heat it to near 70°F without adding moisture, the relative humidity drops very low. Conversely, summer air usually holds an abundance of moisture because of its high temperature. If we cool the air, thus reducing its capacity to hold moisture, the relative humidity (which may be high to begin with) rises even higher.

Free water and bound water

The liquid content of the living tree, called **sap**, is primarily water, but also contains dissolved minerals, nutrients from the soil and carbohydrates manufactured by the foliage. For our purposes we can consider moisture or water in wood to mean either the original sap of the tree or water from other sources that is subsequently picked up by dry wood. Water can return to wood from countless sources, ranging from rain to the moisture in humid air.

To visualize the condition of moisture in the wood of a standing tree, imagine a sopping-wet sponge just pulled from a pail of water. The sponge is analogous to growing wood in that the cell walls are *fully saturated* and swollen and the cell cavities are partially to completely filled with water. If we squeeze the sponge the water pours forth. Similarly the water in wood cell cavities, called **free water**, can be squeezed from wood (see Figure *1*, page 66). If you hit a piece of green lumber with a hammer you may even see the water spurt out.

Now imagine thoroughly wringing out a wet sponge until no further water is evident. The sponge remains full-sized, flexible and damp to the touch. In wood, the comparable condition is called the **fiber saturation point (FSP)**. In this state, the cell cavities are emptied of free water, but the cell walls are still saturated and thus still in their weakest condition. Only when water leaves the cell walls does the wood begin to shrink and increase in strength.

This water remaining in the cell walls is called **bound water**. In contrast to free water, which is held in cell cavities like water in a tumbler, the bound water is held by physical forces of attraction within the cell walls. Just as a sponge must be left to dry—and shrink and harden—so must the bound water be removed by placing the wood in a relatively dry atmosphere. How much of the bound water is lost (and therefore how much shrinkage takes place) will depend on the relative humidity (RH) of the atmosphere. If the air is at 100% RH, no bound water will be lost. To remove all the bound water, the wood would have to be placed in an oven or desiccator, or in a vacuum where the relative humidity is zero. Obviously, we use wood where the relative humidity is somewhere between 100% and zero, so only part of the bound water is lost.

Equilibrium moisture content

Wood always remains hygroscopic—it responds to changes in atmospheric humidity and loses bound water as the RH drops, regaining bound water as the RH increases. For a given RH level, a balance is eventually reached at which the wood is no longer gaining or losing moisture. When this balance of moisture exchange is established, the amount of bound water eventually contained in a piece of wood is called the **equilibrium moisture content (EMC)** of the wood.

The relationship between the amount of bound water in wood and relative humidity is shown in Figure *1*. In my estimation this is the most important item in this book. Although I'm not much for memorizing, I think every woodworker should have a few basic points clearly in mind. A good starting point is to remember that 50% RH gives an approximate 9% equilibrium moisture content (EMC). Then note that 25% RH gives about 5% EMC and 75% RH gives about 14% EMC and you've about got it. The lower end point always originates at 0% RH and 0% EMC, and 100% RH always gives total fiber saturation. Reproduce this graph poster size and hang it on your shop wall. Look at it every day. It's *that* important.

Everything, of course, has its qualifications, a few of which must be mentioned here regarding Figure *1*. First, the curve is for white spruce, a typical species, shown as having a fiber saturation point (FSP) of about 30% moisture content. The FSP may vary among different species. In species having a high extractive content (for example, redwood and mahogany) the FSP will be noticeably lower, around 22% to 24%. For those low in extractives, such as birch, the FSP might range as high as 35%.

Temperature also has an effect upon equilibrium moisture content (EMC). The curve shown in Figure *1* is for 70°F, but at intermediate levels the EMC would be about one percentage point lower for every 25° to 30°F elevation in temperature. The EMC curves always converge at 0% relative humidity and 0% EMC, so variation due to extractives or temperature will therefore be most pronounced toward the fiber-saturation-point end of the curve.

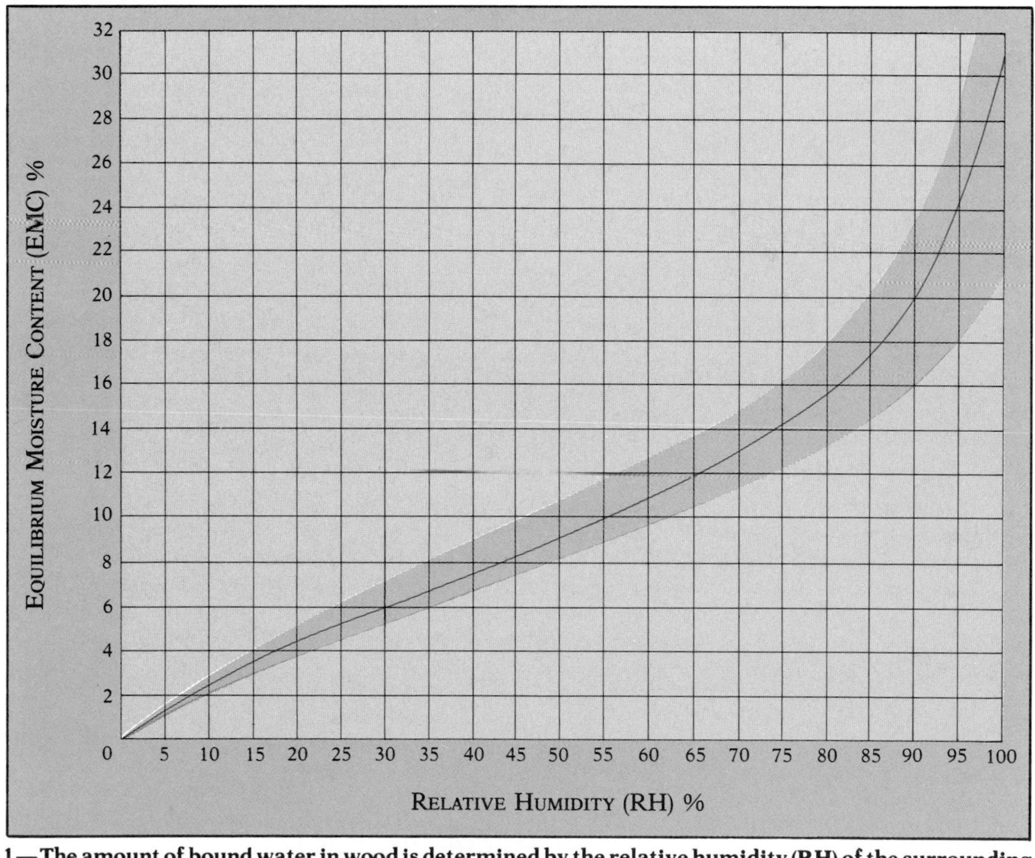

1—The amount of bound water in wood is determined by the relative humidity (RH) of the surrounding atmosphere; the amount of bound water changes (albeit slowly) as the relative humidity changes. The moisture content of wood, when a balance is established at a given relative humidity, is its equilibrium moisture content (EMC). The solid line represents the curve for white spruce, a typical species with fiber saturation point (FSP) around 30% EMC. For species with a high extractive content, such as mahogany, FSP is around 24%, and for those with low extractive content, such as birch, FSP may be as high as 35%. Although a precise curve cannot be drawn for each species, most will fall within the color band.

In addition, when wood is losing moisture (desorbing) the EMC curve is slightly higher than when the wood is picking up moisture (adsorbing). This is called the **hysteresis** effect. Under usual room conditions of slightly fluctuating relative humidity, however, the average or **oscillating** curve is applicable most of the time.

Depending on the degree of environmental control, especially the extent to which we heat during the winter, humidity can vary widely indoors. In summer, with doors and windows open, interiors may approach outdoor conditions. In winter, when buildings are heated, we reach the low extreme. For example, imagine a night in late January when the thermometer outside drops to 0°F. Figure *2* on page 67 shows that nature is supplying air that can carry a maximum of ¼ grain of moisture per cu. ft. As this air diffuses into homes, it is heated to about 70°F, at which temperature it *could* hold 8 grains. Its RH drops to 6%! Checking Figure *1* on page 69, we see that surfaces of unprotected wood or thin veneers would drop to below 2% moisture content. Undoubtedly, humidifiers and the domestic activities of cooking, washing and even breathing add some moisture to the air, not to mention the moisture being released by the wood itself. But under average living conditions, without intentional humidification, it is not uncommon during January and February for the conditions in my shop to hover below 25% RH.

In summer the situation is reversed. In July and August the temperature may reach the 90s. With late afternoon thundershowers, the relative humidity may stay above 85% for days. We are talking about air with 12 or more grains of moisture per cu. ft. In my house, the foundation rests on bedrock, so the cellar rarely gets above 70°F in the summer. Thus for every cu. ft. of muggy August air that enters my cellar, four grains of moisture will be lost either as condensation onto cool surfaces or in raising the interior humidity even higher. In time, the humidity could approach 100%. (I am amused to remember how, when I was a child, we used to open the cellar in such summer weather "to dry it out." It's little wonder we had a damp cellar.) By actual measurements on spruce wafers, I have recorded a moisture content of 23% in August in the same cellar location where I found 5% to 6% readings in January. And this situation is probably typical of many areas of the country where summers are warm and humid and winters are bitter cold. *It is important to realize that if the absolute humidity of air is unchanged, lowering the temperature of the air raises the relative humidity and heating the air lowers the relative humidity.*

As will be emphasized again in discussing shrinkage and finishing, such seasonal extremes must be averaged. The low moisture conditions associated with winter, spring and fall weather seem to outweigh the effects of short-term, high-humidity summer extremes. Thus 7.5% to 8% moisture content is an appropriate average for this kind of area. To bring wood to such low levels it must either be stored indoors or dried in a kiln. Since the latter is the usual practice, the term **kiln-dried** usually means dried to a level appropriate for interior use. To the cabinetmaker, then, kiln-dried suggests a moisture content of below 10%.

In structural lumber, however, **air-dried** levels of moisture content are considered adequate. In this context, kiln-dried may mean 19% or less. In some cases, structural lumber is kiln-dried mainly to reduce its weight for more economical shipping, to kill fungi or other wood-destroying organisms, or simply to speed up the drying process, even though the final moisture content may be scarcely below the fiber saturation point. So the term kiln-dried alone should not be blindly interpreted to indicate any particular moisture content.

One of the more unfortunate yet common fallacies is that kiln-drying leaves wood irreversibly dry, and that once dried the wood somehow becomes dimensionally stable. In reality, if dry wood is stored under relatively moist conditions, bound water will be readsorbed to the equilibrium moisture condition. Tests have shown that even larger packages of 1-in. pine kiln-dried to 7% to 8% moisture content reattain a moisture content of 13% to 14% after sheltered outdoor storage for six months. More rapid gain of moisture to even higher moisture content can result from exposure to weather or from storage in damp indoor conditions.

When we deal with modified wood products, such as particle board, hardboard or decorative laminates, the adhesives and other additives involved as well as the heat applied in manufacture may influence the equilibrium moisture content (EMC) considerably. For example, at a relative humidity of 40%, where wood might come to an EMC of about 7.5%, particle board might average 7%, hardboards 5% and decorative laminates as low as 3.5%. Even among different species, the EMC at 40% RH may vary from 6.5% to 8.5% moisture content. Someone once suggested the concept of thinking of an **equilibrium relative humidity (ERH)** rather than an **equilibrium moisture content (EMC)**, to emphasize the fact that relative humidity determines EMC, not the other way around. It is prudent to think that "my lumber should be at equilibrium with 40% relative humidity" rather than that "my lumber should be at 7.5% moisture content."

Green vs. air-dried vs. kiln-dried

Let's elaborate on these fundamentals of moisture content, because we will encounter them in our work. The amount of water in the living tree is quite variable. *Table 5* gives some average values of moisture content in trees for a number of species. In general, the hardwoods have initial moisture contents in the 60% to 100% range. Notable exceptions are white ash, on the low side, and cottonwood, on the high side. In general, heartwood and sapwood have similar levels of moisture content but sometimes sapwood is higher, sometimes heartwood is higher. In softwood species, the general case seems to be that the heartwood has a fairly low moisture content, often scarcely above the fiber saturation point, while the sapwood is considerably higher. Among the lower-density species such as balsa or even pine, the sapwood moisture content often exceeds 200%.

In some species moisture content shows seasonal variation. A pronounced rise in moisture content of 30% or more about the time of leaf emergence and lasting a month or two has been noted in birch and ash. In many species, moisture levels are low in late summer, then rise again when the leaves fall. In other species, such as beech, there is little year-round variation. Fluctuations in conifers also differ among species. In all cases, however, water content varies principally in the sapwood. Because of the inconsistent behavior among species, there are no general recommendations regarding the best time of year to cut wood to facilitate drying; each species must be considered individually.

There is considerable confusion over the meaning of the word **green** in reference to wood. It is often used to indicate the condition of **freshly cut** wood from a living tree. But because most properties of wood are unchanged regardless of the amount of free water it contains, we consider any wood above the fiber saturation point (FSP) as green, even when the condition has been restored by wetting previously dried wood.

Table 5—Average green moisture content of common wood species.

	Moisture content	
Hardwoods	Heartwood %	Sapwood %
Apple	81	74
Ash, white	46	44
Aspen	96	113
Basswood, American	81	133
Beech, American	55	72
Birch, yellow	74	72
Cherry, black	58	—
Cottonwood, black	162	146
Hackberry	61	65
Hickory, pecan, bitternut	80	54
Magnolia	80	104
Maple, sugar	65	72
Oak, California black	76	75
Oak, northern red	80	69
Oak, southern red	83	75
Oak, white	64	78
Sweetgum	79	137
Sycamore, American	114	130
Tupelo, black	87	115
Walnut, black	90	73
Yellow-poplar	83	106
Softwoods		
Cedar, Alaska	32	166
Cedar, eastern red	33	—
Cedar, incense	40	213
Cedar, Port Orford	50	98
Cedar, western red	58	249
Douglas fir, coast type	37	115
Fir, grand	91	136
Fir, white	98	160
Hemlock, eastern	97	119
Hemlock, western	85	170
Pine, loblolly	33	110
Pine, lodgepole	41	120
Pine, longleaf	31	106
Pine, ponderosa	40	148
Pine, red	32	134
Pine, shortleaf	32	122
Pine, sugar	98	219
Pine, western white	62	148
Redwood, old-growth	86	210
Spruce, eastern	34	128
Spruce, Engelmann	51	173
Spruce, Sitka	41	142

Most people underestimate the amount of water that can be contained in a piece of wood, and especially the amount of free water that must be removed before shrinkage begins to take place. Figure *1* shows the surprising amount of free water that can be contained in an average piece of wood.

Exposed to outdoor conditions, wood will lose its free water and eventually become **air-dry**. This term is used in many confusing ways, but should generally be taken to mean that the moisture content is in equilibrium with the outdoor atmosphere of a particular area. The amount of time to air-dry of course depends on the species, the thickness, the weather conditions and so forth. In New England, for example, where the relative humidity (RH) averages in the 70% to 80% range, lumber stored outdoors will air-dry to about 12% to 15% moisture content. Every woodworker should know about local conditions, and a call to the nearest weather bureau will usually provide information about relative humidity. Outdoors, the average equilibrium moisture content (EMC) is typically rather uniform, subject to only a moderate fluctuation in seasonal conditions, such as spring winds or late summer high humidity.

Recently I was curious about the progress of a couple of houses being built on a new road nearby. It was a typical rainy summer day, wet but pleasant for a walk under an umbrella. The first house I came to had the foundation capped off and the framing partially completed. Framing lumber had been unloaded carelessly along one side of the driveway, which had become a veritable lake. A pile of 2x8s had toppled over so the ends of a third of them were dunked in the big puddle. As the rain pelted down between the pieces in the loosened pile, I noticed the proud grade stamp on each piece boasting "KILN-DRIED." It was like having the delivery boy hang your dry-cleaning in the apple tree next to the lawn sprinkler. Ironically, two pallets of bricks were protected from the weather by a polyethylene covering; I supposed the mason would have to hose them down before he could lay up the fireplace.

In summary, two points of caution are in order here. First, structural lumber is dried in a kiln for efficiency, to reduce its shipping weight and to bring it down to levels that are suitable for construction, say 15% to 19%. Furthermore, even though lumber is kiln-dried, it can still gain or lose moisture depending on how it is stored. *Table 6* summarizes the pertinent moisture-content values.

1—This catalpa block had a moisture content of 114% and weighed almost 60 lb. when cut. When dried to 8% moisture content (for carving), it weighed only 30 lb. The gallon jugs show the amount of free water (F) and bound water (B) that was lost in drying. Some bound water (equivalent to B′) still remains in the wood at 8% moisture content.

Table 6—Moisture-content values.

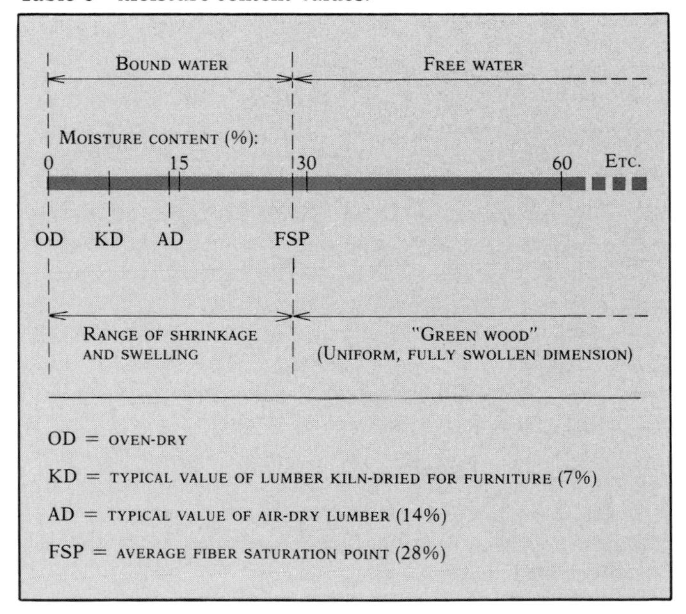

Dimensional change in wood

Shrinkage and **swelling** need little introduction for woodworkers. Wood shrinks or swells due to loss or gain of bound water from the cell walls. The amount of movement varies according to the orientation of the wood cells and is usually measured separately in the three principal directions: tangential, radial and longitudinal. The total amount of linear shrinkage that takes place in a given direction from the green to the oven-dry condition is customarily expressed as a percentage of the green dimensions. This total shrinkage is figured as follows:

$$S = \frac{(D_g - D_{od})}{D_g} \times 100,$$

where S = the total shrinkage, in percent (S_t = tangential shrinkage, S_r = radial shrinkage, S_l = longitudinal shrinkage); D_g = green dimension, D_{od} = oven-dry dimension.

Tangential shrinkage, for example, is expressed by the formula:

$$S_t = \frac{(D_g - D_{od})}{D_g} \times 100.$$

Figure *2* illustrates this formula for tangential shrinkage in a flatsawn board.

The orientation of the long-chain cellulosic structure in the cell wall is nearly parallel to the long axis of the cells. As water molecules enter and leave the cell walls, the resulting swelling or shrinkage is mainly perpendicular to the cell walls and does not influence their length. Similarly, pushing marbles into a straw broom would make the broom head wider, but would have little effect on the overall length of the broom head.

Total shrinkage of wood along the grain is normally only about 0.1%. An 8-ft. wall stud that is installed green and allowed to dry to an average 8% moisture content would shrink only about 1/16 in. In normal wood, longitudinal shrinkage is considered negligible. In juvenile wood or in reaction wood, longitudinal shrinkage can be as much as 2%, about 20 times that of normal wood. Abnormal wood usually develops unevenly in severity and distribution, and the resulting uneven longitudinal shrinkage may cause severe warping. In practice, however, we usually can forget about the longitudinal shrinkage of normal wood.

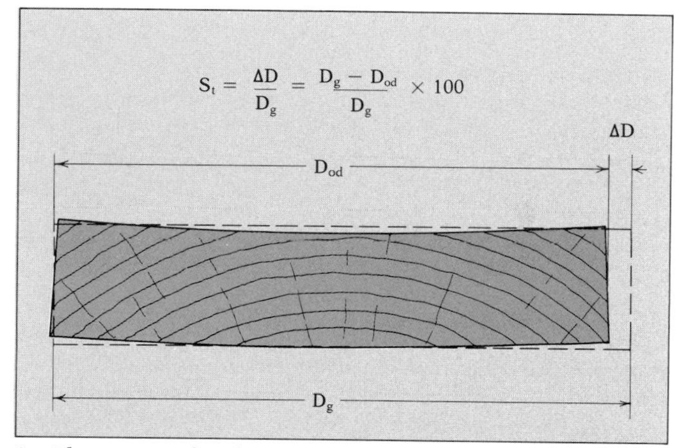

2—The percent shrinkage (in this case tangential shrinkage, S_t) is the change in dimension after oven-drying divided by the original dimension (D_g).

Transverse shrinkage, on the other hand, is significant. The shrinkage values in *Table 7* show considerable difference among different species. Tangential shrinkage (perpendicular to the grain and parallel to the growth rings) is always greater than radial (perpendicular to growth rings). Tangential shrinkage ranges from 4% in teak to 12.7% in overcup oak, with an overall

Table 7—Approximate shrinkage as a percent of green dimension, from green to oven-dry moisture content.

Hardwoods	Tangential	Radial	T/R
Alder, red	7.3	4.4	1.7
Ash, black	7.8	5.0	1.6
Ash, white	7.8	4.9	1.6
Aspen, quaking	6.7	3.5	1.9
Basswood, American	9.3	6.6	1.4
Beech, American	11.9	5.5	2.2
Birch, paper	8.6	6.3	1.4
Birch, yellow	9.2	7.2	1.3
Buckeye, yellow	8.1	3.6	2.2
Butternut	6.4	3.4	1.9
Catalpa	4.9	2.5	2.0
Cherry, black	7.1	3.7	1.9
Chestnut	6.7	3.4	2.0
Chinkapin, golden	7.4	4.6	1.6
Cottonwood, eastern	9.2	3.9	2.4
Dogwood, flowering	11.8	7.4	1.6
Elm, American	9.5	4.2	2.3
Elm, rock	8.1	4.8	1.7
Hackberry	8.9	4.8	1.9
Hickory, pecan	8.9	4.9	1.8
Hickory, shagbark	10.5	7.0	1.4
Holly, American	9.9	4.8	2.1
Honeylocust	6.6	4.2	1.6
Hophornbeam, eastern	10.0	8.5	1.2
Hornbeam, American	11.4	5.7	2.0
Locust, black	7.2	4.6	1.6
Madrone, Pacific	12.4	5.6	2.2
Magnolia, southern	6.6	5.4	1.2
Maple, red	8.2	4.0	2.0
Maple, sugar	9.9	4.8	2.1
Oak, black	11.1	4.4	2.5
Oak, live	9.5	6.6	1.4
Oak, northern red	8.6	4.0	2.2
Oak, overcup	12.7	5.3	2.4
Oak, red	8.9	4.2	2.1
Oak, southern red	11.3	4.7	2.4
Oak, white	10.5	5.6	1.8
Persimmon	11.2	7.9	1.5
Sassafras	6.2	4.0	1.6
Sweetgum	10.2	5.3	1.9
Sycamore, American	8.4	5.0	1.7
Tupelo, black	8.7	5.1	1.7
Walnut, black	7.8	5.5	1.4
Willow, black	8.7	3.3	2.6
Yellow poplar	8.2	4.6	1.8

Softwoods	Tangential	Radial	T/R
Baldcypress	6.2	3.8	1.6
Cedar, Alaska	6.0	2.8	2.1
Cedar, eastern red	4.7	3.1	1.5
Cedar, incense	5.2	3.3	1.6
Cedar, northern white	4.9	2.2	2.2
Cedar, western red	5.0	2.4	2.1
Douglas fir (coastal)	7.8	5.0	1.6
Douglas fir (inland)	7.6	4.1	1.9
Fir, balsam	6.9	2.9	2.4
Fir, white	7.1	3.2	2.2
Hemlock, eastern	6.8	3.0	2.3
Hemlock, western	7.9	4.3	1.8
Larch, western	9.1	4.5	2.0
Pine, eastern white	6.1	2.1	2.9
Pine, loblolly	7.4	4.8	1.5
Pine, lodgepole	6.7	4.3	1.6
Pine, longleaf	7.5	5.1	1.5
Pine, pitch	7.1	4.0	1.8
Pine, ponderosa	6.2	3.9	1.6
Pine, red	7.2	3.8	1.9
Pine, shortleaf	7.7	4.6	1.7
Pine, slash	7.6	5.4	1.4
Pine, sugar	5.6	2.9	1.9
Pine, western white	7.4	4.1	1.8
Redwood, old growth	4.4	2.6	1.7
Redwood, young growth	4.9	2.2	2.2
Spruce, Engelmann	7.1	3.8	1.9
Spruce, red	7.8	3.8	2.1
Spruce, Sitka	7.5	4.3	1.7
Tamarack	7.4	3.7	2.0
Yew, Pacific	5.4	4.0	1.4
Imported Woods			
Apitong	10.9	5.2	2.1
Avodire	6.5	3.7	1.8
Balsa	7.6	3.0	2.5
Banak	8.8	4.6	1.9
Cativo	5.3	2.3	2.3
Greenheart	9.0	8.2	1.1
Khaya	5.8	4.1	1.4
Lauan	8.0	3.8	2.1
Limba	5.4	4.4	1.2
Mahogany	5.1	3.7	1.4
Obeche	5.3	3.1	1.7
Parana pine	7.9	4.0	2.0
Primavera	5.2	3.1	1.7
Ramin	8.7	3.9	2.2
Spanish cedar	6.3	4.1	1.5
Teak	4.0	2.2	1.8
Walnut, European	6.4	4.3	1.5

average of 7.95%. Radially, shrinkage values range from 2.2% for teak or redwood to 8.5% for eastern hophornbeam, averaging 4.39%. It is reasonable to think of wood as having roughly 8% tangential shrinkage and 4% radial shrinkage.

This listing of shrinkage values suggests why some species, such as teak, mahogany, northern white cedar, redwood and catalpa, have a reputation for being woods that are quite stable. The values also suggest that others, such as beech, certain oaks and hickories, can be potentially troublesome.

The difference between tangential and radial shrinkage is caused by anatomical structure, principally the restraining effect of the wood rays, whose long axes are radially oriented. The **magnitude** of the differential shrinkage is critical to development of certain forms of warp and defects. The woodworker soon learns the importance of the **ratio** of tangential to radial shrinkage. *Table 7* reveals that the tangential shrinkage/radial shrinkage ratio averages 1.8, but individual species range from 1.1 (greenheart) to 2.9 (eastern white pine).

Over the entire range of moisture content—from fiber saturation point to oven-dry—shrinkage is approximately proportional to moisture loss. Typical relationships between tangential, radial and longitudinal shrinkage and moisture loss are shown in Figure *1* and in the photos *(2)* of red oak, red pine and beech.

1—Shrinkage vs. moisture content in northern red oak, a typical species.

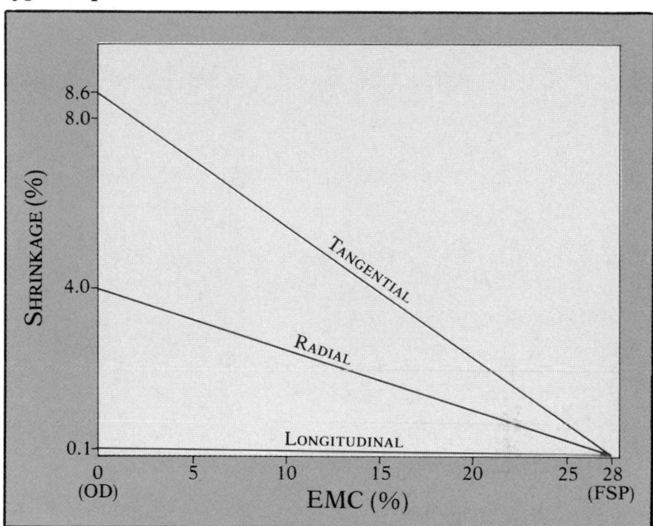

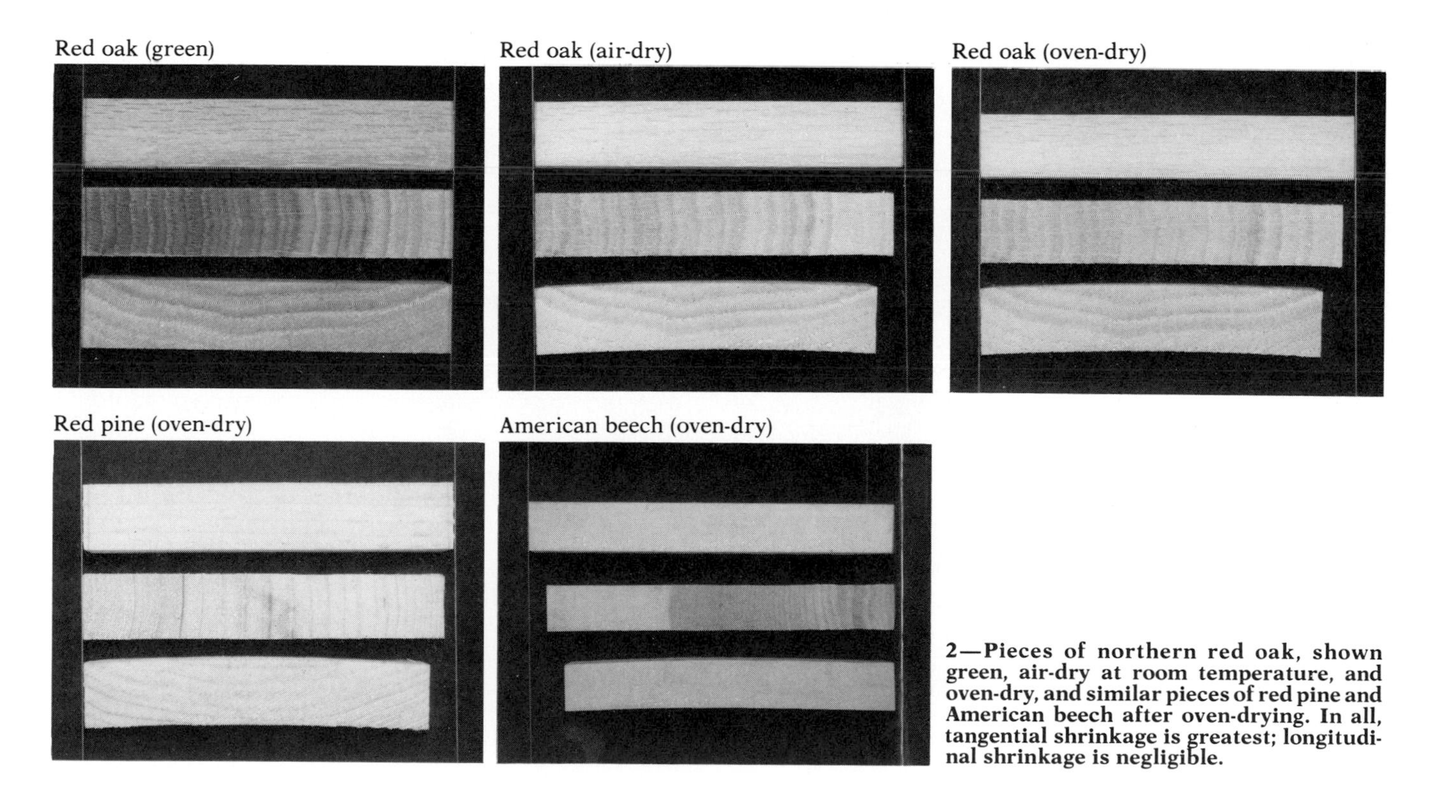

2—Pieces of northern red oak, shown green, air-dry at room temperature, and oven-dry, and similar pieces of red pine and American beech after oven-drying. In all, tangential shrinkage is greatest; longitudinal shrinkage is negligible.

Estimating shrinkage and swelling

There are many ways in which dimensional change in wood can affect the woodworker, and the discussion below considers some typical examples. In most cases, the mere change in dimension is noteworthy or troublesome. In others, uneven or unequal shrinkage is responsible for the problem.

Let's look at shrinkage from a quantitative point of view. Can we estimate how much a piece of wood will shrink under a given set of circumstances? Assume that a rough workbench was constructed out of freshly cut northern red oak *(1)*. Suppose the top were formed by spiking down flatsawn planks, edge-to-edge, that were originally 10 in. wide, as in the diagram. If the bench were left in an unheated garage, how wide would the cracks between the planks eventually become? In other words, how much would each plank shrink across its width as the wood dried out?

First, recall what we considered in Figure *1* on page 69 and Figure *1* on page 75. In the curve relating EMC to shrinkage, the tangential shrinkage is given as 8.6%. We can now construct a composite graph for northern red oak *(2)* that will relate relative humidity to moisture content and moisture content to shrinkage.

If the average relative humidity in the garage (unheated, but generally open to the outside air and sheltered from the direct sun and rain) is about 75%, we can follow the broken lines on our chart and see that a relative humidity (RH) of 75% will give an equilibrium moisture content (EMC) of about 13%, resulting in about 4.6% shrinkage. On a 10-in. board, this means 0.46 in., so each of the planks would be expected to shrink almost ½ in. in width.

Another way to estimate this shrinkage, once we know the equilibrium moisture content, is by the formula:

$$\Delta D = D\,S\left(\frac{\Delta MC}{fsp}\right)$$

where ΔD = change in dimension due to shrinkage

D_i = initial dimension

S = total shrinkage percentage. From *Table 7* use S_t (tangential shrinkage) for flatsawn, S_r (radial shrinkage) for edge-grain lumber

ΔMC = change in moisture content*

fsp = fiber saturation point (average value = 28%).

(In using the formula, remember that 8.6% means $\frac{8.6}{100} = 0.086$; 28% is similarly written as 0.28, etc.)

For our example, then,

$$\Delta D = (10'')(0.086)\left(\frac{0.28 - 0.13}{0.28}\right) = 0.46''.$$

* Since the formula applies only to moisture gain or loss below fiber saturation point, no values above fiber saturation point should be considered. For example, if the original moisture content of a board were 86% and it dried to 9%, ΔMC would be computed as 0.28 to 0.09 = 0.21, since no shrinkage would take place in drying from 86% down to the fiber saturation point, 28%.

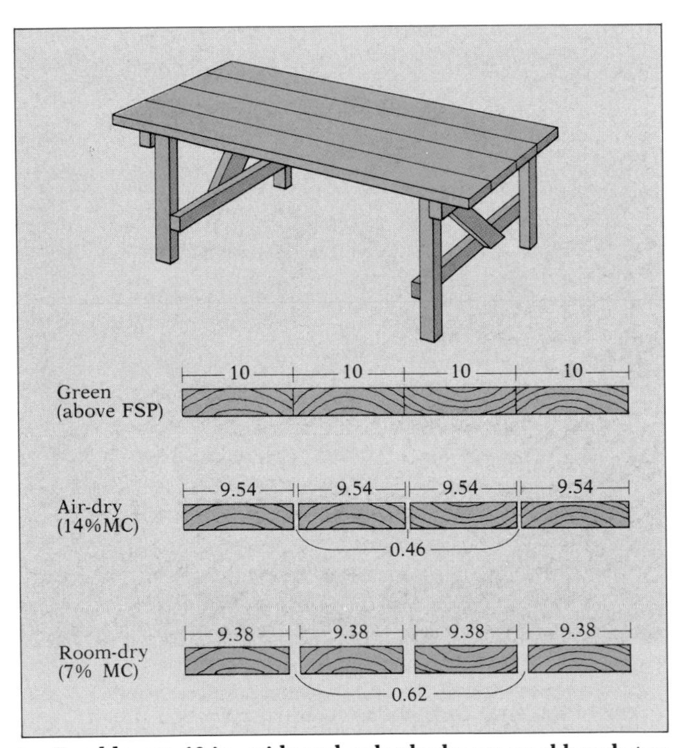

1—Freshly cut 10-in. wide red oak planks on workbench top would shrink as wood dried, leaving cracks.

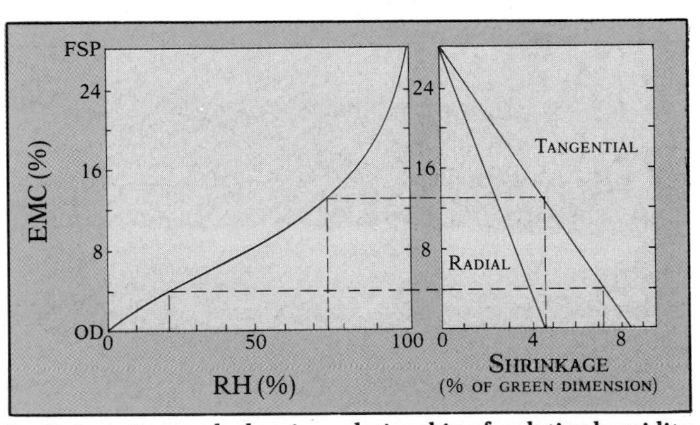

2—Composite graph showing relationship of relative humidity to moisture content and moisture content to radial and tangential shrinkage in northern red oak. At 75% relative humidity, the wood will shrink 4.6% in the tangential direction. Dimensional change in wood with change in atmospheric humidity is of utmost importance to the woodworker.

The boards are now 9.54 in. wide. Now suppose we move our bench into the house where the relative humidity is 40%, giving an equilibrium moisture content (EMC) of about 7.5%. How much more will the boards shrink? We can get an estimate* by again using the formula:

$$\Delta D = (9.54'')(0.086)\left(\frac{0.13 - 0.075}{0.28}\right)$$

$$= 0.1612''.$$

Each board will shrink another 0.16 in. to an estimated final width of 9.38 in.

Calculations using the above formulas are extremely valuable in accommodating dimensional change to allow swelling space for panels or clearance allowance for doors and drawers. A typical shop problem might be as follows: You are constructing a chest with drawers such as the one shown in Figure *3* using flatsawn sugar maple. How much allowance will ensure clearance of each drawer front in the frame?

Table 7 (page 74) indicates that the tangential shrinkage (S_t) for sugar maple is 9.9%. Assume that shop conditions have your lumber at an equilibrium moisture content (EMC) of 8%, but your experience has determined that with the light oil finish you plan to use, the EMC of the finished piece could go as high as 12% in summer weather. The drawer front should be no greater in width (F) at 12% than the drawer opening (D/O). The question becomes "how much smaller should it be at the current moisture content of 8%?" An estimate would be:

$$\Delta D = (D/O)(S_t)\left(\frac{\Delta MC}{fsp}\right)$$

$$= (D/O)(0.099)\left(\frac{0.12 - 0.08}{0.28}\right)$$

$$= (D/O)(0.0141).$$

For a 9-in. drawer front

$$\Delta D = 0.1272'', \text{about } 1/8''.$$

In the dead of winter, if the EMC drops to 5%, the gap for a 9-in. drawer will be

$$\Delta D = 9''\ (0.099)\left(\frac{0.12 - 0.05}{0.28}\right)$$

$$= 0.223'', \text{ over } 7/32''!$$

* Because the shrinkage percentages are based upon shrinkage from the green condition, the above formula is accurate only for shrinkages starting from the green condition, as in the original computation. For shrinkage of wood starting at a partially dry condition, the above formula will introduce an average error of about 5% of the calculated change in dimension. For most purposes such error is insignificant in view of other inherent sources of error. However, where a more refined estimate is desirable, the following formula should be used:

$$\Delta D = \frac{D_i\,(MC_i - MC_f)}{\frac{fsp}{S} - fsp + MC_i}$$

where

MC_i = initial moisture content
MC_f = final moisture content

(as noted above, neither MC_i nor MC_f can be greater than the fsp).

Thus, our last example would be computed as follows:

$$\Delta D = \frac{9.54''\ (0.13 - 0.075)}{\frac{0.28}{0.086} - 0.28 + 0.13}$$

$$= 0.1689''.$$

Thus our previous estimate of 0.1612 in. was off by 4.6%.

The formula can be used to estimate swelling that would be indicated by a negative value of ΔD.

Note that when the initial moisture content, MC_i, is at or above the fiber saturation point, the formula becomes

$$\Delta D = \frac{D_i\,(fsp - MC_f)}{\frac{fsp}{S} - fsp + fsp}$$

$$= \frac{D_i S\,(fsp - MC_f)}{fsp}$$

$$= D_i S\left(\frac{\Delta MC}{fsp}\right), \text{ as originally introduced.}$$

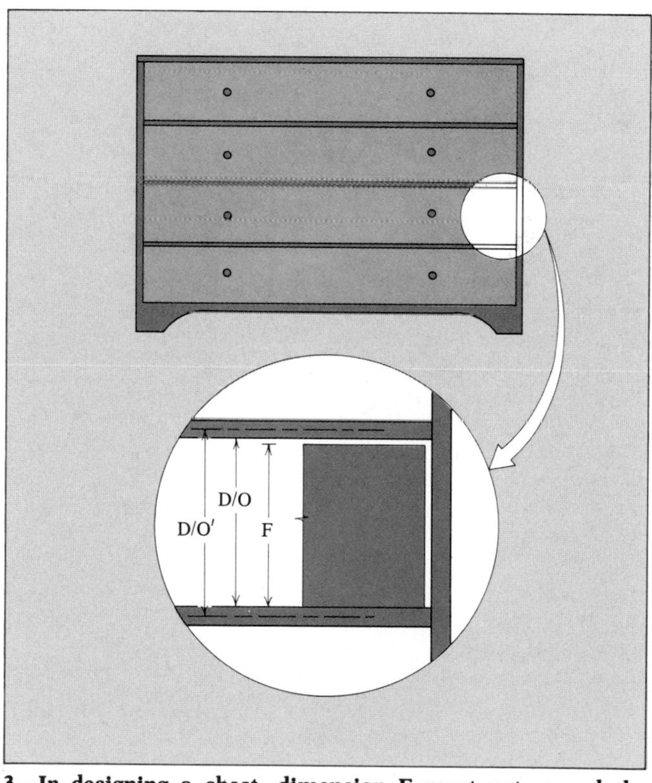

3—In designing a chest, dimension F must not exceed the drawer opening (D/O) when the wood swells in summer humidity. Dimension D/O′ is the centerline spacing of the drawer dividers.

If the dividers between drawers are large enough to be taken into account, D/O′, the centerline spacing of the dividers, should be used: the shrinkage, ΔD, will therefore account for the dividers as well.

A point usually overlooked in drawer design is the difference in future wood movement between two species worked at about the same equilibrium moisture content. Suppose we want to make a drawer 10 in. deep by lap-dovetailing flatsawn beech sides to quartersawn mahogany fronts. Can we expect the sides and the front to stay the same width while the relative humidity varies with the seasons? Assume that the drawer is made during the summer, with both species at an EMC of 10%, which drops to 6% in winter. The calculation would be the same the other way around, from winter to summer.

For a drawer front, S_r = 3.7% for mahogany.

$$\Delta D = (10)(0.037)\left(\frac{0.10 - 0.06}{0.28}\right)$$

$$= 0.052 \text{ in., or about } 1/20''.$$

For the drawer side, S_t = 11.9% for beech.

$$\Delta D = (10)(0.119)\left(\frac{0.10 - 0.06}{0.28}\right)$$

$$= 0.170 \text{ in., or about } 1/6''.$$

Thus the width of the drawer front changes a mere 1/20 in., while the side changes almost 1/6 in., three times as much. This change could have been predicted from the ratio between shrinkage coefficients, S_t of 11.9% for the beech and S_r of 3.7% for the mahogany. The choice of woods is extreme for the sake of illustration, but the amount of seasonal variation is not.

The importance of dimensional change in making furniture is generally known, but we may be tempted to assume it has no consequence in carpentry. Two examples will illustrate that even in building construction, dimensional change must be seriously considered.

The first instance came to my attention while visiting a partially completed home. The house was closed in, rough floors were down, and the interior sheetrock was installed. In the kitchen, 2x4s had been placed on the floor, approximately where they were to be used as base framing for the counters. They were sad-looking specimens of lumber, badly weathered and warped. Nudging one piece with my foot, I noticed that the floor underneath showed a wet spot, indicating that the moisture content was probably near or above the fiber saturation point. Sensing my concern over the quality of the material, the builder assured me that he used such pieces only where they would be nailed down securely, as in this case, so that the condition of the material was of no consequence. I did not pursue the matter, but I immediately recognized the ingredients of a potential problem.

It is customary to frame cabinets in place as shown in Figure *1*. The rear edge of the countertop is supported by a cleat anchored to the vertical studs of the wall. The forward edge is supported by the cabinet front, which in turn is supported by the base frame. Any shrinkage in the base frame will therefore be reflected by a drop in the front edge of the counter.

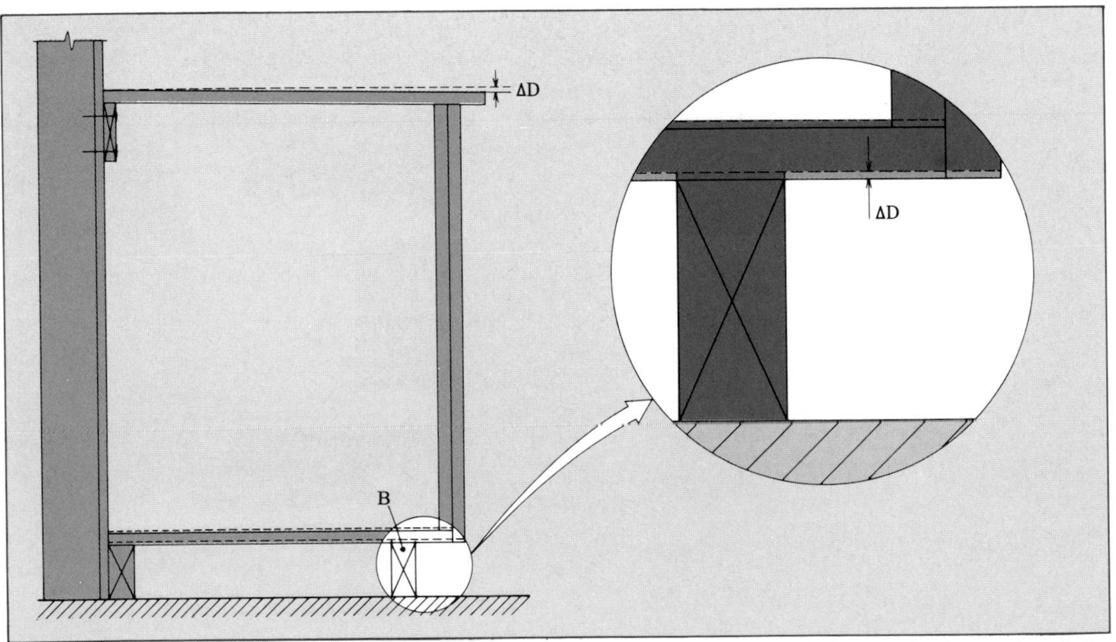

1—In typical counter construction, shrinkage in the base framing (B) eventually lowers the front edge of the counter. If this shrinkage is accounted for in the design stage, the countertop will be level when the equilibrium moisture content is reached.

How much? Let's consider the extreme. The piece in question had tangential grain across its 3½-in. dimension. It was probably red spruce, for which S_t (tangential shrinkage) is 7.8%. Assuming it was originally green but would eventually dry to 7.5% moisture content, $\Delta MC = 28 - 7.5 = 20.5\%$. Using our formula to estimate shrinkage:

$$\Delta D = 3.5'' \; (0.078) \left(\frac{0.205}{0.28}\right)$$

$$= 0.200'', \text{ a shade over } 3/16''.$$

I had occasion to check the counter a few years later, and the front appeared to have just about that amount of pitch, enough so that an egg would roll along the Formica and over the edge. I know the builder's work well enough to believe that it was level when he finished the job. I have since checked other counters and find the problem to be common. (I secretly check the roll of a marble or ping-pong ball when I don't have the nerve to ask to measure with a level.) I've never found a counter that slopes back to the wall.

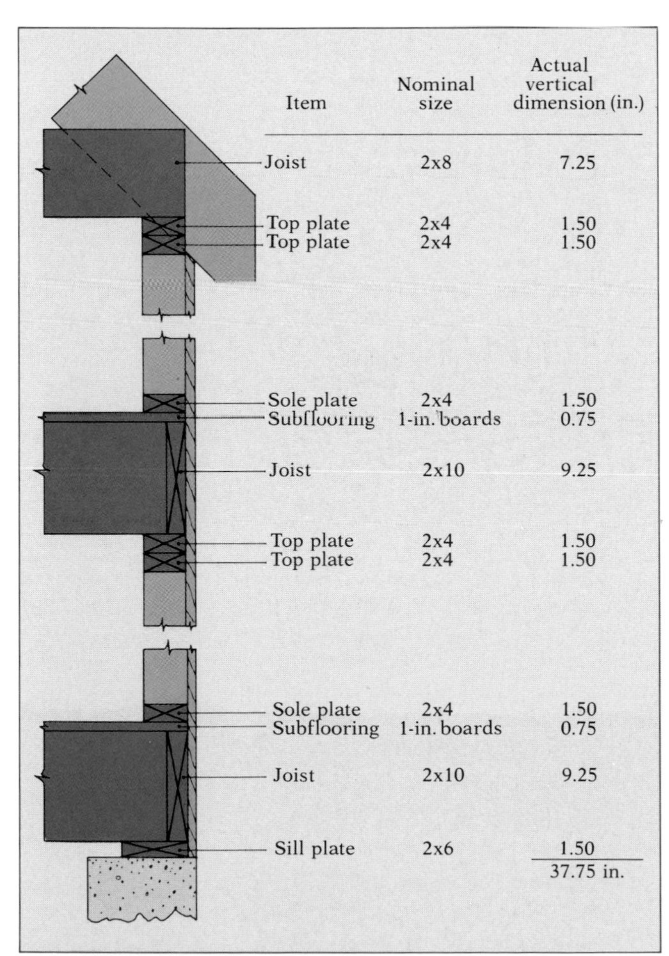

2—Wall-framing diagram of platform construction for a two-story dwelling shows 37.75 in. of framing members stacked vertically. From this figure the potential vertical shrinkage of the house can be calculated.

How does shrinkage affect an entire structure? Let's consider as an example a two-story house of typical platform construction, such as the one that is shown in Figure *2*. It shows a stacking of some 37.75 in. of framing members whose transverse shrinkage would take place vertically. What is the effect in the framing of moisture-content change that takes place after the structure is occupied and heated?

To assess the original moisture content in such cases, I have made extensive moisture measurements on framing lumber, both in lumberyards and in houses under construction. Readings can range between 11% and 26%, but the average has usually been about 18%.

Using moisture meters with insulated pin electrodes, I have also measured the moisture content of comparable interior framing lumber in a number of houses that had been occupied for at least three years. These readings were then checked against spruce wafers placed in indoor locations and allowed to reach equilibrium. Accounting for variations due to seasons of the year and to location, an overall average would be about 8%. The average moisture content change (ΔMC) then would be approximately $18\% - 8\% = 10\%$.

The most common woods for framing are Engelmann and eastern spruce, western hemlock and western fir. Because pieces vary from quartersawn to flat-grained, we arrived at the compromise species/growth-ring-orientation shrinkage value for the total shrinkage (S) of 5.8%. Applying the shrinkage formula once again:

$$\Delta D = 37.75'' \; (0.058) \left(\frac{0.10}{0.28}\right) = 0.782''.^*$$

The top of the house would be expected to drop by something over ¾ in. due to shrinkage. (If green lumber were used, the shrinkage would be about twice as great.) If the entire house were to shrink all over, the result would doubtless be quite insignificant. However, other problems may be forthcoming where rigid vertical plumbing or masonry structures are involved. I have seen countless broken joints along roof-chimney junctions apparently due to the shrinkage of the building. When framing is anchored to the chimney system, sloping floors, racked walls, and separated corner and ceiling joints may result.

These consequences are often referred to as the "settling" of a house. To me, this is unfair to the builder, because it implies that the building components were so loosely assembled that they eventually settled more compactly together, like so many crackers in a box. We should recognize "settling" as just another manifestation of the shrinkage of wood.

* If we also consider the longitudinal shrinkage ($S_l = 0.1\%$) in the upper and lower studs, for each 7′ 6″ or 90″ in length, $\Delta D = (2x90)\,(0.001)\,(0.10/0.28) = 0.064''$, an additional 1/16″. Some additional component for the roof rafters would also be present, depending on the roof pitch, etc.

Uneven shrinkage and swelling

Change in dimension is only one consequence of shrinkage or swelling. Even more serious effects may result when shrinkage or swelling is uneven throughout the piece, even though it is very small in magnitude. **Warp**, which is the distortion of a piece from its desired or intended shape, usually results from variable shrinkage in different directions, or from uneven shrinkage that causes stress in the piece *(1)*. **Cup** is a form of warp that is characterized by deviation from flatness across the width of a board. **Bow** is deviation from lengthwise flatness in a board. **Crook** is departure in end-to-end straightness along the edge of a board. **Twist** signifies that the four corners of a flat face do not lie in the same plane. **Kink** describes a localized crook, due to a knot. *Table 8* lists a number of woods by their tendency to warp during seasoning.

When uneven shrinkage causes stress that exceeds the perpendicular-to-grain strength of the wood, separation of cells occurs along the grain. Such failures are termed **checks**. Although most common on the surfaces and ends of pieces, they may also occur internally.

Tangential vs. radial shrinkage—Figure *4* illustrates cupping of flatsawn boards, with concavity away from the pith, the result of greater tangential than radial shrinkage. The severity of this effect is greater in boards with one face intersecting the pith. Since this face is radial, the opposite will be tangential, at least in one portion, and therefore will shrink twice as much. Some woodworkers say that the "rings tend to flatten out," which is an easy way to remember which way the

1—Principal forms of warp in boards.

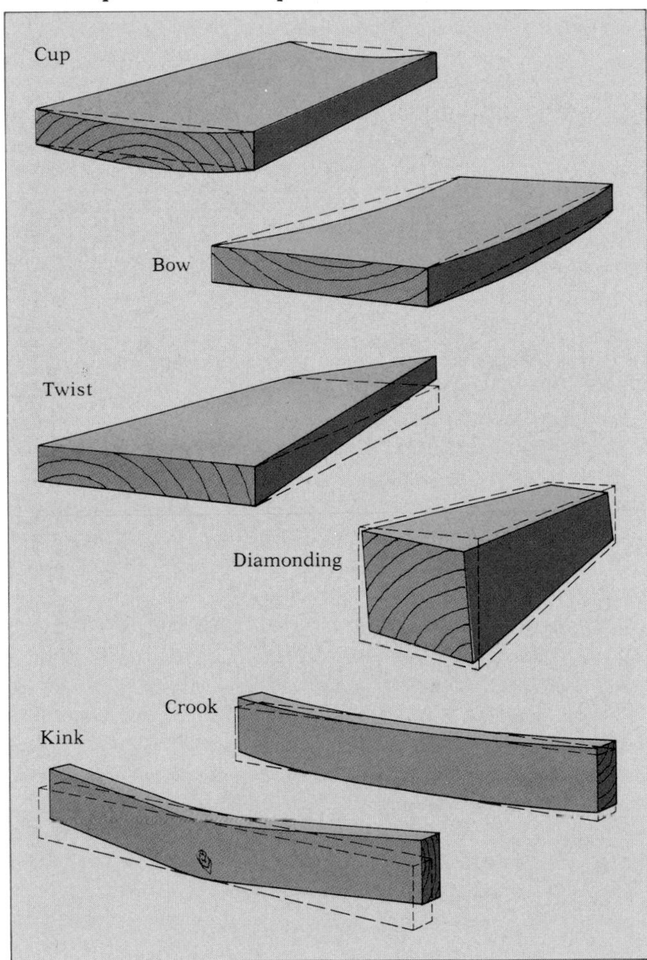

Table 8—Tendency to warp during seasoning of various woods.

Low	Intermediate	High
	Softwoods	
Cedars	Baldcypress	
Pine, ponderosa	Douglas fir	
Pine, sugar	Firs, true	
Pine, white	Hemlocks	
Redwood	Larch, western	
Spruce	Pine, jack	
	Pine, lodgepole	
	Pine, red	
	Pine, southern	
	Hardwoods	
Alder	Ash	Beech
Aspen	Basswood	Cottonwood
Birch, paper and sweet	Birch, yellow	Elm, American
Butternut	Elm, rock	Sweetgum
Cherry	Hackberry	Sycamore
Walnut	Hickory	Tanoak
Yellow-poplar	Locust	Tupelo
	Magnolia, southern	
	Maples	
	Oaks	
	Pecan	
	Willow	

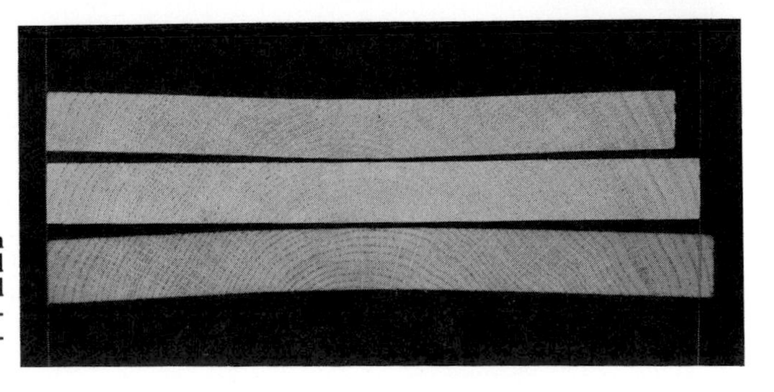

2—Strips in photo at right were cut in sequence from the end of an air-dry red oak board. As shown by the middle strip, it measured 9½ in. wide at 14% moisture content. The top strip was then dried to below 4% moisture content—it both shrinks and cups. The bottom strip was allowed to readsorb moisture to over 20% MC. It expands, and cups in the opposite direction.

board will cup, but has little to do with the reason why. Cup is reversible upon swelling *(2)*. If lumber cut against the pith is restrained from cupping, it may crack itself open. Furthermore, any attempts to flatten by force a board that has cupped in drying will usually produce a crack *(3)*.

A square or rectangular piece with diagonally oriented growth rings will shrink twice as much across one diagonal than the other, distorting the board into a diamond—the term **diamonding** designates the effect *(4)*. This has its counterpart in turnings, since the greater tangential shrinkage will turn a round cross section into an oval. Holes bored lengthwise in wood also show this behavior, since a hole in a piece of wood reacts the same as a piece of solid wood of the same shape.

The classic problem of non-uniform thickness change in boards is often traceable, at least in part, to the difference between radial and tangential shrinkage. The typical case in point emerges in quartersawn boards. If one edge of a board is located near the pith in a tree, the growth rings within it will have sharp curvature, and shrinkage in its thickness will be essentially radial *(r)*, as at *A* in Figure 5. At the opposite edge, the growth-ring orientation is more truly edge-grained and will have tangential shrinkage *(t)* through the thickness of the board along the edge toward the bark. Although slight, such differences may be quite noticeable along a glue joint if adjacent edges have bark edges butted to pith edges. The problem can be minimized by matching bark edges to bark edges and pith edges to pith edges, as in *B*.

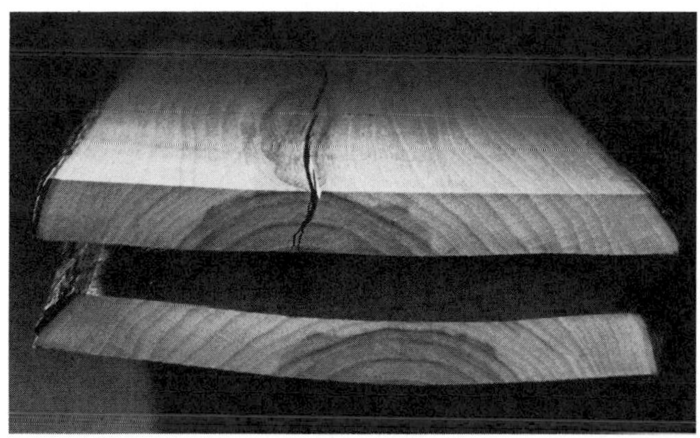

3—Half of a cupped elm board (top) was flattened and cracked by roller pressure in the thickness planer. The other half (above) exhibits the original cup.

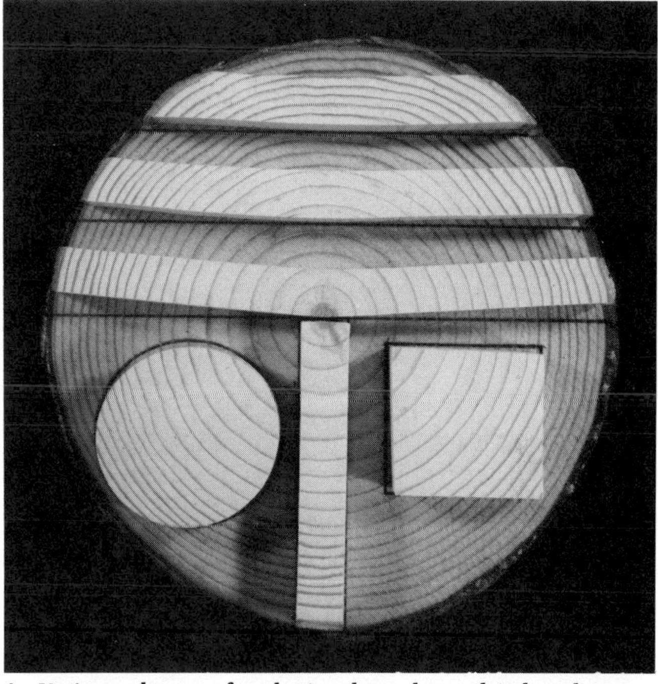

4—Various shapes of red pine have been dried and superimposed on their original positions on an adjacent log section. The greater tangential than radial shrinkage causes squares to become diamond-shaped, cylinders to become oval. Quartersawn boards seldom warp, but flatsawn boards cup away from the pith.

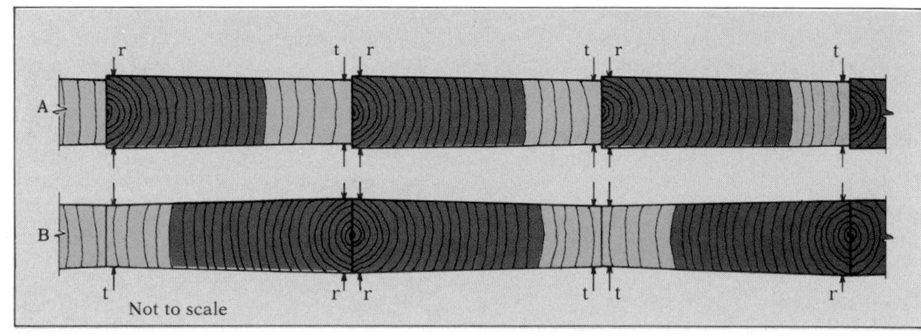

5—The difference between radial and tangential shrinkage can be quite noticeable along a glue joint (A), where quartersawn boards are butted bark side to pith side. The problem can be minimized by matching bark to bark and pith to pith (B).

A

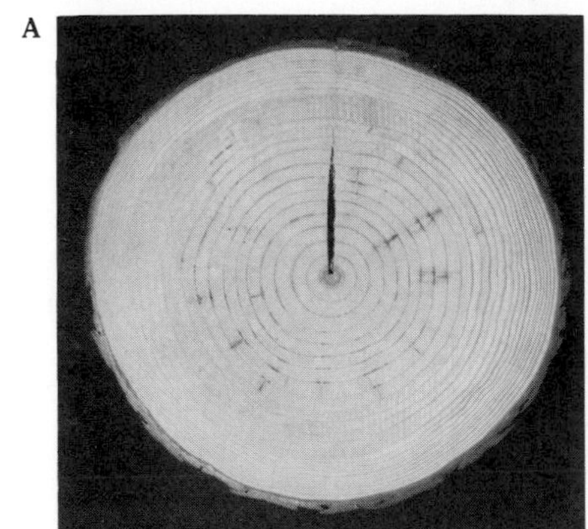

B

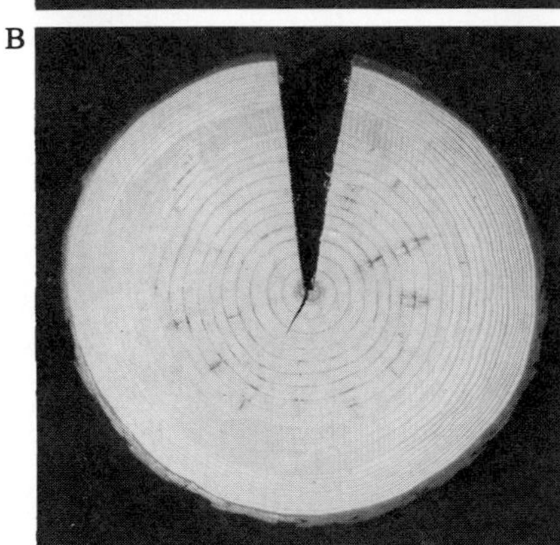

C

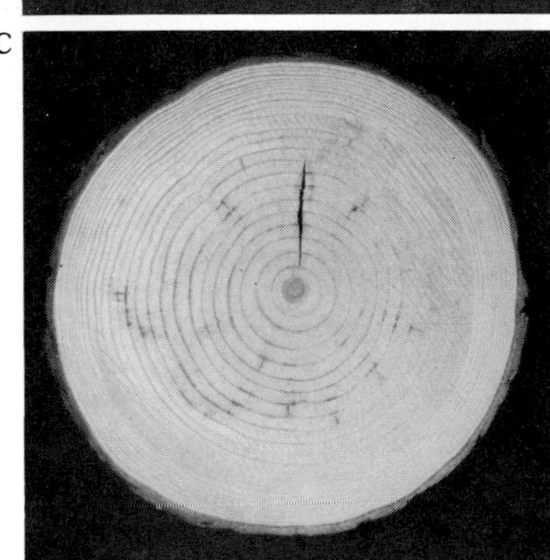

D

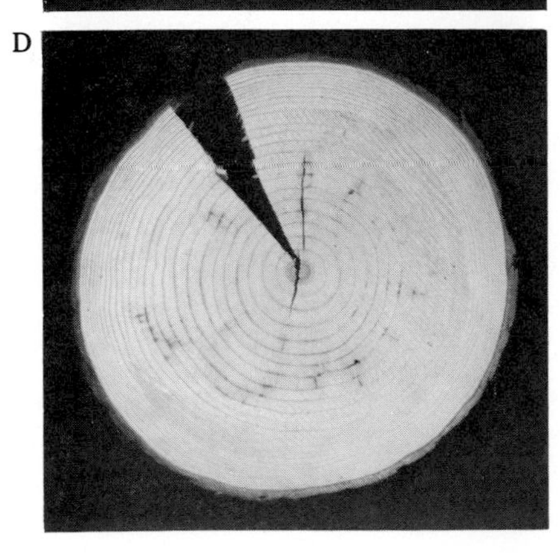

1—Results of drying cross-sectional discs of red pine. A disc with a radial slot sawn in it begins to dry, (A). The heartwood has a lower moisture content and shrinks first, opening the slot near the pith. Later the sapwood also dries, and the greater tangential shrinkage opens the slot wide, (B). An unslotted disc begins to dry, (C), and a crack opens first near the pith. As the disc dries further, the stress of greater tangential shrinkage exceeds the strength of the wood, and the disc cracks radially, (D). The crack is not as wide as in the first disc (B), where the strain was relieved by slotting.

Probably the most familiar—or notorious—manifestation of shrinkage is the radial **cracking*** of logs or log sections caused by the stress resulting from greater tangential shrinkage, which cannot be accommodated by distortion alone. The stress eventually becomes great enough to crack the wood radially. It's interesting to compare the effect of bandsawing a radial cut into discs before drying begins *(1, 3)*. In the red pine discs used, the heartwood begins to shrink first, doubtless because of its lower moisture content (41% in contrast to 206% for the sapwood). In oak, the bandsawn piece begins to open before any dominant crack occurs in the unsawn disc, suggesting that stress is developing in the unsawn one while it is being relieved by the saw slot in the other. In each species the crack in the sawn disc is obviously wider than in the uncut disc.

Sometimes a disc can be dried without cracking. Success is favored by a number of factors. Species with low shrinkage percentages and low tangential-to-radial shrinkage ratios are better prospects. Catalpa, for example, develops only a slight tendency to crack because of its low shrinkage percentage *(2)*. I have dried many catalpa logs and discs up to about a foot in diameter merely by drying slowly. But the slowest drying in the world wouldn't bring a red pine disc through without a crack. Every once in a while I get cocky and think I have succeeded, but the fact remains that the stress is there and usually exhibits itself by eventual cracking. I had a disc of unfinished red pine cut near a branch whorl, giving it an attractive star effect. It sat around my office for three years without defect. Finally one day during a bitter cold spell the humidity bottomed out and did the job. The result is shown in Figure *4*.

A friend of mine once turned a pedestal out of a green

2—Discs of red pine (left) and catalpa (right) after drying to 6% moisture content. Radial slits were sawn into green discs; crack width indicates the relative stability of these species.

* The large radial failures in pieces containing the pith due to dominant tangential shrinkage are often called **cracks** rather than checks.

cherry log. He took it into the kitchen to admire during dinner and *listened* to it crack.

On one occasion I was asked to give a talk, on rather short notice, on the subject of stabilizing wood with polyethylene glycol-1000 (PEG). I found I had no samples available to illustrate the results. With only five days to get ready, I sawed two red oak discs about 9 in. in diameter. For four days I soaked one disc in PEG, the other in water. The day before the lecture I popped them both in the oven, sure that the untreated one would crack. The next day, about two hours before the lecture, I opened the oven. To my horror neither had a crack. I knew the untreated disc must be loaded with shrinkage stress, but who can see stress? In disgust, I slammed the discs down on a bench top. A crack suddenly appeared in the untreated disc, and within a minute or two opened up to over ¼ in. I had my lecture props on time.

Low-density woods seem favorable to disc drying, apparently because of their ability to deform internally to relieve the shrinkage stress. Here again catalpa is a winner because of its moderately low density.

Size also has an important effect on drying. I've been able to dry only a few species of wood in disc diameters over 8 in., mostly low-density species. However, I have a set of coasters of 27 different species which are cross-sectional discs of 3-in. to 3¼-in. diameter. To make them, I cut lengths of tree stem about 18 in. long in the winter (so the bark will be tight), coat both ends with paraffin and put them aside. If the logs crack, they join my fuel supply. If not, I bandsaw them into discs ⅜ in. thick after the logs have stopped losing weight. Then I wait another several months, sand them smooth and apply a few rubbed coats of urethane. Of the hundreds I have made, I know of only one that cracked in use. But I know they all have some stress because of greater tangential than radial shrinkage.

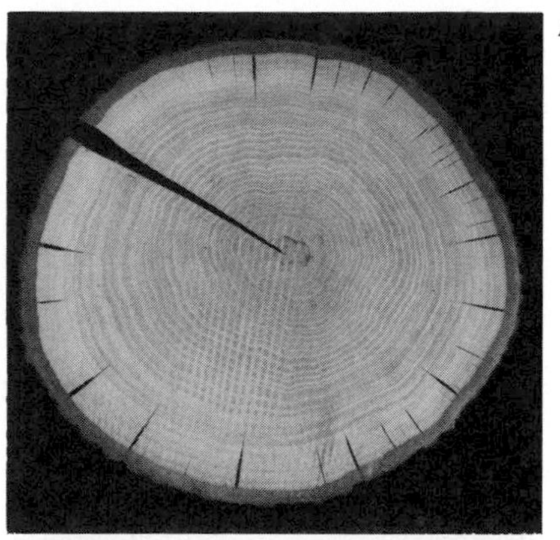

A

B

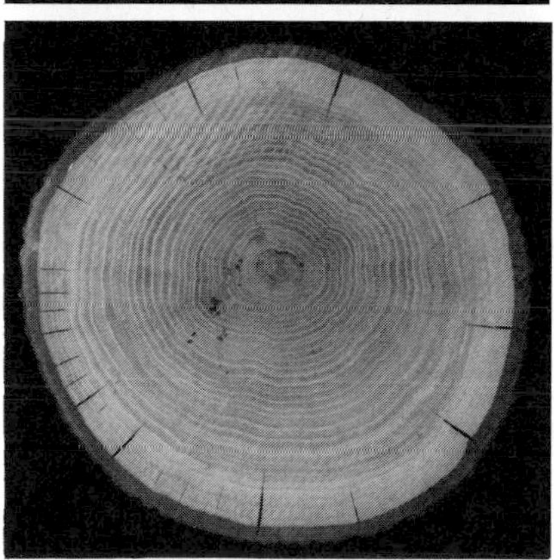

C

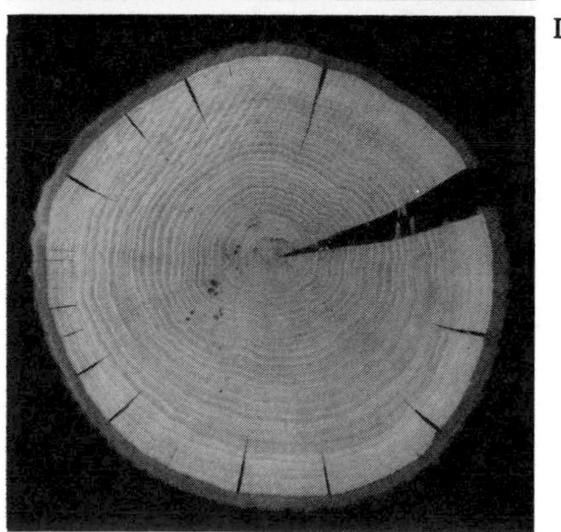

D

3—Results of drying cross-sectional discs of northern red oak. This disc had a radial slot sawn in it, (A). Because of heartwood extractives, the sapwood dries first and shrinks more, resulting in radial sapwood checks. When the heartwood dries, the slot opens wide, (B). A disc without a slot, (C), also forms sapwood checks in early stages of drying. Eventual shrinkage stress opens a radial check, (D).

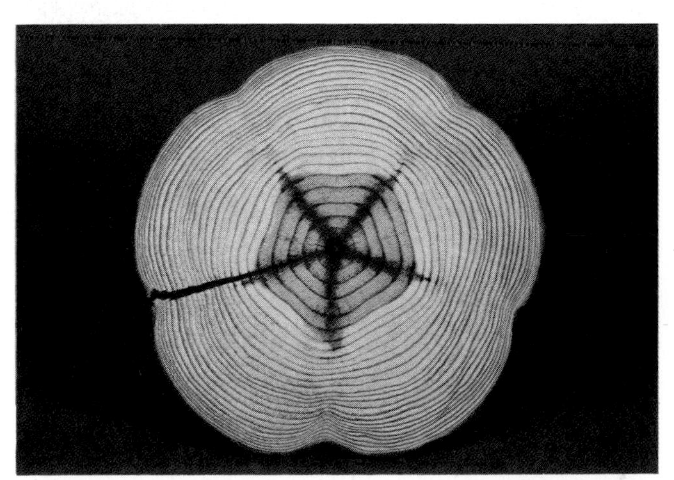

4—This star-patterned red pine disc remained crack-free for three years, until an extremely dry period finally caused the radial crack.

Perpendicular-to-grain vs. parallel-to-grain—A second type of dimensional behavior that causes problems is the wide discrepancy between perpendicular-to-grain dimensional movement and the nearly negligible longitudinal instability. A classic problem is the mitered joint, which is shown in Figure *1*. The joint opens on the outside in summer humidity, and on the inside in winter dryness. This one is familiar because it is out in plain sight. But a hidden example of problems caused by differences in longitudinal and transverse shrinkage is the mortise-and-tenon joint. It somehow seems absurd that in an era in which color television is commonplace, men walk on the moon and heart transplants succeed, we still have wobbly chairs.

Wobbly chairs are caused principally by the difference between the dimensional change of a mortise and the dimensional change of a tenon. The simplest joint is a round tenon in a drilled hole or mortise, as in the insertion of a chair rung into a chair leg. Perpendicular to the grain direction of the chair leg, the tenon and the hole shrink and swell by about the same amount in diameter. In the direction of the leg the hole is virtually stable; the rung, however, will have pronounced dimensional response, especially if the growth ring orientation of the rung is vertical in the joint. As will be discussed in Chapter 6, the allowable elastic compression and tension strains are smaller than the amounts of swelling and shrinkage that develop in response to seasonal humidity fluctuation. The result will be overcompression of the rung when it tries to expand in the vertical direction. Upon redrying to its original moisture content, it will shrink to a smaller diameter than the original. Glue is only partially successful in preventing this compression-set loosening. With or without glue, extreme moisture variation can cause looseness of the joint *(2)*.

1—These red oak frame corners were tightly mitered when originally assembled. The upper one was dried, and the lower one dampened. Because wood is stable along the grain but shrinks and swells across the grain, the joints open as shown.

Uneven drying—A third cause of dimensional troubles is uneven shrinkage due to uneven drying. A familiar case is when a pile of air-dry lumber is brought into a heated building. Cupping soon develops on the top boards as the exposed faces dry and shrink first. The cupping back and forth of a tabletop finished only on the upper face is another common example. But perhaps the most universal problem is end-checking. Water moves longitudinally through wood 10 to 15 times faster than it moves perpendicular to the grain. Therefore, end-grain surfaces rapidly lose their moisture and will be first to drop to below fiber saturation point and begin to shrink. If the shrinkage exceeds about 1.5%, tension failures in the form of end-checking may occur *(3, 4)*.

Here's another way to look at it: Let's assume that moisture moves, on the average, twelve times faster along the grain than across it. Suppose a board is 1 in. thick. Up to 6 in. from either end, water molecules at the mid-thickness of the board have a better chance to escape through the end-grain surface than through the

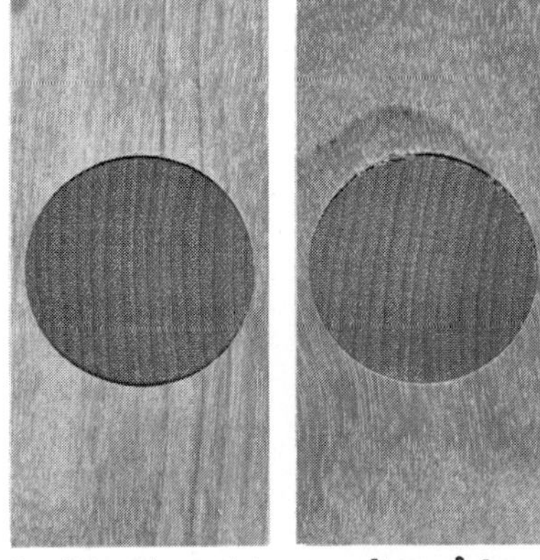

2—Drastic moisture cycling of dowel joints with vertical growth-ring orientation results in compression set and loosening of an unglued joint, left. Glue reduces the extent of loosening, right. Note glue failure at top of joint.

3—The ends of logs dry first, and the resulting shrinkage causes end-grain checks, as shown here in northern red oak.

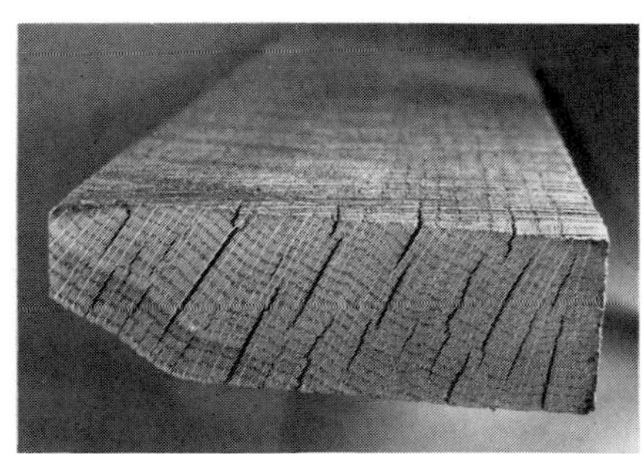

4—Because drying is more rapid along the grain than across, boards dry first near the ends, shrink and develop end checks.

side-grain surface. Except for the end 6 in., drying from the board should be uniformly slow because most molecules will escape through the side grain. The objective of end-coating boards with sealers is to prevent rapid end-drying and create uniform side-grain drying right to the end of the board.

Stresses are ever-present in drying, because there must be differential drying in a piece of wood to make the moisture move. If the moisture gradients are great enough, serious defects will develop.

Variation in shrinkage properties—A fourth category of troublesome uneven shrinkage results when shrinkage properties vary within a given piece of wood—a characteristic of juvenile and reaction wood *(5)*. These abnormal woods shrink more along the grain than normal wood. That might not be a problem if the degree of abnormality were uniform, but it never is. Typically, the severity of reaction wood varies within a given piece, or it may even be combined in the piece with normal wood. Bow and crook are commonly traceable to such variable longitudinal shrinkage.

Twist is sometimes the result of uneven reaction wood formation but most pronounced twist is usually associated with spiral grain. Those boards that form veritable propellers are usually caused by spiral grain.

In some cases, the reaction wood may shrink so powerfully that it crushes adjacent normal wood in compression parallel to the grain (*6*, left). At the same time, if the reaction wood is a small portion of the board, it may pull itself apart by failing in tension parallel to the grain (*6*, right). A house we once lived in was framed in native hemlock, which runs heavy to reaction wood. One night I heard what sounded like a gunshot coming from the cellar. Racing down the stairs, I found nothing alarming. However, in one corner there was a concentration of dust particles in the air. Looking up, I discovered the source of the noise. The lower face of one of the hemlock joists, with obvious reaction wood, had a gaping cross-break. The dryness of furnace heat was the final increment of unbearable shrinkage tension, causing the beam to fail abruptly.

In some species, extractives may significantly reduce shrinkage of heartwood as compared to sapwood. Such differences between dimensional change as noted in Figure *3* on page 83 may also produce troublesome results in boards. The sapwood/heartwood shrinkage difference is also doubtless responsible in part for the uneven thickness variation noted in Figure *5* on page 81. In any case, the practice of matching sapwood to sapwood and heartwood to heartwood is a logical one.

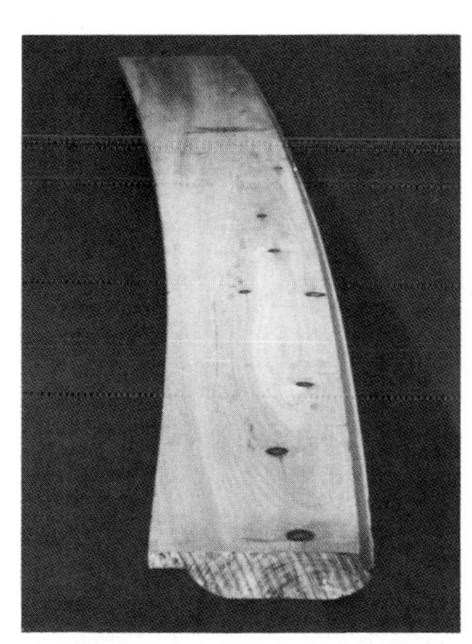

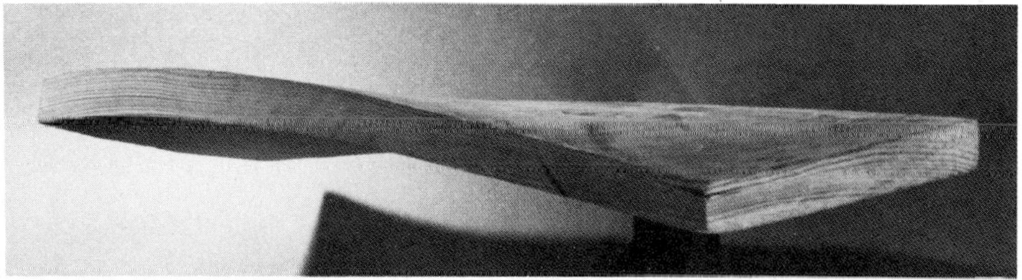

5—Reaction wood is usually to blame when boards bow or crook, like this piece of pine siding (left). Reaction wood may also cause twisting, as shown in the Japanese fir sample above.

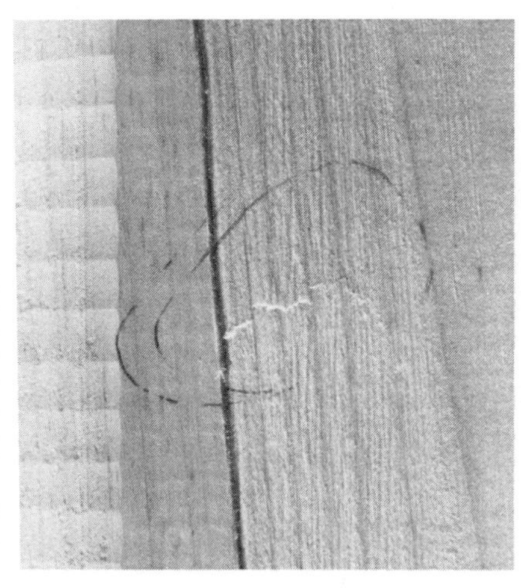

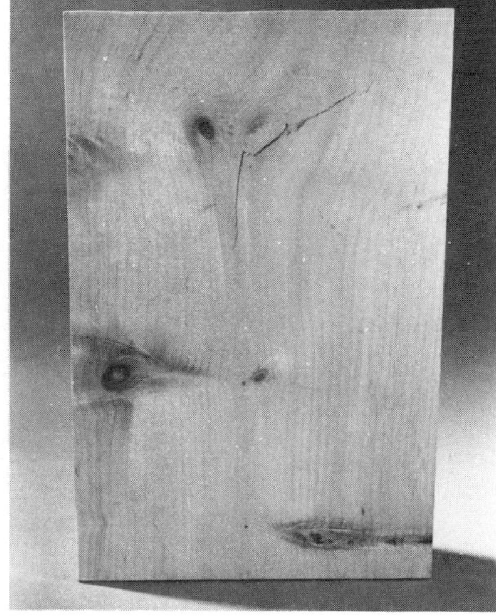

6—Results of differential shrinkage in eastern white pine boards containing reaction wood. In the board at right, compression wood predominates, and its greater longitudinal shrinkage caused compression failures in the adjacent normal wood. The ridge indicates reaction wood that has shrunk less across the grain than the normal wood did. In the board at far right, normal wood predominates, and the greater longitudinal shrinkage of reaction wood has pulled it apart.

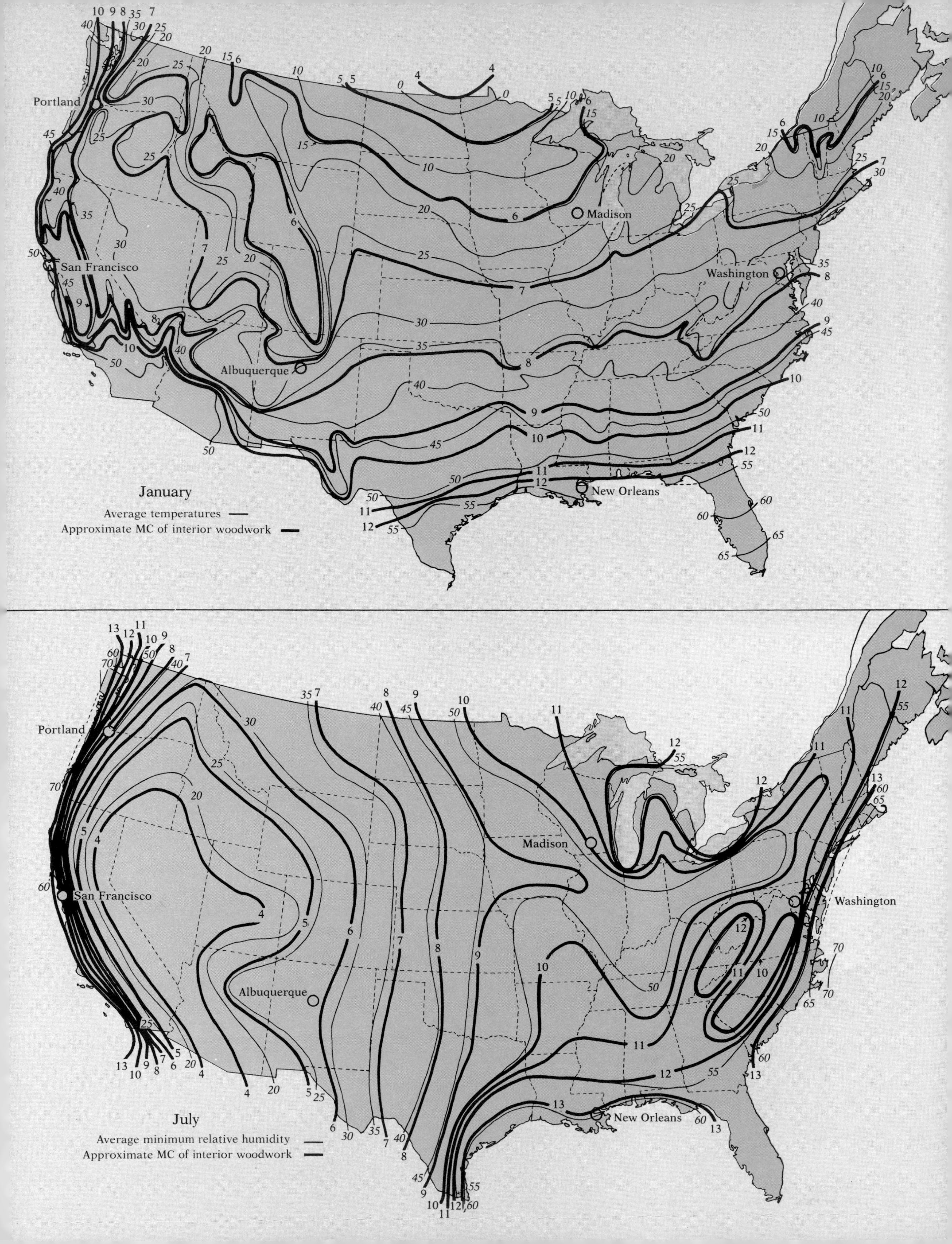

Portland
San Francisco
Albuquerque
Madison
Washington
New Orleans
January
Average temperatures
Approximate MC of interior woodwork
Portland
San Francisco
Albuquerque
Madison
Washington
New Orleans
July
Average minimum relative humidity
Approximate MC of interior woodwork

Coping with Wood Movement

The examples mentioned in the previous chapter describing the ways that moisture-related instability of wood can cause problems suggest that the woodworker must not only understand basic wood-moisture relationships, but must also know the alternatives that can be considered in dealing with dimensional instability.

As I see it, there are five fundamental approaches to coping with dimensional change in wood: preshrinking by drying prior to use, control of humidity, mechanical restraint, chemical stabilization and design. Alone or in combination, each has its individual merits in dealing with a particular woodworking situation.

Preshrinking wood by seasoning, as obvious as it is, is too important to pass over lightly. Although wood is dried for many other reasons (to reduce weight, to prevent deterioration by fungi, to increase strength, to permit gluing and finishing) the principal objective is to have shrinkage take place *before* rather than *after* the final product is completed. The key to this approach is drying the wood to a moisture content consistent with the average relative humidity in which the finished piece will be used. The ideal, of course, would be wood that subsequently would never shrink or swell. However, this ideal is not totally realistic. It is difficult to predict the appropriate equilibrium moisture content perfectly, and environments seldom remain stable.

The target moisture content depends upon a number of climatic and environmental factors, such as the local extreme and average levels of humidity, whether the finished item will be used indoors or out, and the extent of winter heating. In the Northeast, where the average annual relative humidity is typically in the 70% range, a moisture content of 12% is appropriate for outdoor items. For interior work, however, where central heating drives the relative humidity to 25% or less, a moisture content of 7% or 8% is more appropriate. The moisture content would be understandably higher in humid areas such as the Gulf states and lower in arid regions like the Southwest. I recall a day in early June where the outdoor temperatures in central Massachusetts were in the mid-60°s F and the relative humidity was in the 60% to 80% range. Flying to Phoenix, Arizona, I discovered the day's weather featured a high of 113° F and a low of 84° F with a 6% to 24% range in relative humidity. Since this would represent the immediate transition from an EMC of 13% to 14% to one of 3% to 4%, I was glad I had not brought along a half-finished piece of sculpture to work on. Imagine the problems encountered by anyone moving furniture (or any wooden items) from one such extreme to another. Figure *1* shows the average January and July levels of moisture content for interior woodwork in all areas of the United States. Drastic variation can also exist within a rather limited geographic region—the summer moisture content is 13% along the coast of California while eastern California falls within the 4% isotherm. These data stress the need for becoming familiar with local weather history and for keeping track of current weather conditions.

Preshrinking wood is one thing: keeping it there is another. Careful attention must be given to the second basic consideration—**atmospheric control**. Air conditioning is effective, but not always possible or even sensible, except for priceless museum objects and the like. Another approach is to control humidity through **isolation**, by keeping the wood in a reasonably airtight container. This may be a small display box, a glass jar, a plastic bag or simply a coat of finish *(2)*.

If there were such a thing as a coating system for wood that was totally impervious to moisture, the problem would be solved. All subsequent loss or gain of moisture would be arrested and the original moisture content would not be very critical. In reality, however, no finish can block the passage of all moisture. Finishes are merely obstacles to moisture passage or buffers to curb the extremes. This point can not be emphasized enough. As a companion to proper preshrinking, an effective finish is the most relied-upon approach to minimizing dimensional response in our variable atmo-

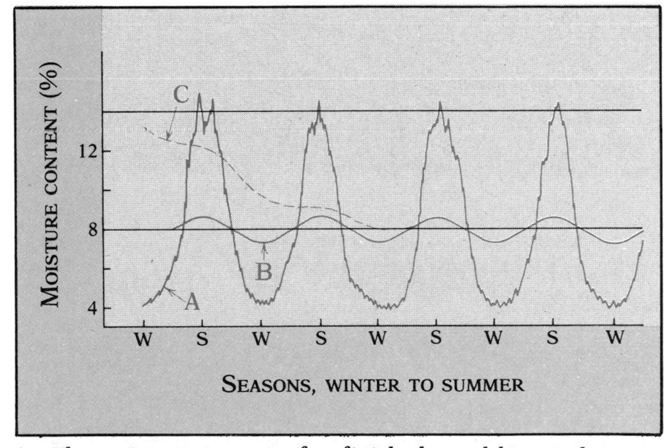

2—The moisture content of unfinished wood kept indoors in most parts of the United States fluctuates seasonally from 4% to about 14% (A). The moisture content of wood that has been kiln-dried and well finished with lacquer or varnish oscillates in a much narrower range, around 8% (B), while wood that has been air-dried and coated with finish continues to dry gradually, then oscillates in the same range (C).

1—Average January and July levels of moisture content for interior woodwork in areas of the United States.

sphere, but even as we must appreciate the important potential role of finishes as moisture barriers, we must not expect too much of them. In Chapter 10 we will return to this important subject and consider the relative effectiveness of various finishing materials in retarding moisture exchange with the atmosphere.

In some cases the above measures may not be adequate to ensure the desired dimensional control. It may then be advantageous to prevent wood from changing its dimension through other means. One such approach is **mechanical restraint**.

A stable material such as metal, whose strength is clearly greater than that of wood, can be used to limit swelling or shrinkage of the wood. Metal straps or long bolts can be used to some extent, as can plastics and other synthetics. More commonly, wood itself is used by taking advantage of its superior strength and stability in the grain direction. This is demonstrated in the cross-ply construction of plywood. Douglas fir, for example, has an average tangential shrinkage of 7.6% to 7.8% and a longitudinal shrinkage of about 0.1%. When Douglas fir plywood is made having equal amounts of wood in both directions, providing mutual restraint, the panel has identical shrinkage levels in both directions of only about 0.5%. With only a slight sacrifice of stability along the grain, the normal tangential shrinkage is reduced to one-fifteenth of its original value. Various types of composite boards (referred to generically as particle board or flake board) have stability similar to that of plywood because of the mutual stabilization of the particles, chips or flakes, plus the adhesive used.

Crossbanding solid lumber panels with thin veneers yields dramatic results. An interesting experiment is to make up panels of pine or basswood with and without hardwood plywood crossbanding on face and back, trim the panels to the same size, leave them unfinished and compare their dimensions over a period of time.

For even more revealing results, add a third matched panel that has a crossband on only one face. This will show the importance of **balanced construction** (symmetry on either side of the central plane with regard to thickness, species properties, etc.). Otherwise the uneven restraint will result in unbelievable warp *(1)*. That is why plywood is made in odd numbers of plies, so the face and back plies will always have parallel grain direction, thus preserving symmetry and balance of construction. A warped panel of plywood may be the result of uneven sanding of the opposite face plies, thereby "unbalancing" the panel.

Cleats or other cross members are used in countless ways to restrain dimensional movement in wood, but such constructions are successful only if the restraint is uniformly transmitted and distributed. Adhesives offer the most uniformity, in contrast to the point attachment of mechanical fasteners. As in the face plies of plywood, restraint can be well distributed in thin layers by attachment along only one face. When lumber thicknesses are restrained on only one surface, the free surface may develop stress concentrations, resulting in checking. Also, when panels are restrained only at their ends, strain concentration may cause failure in the unrestrained central area.

Chemical treatment also helps prevent dimensional movement. One such **chemical stabilizing** technique is impregnation of the wood with a monomer (single molecular form of chemical) and polymerizing it (forming it into a long-chain or complex molecular form) with heat or radiation, thus forming a "wood-plastic composite (WPC)." One of the most commonly used monomers is methyl methacrylate. Most of us know this material in its polymerized form, commercially produced as Plexiglas and Lucite. Since the plastic is in the wood, not on the wood, the wood can be polished without further application of finishing materials. Because methyl methacrylate is colorless, wood treated with it is similar in appearance to wood finished with conventional materials. Stability is drastically improved as well. Such wood has been used for novelties and other items in which hard-

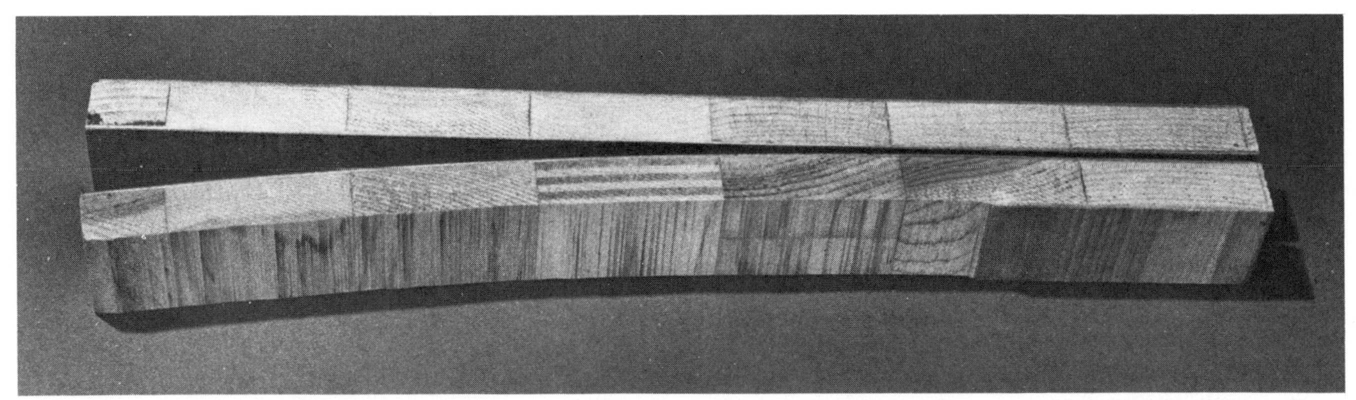

1—The consequences of unbalanced construction are evident here, where crossband and face veneer have been bandsawn off one side of a strip of lumber-core panel.

ness and stability are advantages that justify the cost of the treatment. A greater degree of stabilization can be obtained by using phenol-formaldehyde rather than methyl methacrylate, but the dark color it imparts greatly changes the appearance of the wood.

Another popular stabilizing treatment for wood, discussed in Chapter 11, is impregnation with polyethylene glycol (PEG) with an average molecular weight of 1000 (PEG-1000). If freshly cut wood having an abundance of free water is soaked in a 30% to 50% solution of PEG-1000, the chemical diffuses into the wood, replaces the bound water and thus **bulks** the cell walls. The PEG prevents shrinkage when the wood is dried. The treatment depends on the rather slow diffusion of the chemical into the wood. Since end-grain diffusion is 12 to 15 times greater than side-grain diffusion, the material diffuses rather well into cross-sectional discs and has been used to prevent crack formation in such discs due to differential tangential and radial shrinkage. PEG-1000 also has been used as a surface treatment to prevent checks on objects such as woodcarvings.

But let's not forget what is perhaps the most important consideration of all—**design**. With all the scientific innovations and materials that allow us to seal or stabilize wood, it is tempting to forget the value of intelligent design, which allows dimensional change to take place if it must. Traditional woodworking technique reveals how essential logical design can be. Many wooden objects have remained intact for centuries only because they were designed to accommodate natural dimensional changes when chemical treatments, synthetic finishes and environmental control were unheard of.

Framed panel construction exemplifies the idea of capitalizing on a functional design requirement and featuring it aesthetically. Centuries ago, woodworkers learned that a rectangular area could be well defined and maintained by framing it in longitudinal members whose dimensions would be faithfully held. The enclosed surface was filled with a panel whose perpendicular-to-grain dimension was sure to change considerably with the seasons. However, by beveling the panel to a **feather edge** and fitting it into slots in the interior of the frame, the panel could swell and shrink freely. Because the bevel is made quite wide, the eye does not detect the amount of change, which would show up if the panel were tightly fitted with a simple tongue and groove joint. The basic design has infinite variations, from wall paneling and architectural woodwork to furniture and cabinet doors. We are all familiar with the six-panel door, another traditional application.

In the classic framed panel *(2)*, it was customary to fasten the panel in the frame only at the center of the upper and lower stiles so that movement of the panel would be approximately equal within each side rail. Early woodworkers undoubtedly learned by experience the necessary amount of clearance to leave between the beveled panel edges and the bottom of the grooves in the side rails. Until such experience is gained, an intelligent estimate of the necessary clearance can be computed by the shrinkage formula given on page 76.

To illustrate, consider an 18-in. wide paneled cabinet door to be made from eastern white pine. In framed panels it is disastrous to leave too little clearance but quite harmless to leave an excess. It is therefore a good idea to be a little liberal in making an estimate. In cases where the pieces are not strictly either flatsawn or edge-grained, the conservative assumption would be that all material is flatsawn. From Table 7 on page 74, tangential shrinkage (S_t)=6.1%. Thus extreme moisture conditions might well be anticipated.

Let's prepare for a maximum moisture content of 14%, but assume our lumber is at 8% moisture content. Since the panel will be pinned in the center, we can assume symmetrical behavior and look at either half:

$$\Delta D = 9'' \, (0.061) \left(\frac{0.14 - 0.08}{0.28}\right)$$
$$= 0.118'', \text{ approximately } \tfrac{1}{8}''.$$

Therefore at least ⅛ in. of clearance should be allowed for swelling of the panel and frame.

Let's also consider how much shrinkage would occur if the moisture content of the door dropped to 4%:

$$\Delta D = 9''(0.061) \left(\frac{0.08 - 0.04}{0.28}\right) = 0.078'' \text{ or about } \tfrac{5}{64} \text{ in.}$$

Therefore, the final detailing of the joint should allow not only for swelling room, but also for shrinkage.

The total estimated movement of some 0.196 in. at each side is of special concern with regard to the finish-

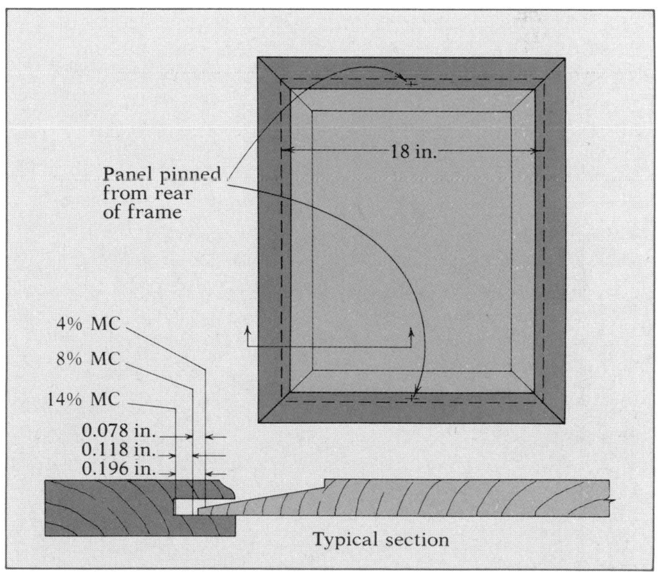

2—In frame-and-panel construction, a panel pinned only at the center top and bottom is free to expand and contract within a rectangular frame, which moves little along the grain with changes in humidity. The cross section shows estimated positions of the panel at different moisture levels.

ing of the door. If the door is to be stained, the bevel of the panel should be stained and given a preliminary finish before assembly. Otherwise, if assembled first and then stained at an intermediate moisture content, an unstained strip of the bevel will be revealed as the panel shrinks and withdraws from the frame.

If some naive woodworker tries to imitate only the appearance of a framed panel and pays no heed to the mechanics involved, the results can be disastrous. A panel tightly fitted into the frame can split itself lengthwise in a dry period or burst the frame apart during a humid spell.

It is amusing to note that a style that originally evolved to accommodate the dimensional change in wood is mimicked today in doors that are stamped out of sheet metal or molded in urethane plastic. For many, the style evidently is disassociated completely from the functional necessity that originally created it.

There are numerous other techniques for accommodating dimensional change by intelligent design. For example, in fastening wide tabletops to the apron by screws from below, slotted holes for the screws will allow the top to change dimension across its width. Drawer bottoms can shrink and swell if held without fastening by grooves in the sides. Attaching support strips to the ends of large panels with longitudinal dovetails provides for dimensional change of the panel across its width.

The shrinkage formula discussed on page 76,

$$\Delta D = D_i S \left(\frac{\Delta MC}{fsp}\right),$$

suggests ways of modifying design to reduce the consequences of shrinkage and swelling. For example, one might reduce the dimensions (D_i) of the members. Narrow flooring strips develop smaller cracks between boards than wide flooring; large-diameter tenons have a greater tendency to loosen than small-diameter ones.

Likewise, choosing a species with a small shrinkage percent (S) obviously helps; for example, mahogany is more stable than beech. Also, using edge-grain rather than flat-grain lumber can take advantage of the smaller radial shrinkage percentage (S_r) compared to tangential shrinkage percentage (S_t). A tangible estimate of the degree of improvement is suggested by the numerical quantities involved, as listed in *Table 7* on page 74.

On page 77 we considered the clearance allowance of a 9-in. maple drawer front, calculated to fit a drawer opening when swollen to an equilibrium moisture content (EMC) of 12%. Our estimates indicated that at an EMC of 5%, a gap of 0.223 in. might be expected. If we respecified quartersawn maple, thus substituting radial shrinkage (S_r)=4.8% for tangential shrinkage (S_t)=9.9%, the gap would be reduced to

$$\Delta D = 9'' \,(0.048)\left(\frac{0.12-0.05}{0.28}\right) = 0.108'',$$

slightly over 1/10 in.

Furthermore, if we were to select quartersawn teak instead of maple, we would be dealing with a radial shrinkage of 2.2%, so our anticipated gap would now be reduced to

$$\Delta D = 9'' \,(0.022)\left(\frac{0.12-0.05}{0.28}\right) = 0.050'',\ 1/20 \text{ in.}$$

If it were possible to redesign the drawer fronts to be only 5 in. high (that is, 5 in. across the grain radially), thereby reducing the dimension factor from 9 in. to 5 in., the dry weather gap would be estimated as only

$$\Delta D = 5\,(0.022)\left(\frac{0.12-0.05}{0.28}\right) = 0.0275 \text{ in.},$$

a mere 1/32 in. Thus, by carefully considering dimension, species and growth-ring orientation, it is possible to shrink the gap from 7/32 in. to 1/32 in.

As I will repeatedly emphasize, reducing moisture variation (ΔMC) by moisture-retarding finishes further reduces the magnitude of dimensional change.

I find that part of my philosophical attitude toward woodworking seems to have come full circle. I began trying to use wood "in the raw," but discovered I didn't know how. Then I became eager to learn ways to "overcome" all the "problems" that wood has, and experimented with wood stabilizers and chemicals, impatient to improve upon nature's product. In time, however, I realized a certain distaste for trying to make wood into something else and for trying to make it do things other materials do better. I'm back to using wood "as is" now, but with a different point of view. I concentrate on learning what wood is, rather than worry about what it isn't; I try to work with it, not against it. Whether this makes me a better woodworker I am not sure, but I am happier, for certainly part of the reward of working with wood is accepting the challenge of understanding it. The dimensional behavior of wood should be looked upon as simply a property of wood to be taken in stride, not as a problem to be corrected.

In summary, live with the dimensional properties of wood by trying to follow these guidelines: First, season lumber before working it to the average moisture content it will have in use. Second, provide the finished work with the most impervious finish consistent with its intended use. Third, design to allow normal dimensional change to take place. (Even if you don't think it necessary, it's a nice "fail-safe.") Finally, think about mechanical restraints and chemical stabilizers only when instability still remains unresolved.

Monitoring moisture

I think it was Charles Dudley Warner who observed that everyone talks about the weather, but nobody does anything about it. I often get the same feeling about moisture in wood. An amazing number of woodworkers can recite informed and intelligent-sounding numbers about the recommended moisture content for such and such a job. Or they may even know three ways to determine moisture content. But they don't actually have a moisture meter, a balance or an oven. Likewise, I've heard museum people discuss the prescribed humidity conditions for safe-keeping valuable wooden objects, but they don't actually use humidity-measuring equipment. I've lost count of the number of woodworking shops whose capital outlay is in six figures, yet who don't have instruments for measuring or monitoring moisture content or humidity. This holds true for the hobbyist or small-scale professional—whose equipment outlay may be many thousands of dollars—where a $100 moisture meter or even a $25 hygrometer could help avoid the problems whose eventual real costs probably total in the thousands. Not equipping to measure moisture, in the air or in the wood, is penny-wise but pound-foolish.

Relative humidity—Various instruments are now available for measuring relative humidity. About the simplest is the dual-bulb thermometer **hygrometer** *(1)*. This instrument uses two liquid-filled glass thermometers—one is a **dry bulb**, the other has a damp cloth sleeve or "wick" over its bulb (therefore called a **wet bulb**). The wick of the wet bulb usually dangles into a little vessel of water, or it may be moistened before each use. Water evaporation from the wick cools the wet bulb below the ambient temperature indicated by the dry bulb. The drier the air, the greater the evaporation rate and the lower the wet-bulb temperature. To make a reading, the wet bulb is fanned vigorously to maximize evaporation. When the wet-bulb temperature stops decreasing, the dry-bulb and wet-bulb temperatures are noted. The relative humidity is determined through the combination of dry-bulb temperature and wet-bulb depression by using **psychrometric** tables or charts *(2)*.

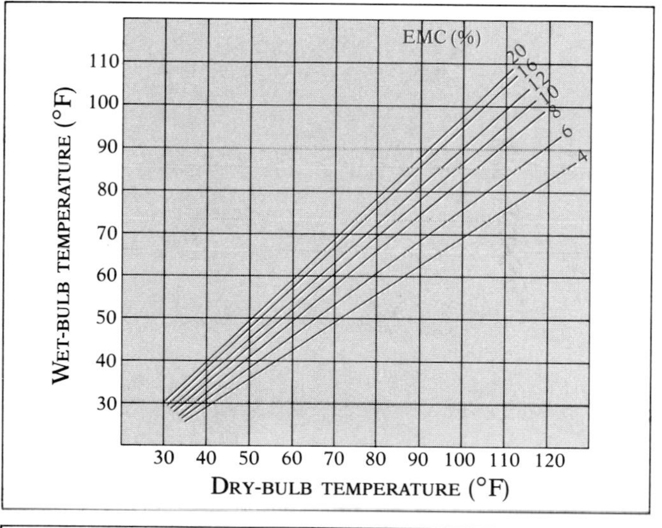

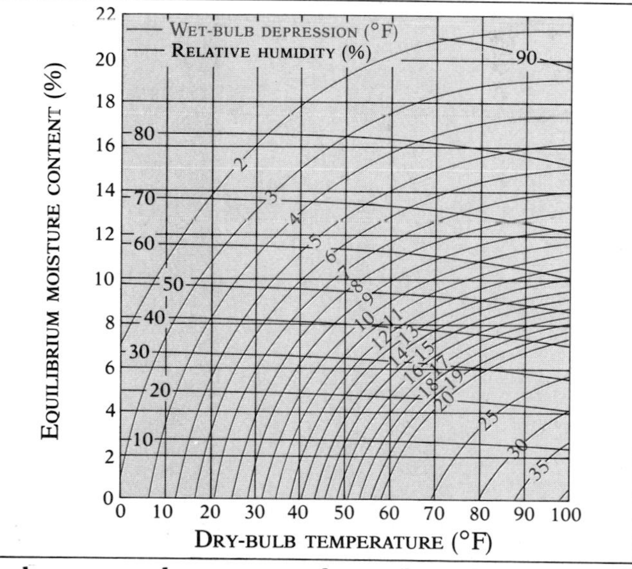

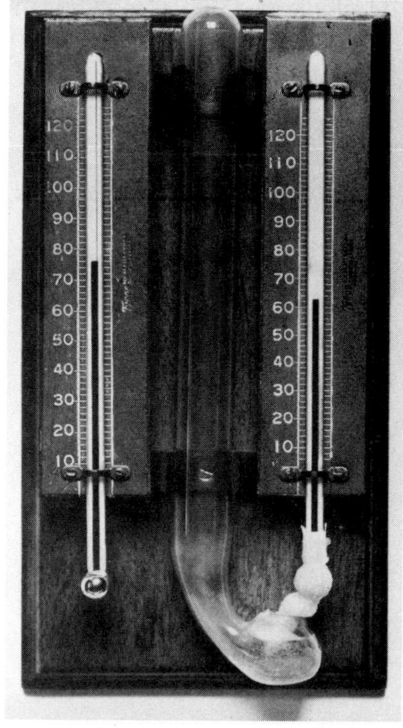

1—Dual-bulb hygrometer. The dry bulb (left) measures normal air temperature. The wet bulb (right) is covered with a moist cotton sleeve. The drier the air, the greater the evaporation rate and the greater the cooling effect on the wet bulb, which is fanned vigorously before taking a reading to develop full evaporation rate.

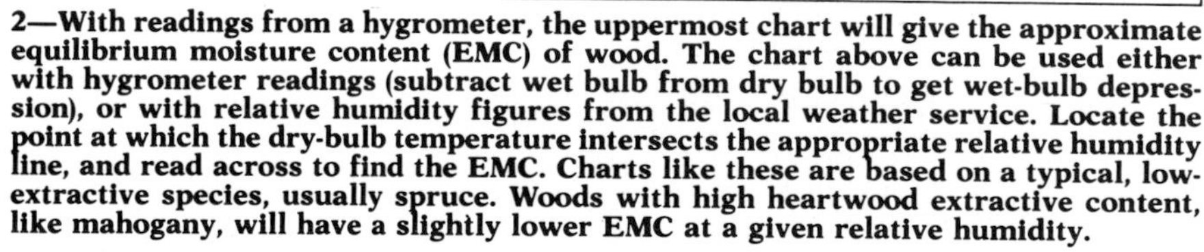

2—With readings from a hygrometer, the uppermost chart will give the approximate equilibrium moisture content (EMC) of wood. The chart above can be used either with hygrometer readings (subtract wet bulb from dry bulb to get wet-bulb depression), or with relative humidity figures from the local weather service. Locate the point at which the dry-bulb temperature intersects the appropriate relative humidity line, and read across to find the EMC. Charts like these are based on a typical, low-extractive species, usually spruce. Woods with high heartwood extractive content, like mahogany, will have a slightly lower EMC at a given relative humidity.

Wall-mounted models like the one shown in Figure *1* on page 91 are quite popular and the least expensive, but a dual-bulb hygrometer can easily be built from a pair of accurate glass thermometers. Despite the simplicity and economy of the dual bulb, it is highly accurate. In fact, it is used to calibrate most other types.

A **sling psychrometer** is a common type of portable dual-bulb hygrometer. It is operated by whirling the pair of thermometers around on the end of a swiveled handle *(1)*.

The dial hygrometer *(2)*, while perhaps not as accurate as the dual-bulb thermometer, offers the convenience of information at a glance. You will be amazed to discover how rapidly the humidity can change.

Moisture content—The traditional standard for measuring the moisture content of wood is the **oven-drying method**. As the term implies, water is driven from the wood by placing a sample in an oven at 212° to 221°F (100° to 105°C) until constant weight is reached. This final weight is the **oven-dry weight**. If samples from boards or other pieces are taken as cross-sectional wafers no longer than one inch along the grain, oven-dry weight is usually attained within 24 hours. The procedure involves weighing the initial sample (W_i), oven-drying, and then determining the oven-dry weight (W_{od}). Moisture content (MC), expressed as a percentage, is calculated by the formula:

$$MC = \left(\frac{W_i - W_{od}}{W_{od}}\right) \times 100.$$

In determining weights, a balance must be available that can weigh samples within at least 0.5% of the sample weight; 0.1% is preferable.

Moisture meters—An easy way to determine moisture content is by using modern moisture meters, which give immediate and highly accurate readings. These magical little meters use the electrical properties of wood, and their development has followed the usual trend in electronics toward portable and miniature units with simplified operation. A wide range of models is now available to suit virtually every situation, from the hobbyist's use to production operations in the shop or in the field.

For typical woodworking applications two principal types of meters are available. One is based on the direct-current electrical resistance of the wood and involves driving small, pin-type electrodes into the wood surface; the other uses the dielectric properties of the wood and requires only surface contact with the board.

The resistance meter takes advantage of the fact that moisture is an excellent conductor of electricity but dry wood is an effective electrical insulator. The meter itself is simply a specialized ohmmeter, which measures electrical resistance. The piece of wood is arranged as an element in an electrical circuit by driving the two-pin electrodes into it. The current (usually supplied by a battery) flows from one electrode through the wood to the other, then back through the ohmmeter. Actually, by simply driving pairs of nails into a piece of wood for electrodes and taking resistance measurements with a standard ohmmeter, readings could be obtained that would indicate relative moisture content. But commercial meters have the resistance translated directly into percent moisture content instead of ohms of resistance, and are therefore more convenient.

Because electricity follows the path of least resis-

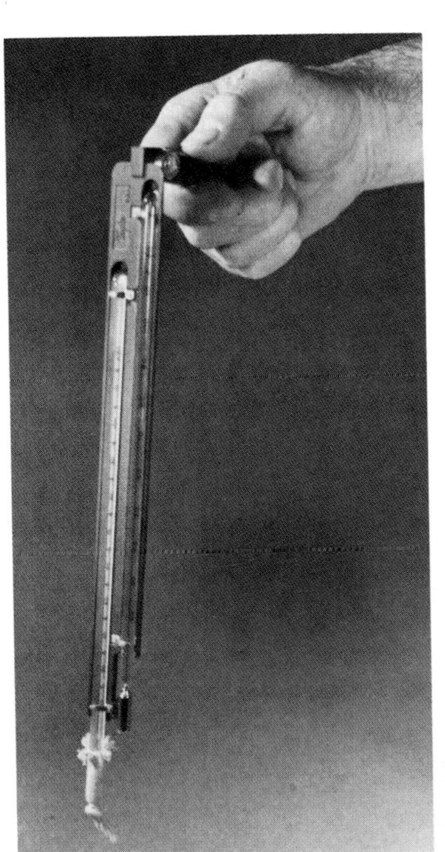

1—To take a reading with a sling psychrometer, the cotton sleeve on the wet bulb is first moistened. Then the instrument is whirled to maximize evaporation on the wick, until the wet-bulb temperature no longer drops.

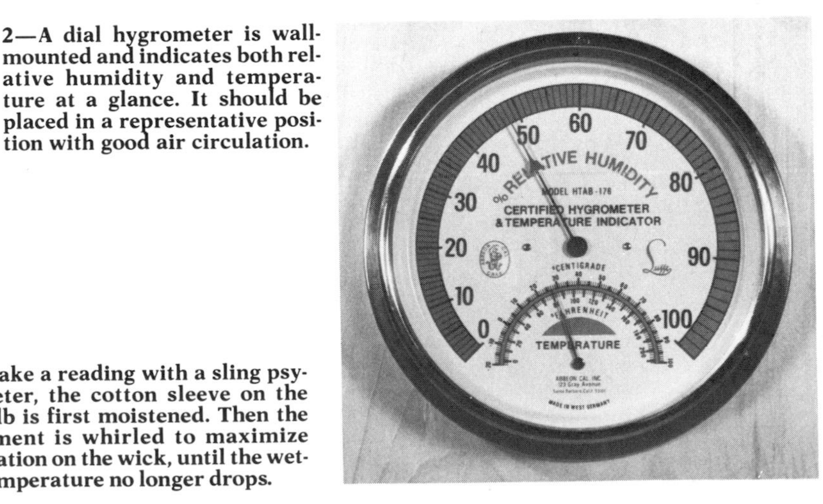

2—A dial hygrometer is wall-mounted and indicates both relative humidity and temperature at a glance. It should be placed in a representative position with good air circulation.

tance, the wettest layer of wood penetrated by the electrodes will be measured. For boards that dry normally, a drying gradient usually develops from the wetter core to the drier surface with an average moisture content about ⅕ or ¼ the board thickness from the surface. Thus for 1-in. lumber, the electrodes should penetrate only ¼ in. to ⅕ in. In some models, the electrodes are a pair of pins extending from one end of the unit that can be pushed into the wood by hand *(3)*. In other models, the electrode pins are mounted in a separate handle attached by a plug-in cord to the meter box *(4)*.

Electrodes of various lengths, up to 2 in. or more, are available for measuring thick material so the same meter can be used for thin veneer and heavy planks. Electrodes should be inserted so current flow is parallel to the grain. Electrical resistance is greater across the grain than parallel to it, although the difference is minor at lower moisture-content levels.

Meters using the dielectric properties of wood have a surface electrode array that generates a radio-frequency field that extends for a prescribed distance when placed against the wood *(5)*. Some meters measure the power-loss effect, which varies according to moisture content, whereas others respond to changes in electrical capacitance. Different models have electrodes designed for field penetration to various depths. Field penetration to about half the stock thickness is usual. Where moisture content is uneven, a more or less average reading will be given.

Green wood may have an extremely high moisture content, but woodworkers are most concerned with moisture measurement of seasoned stock. Fortunately, the electrical properties of wood are most consistent at moisture levels below fiber saturation (25% to 30%), the range of most interest to woodworkers. Dielectric meters can indicate moisture contents down to zero. The electrical resistance of wood becomes extreme at low moisture contents, limiting the lower end of the range of resistance meters to about 5% or 6%. More elaborate meters sometimes have scales extending to 60% or 80% moisture content; however, electrical properties are less consistent above fiber saturation so readings in this range must be considered approximate.

Moisture meters usually give scale readings of percent moisture content that are correct for certain typical species at room temperature. Instruction manuals give correction factors for other species and different temperatures. Since density has little effect on electrical resistance, the species corrections are usually less than two percentage points for resistance meters; correction factors may be greater with power-loss meters. Resistance readings must also be corrected about one percentage point for every 20°F departure from the calibration standard. With dielectric meters the correction

3—Electrical-resistance moisture meter with attached electrodes is pushed directly into the wood.

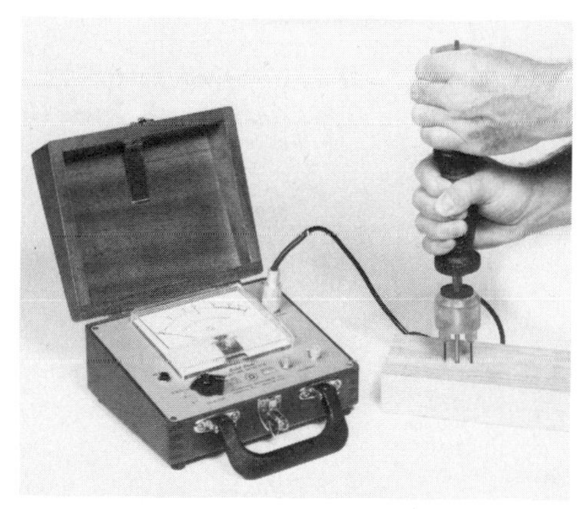

4—A resistance meter with external electrodes, which are driven into the board. The center pin gauges penetration.

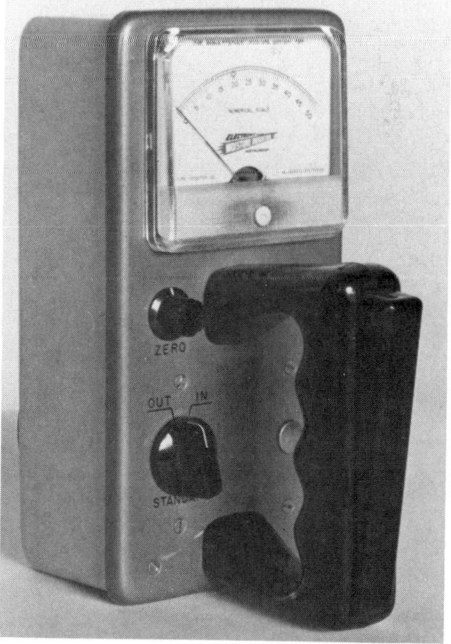

5—The electrode array on the back of a dielectric meter generates a radio-frequency field as it is pressed against the face of a board.

is more complicated, but is well explained in the instruction manuals. For anyone using meters under regular conditions—with one or a few common species and always at room temperature—correction factors either are not applicable or become routine.

The values obtained with a resistance meter can be expected to agree within one-half a percentage point with those obtained by oven-testing for samples in the 6% to 12% range; within one point in the 12% to 20% moisture-content range, and within one to two points in the range from 20% to fiber saturation.

It is important to appreciate that a meter in good condition will faithfully and accurately measure the electrical properties of the wood being sampled. It must be realized that the moisture content in a tree may show considerable variation, which may be reflected in the lumber moisture content until an equilibrium has been reached. The operator must understand the vagaries of wood moisture and interpret accordingly. For example, a new owner of a meter might discover a variation of two or three percentage points up and down a given board. The common reaction is, "the meter is accurate only to within three percent" or, "it gives variable readings." But in fact the meter is properly measuring moisture variations that exist in the board. Thus, one must measure average or typical areas of boards and avoid the ends or cross-grain around knots, which dry most rapidly.

Each type of meter has its strengths and weaknesses. Resistance meters have the disadvantage of leaving small pinholes wherever the electrodes were inserted, which might be unacceptable in exposed furniture parts, gunstocks and the like. On the other hand, a given meter can be used with a variety of electrodes in a wide range of situations. Resistance meters with a 6% to 30% range are available down to pocket size, with both built-in short-pin electrodes and separate cord-attached electrodes. Radio-frequency power-loss meters are available in compact hand-held models, with electrodes for 1-in. field penetration and scaled from 0% to 25% moisture content. Their distinct advantage is the ability to take readings without marring surfaces, thereby allowing measurements of completed items, even after the finish has been applied. These meters are extremely quick to use, but are less versatile because a given electrode style works only for a particular area and depth of field.

The moisture "widgit"

I am convinced that the greatest single impediment to moisture control is simply neglecting (or even refusing) to think about it. More lumber is dried and conditioned by assumptions and wishes than by controlling the atmosphere. An interesting gimmick for reminding oneself about moisture content is to build a simple gadget consisting of a horizontal wand with a wood sample suspended from one end and hung at its balance point *(1)*. As the wood sample picks up and loses moisture in response to changes in humidity, the free end of the wand will float up and down. The device can be made most sensitive by making the wand out of the lightest material possible that is also non-hygroscopic (stiff, thin-walled plastic tubing works well); by making the wood sample a cross-sectional piece so it will have as much end grain exposed as possible; and by making the suspension points as frictionless as possible. (Note: Don't try to rest the wand across a fulcrum. The farther out of balance it gets the less stable it becomes, and it will fall off the fulcrum. It is best to suspend the wand by a short thread or eye screw.) While merely noting the angle of the wand's inclination will be informative, the gadget can be made into a more sophisticated instrument. Hanging the wand parallel and close to a wall enables reference marks to be recorded where the tip of the wand is observed at intervals. If an oven and balance are available for moisture determination, the moisture content of the wood element can be determined by matched "control" material. Once the moisture content is known, the wand can be "calibrated" by hanging known weights (equivalent to the sample at various moisture content values) on the wand and marking the wall.

Once the gadget is installed and operating on a wall of a shop or classroom, it's easy to check it often. After a while, you think more and more about the effects of weather, and pretty soon you find yourself each morning wondering what it will read when you get to the shop that day. It's a great low-cost way to have a continuously reading hygrometer.

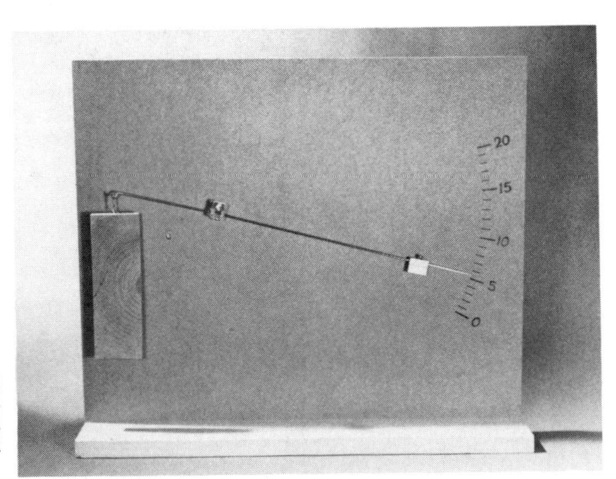

1—This simple homemade gadget indicates equilibrium moisture content. The pointer rises and falls as the wood absorbs and desorbs atmospheric moisture. The wand can be calibrated with moisture-content values derived from matched samples that were oven-dried.

How wood dries

We have discussed many aspects of wood-moisture relationships, from the nature of water in the wood to final equilibrium between bound water and the atmosphere. Now we're ready to tackle the most important and perhaps the trickiest subject of them all—the process of drying wood. It's important for woodworkers to understand the drying procedure for two reasons. First, woodworkers should appreciate and recognize the difference between poorly dried and properly dried lumber. Assuming lumber is well dried because "they said so," or because "it ought to be" won't help when the lumber turns out to have drying defects. Woodworkers should watch for telltale warnings of improper drying. Second, most woodworkers eventually attempt some drying on one scale or another, either for economy or to acquire otherwise unavailable material.

Abundant tree material is available to those who seek it out from such sources as storm-damage cleanup, construction-site clearance, firewood cuttings and even direct purchase from local loggers. With chain saws, wedges, band saws and a measure of ingenuity, chunks and flitches for carving or even lumber can be worked out. Also, it is usually possible to buy green lumber, either hardwood or softwood, at an attractive price from small local sawmills. Many an eager woodworker has produced a supply of wood to the green-board stage, but has been unable to dry it to usable moisture levels without serious degrade or even total loss. Therefore, it's quite appropriate that we look carefully at both the fundamentals and the procedures involved in drying.

It should be emphasized here that the so-called "seasoning" of wood is a water-removal process, rather than the chemical-modification process that is associated with seasoning certain foods or with curing hides. As previously discussed, it involves the removal of the free water and part of the bound water down to the target equilibrium with the average use conditions of the finished item.

One is tempted to wonder why lumber can't simply be placed in an oven to drive off sufficient moisture, as is done when determining moisture content. We would certainly get the moisture out, but the uneven shrinkage, resulting in defects, would totally ruin the lumber. The drying sequence is a continuously changing process, but there are three stages that every piece of wood passes through in drying that will illustrate how stress and defects develop.

Assume we take a plank of green wood and decide to try to dry it under rather drastic conditions. Let's visualize the cross section of our plank well away from the ends, as shown in Figure *2*. Initially, it is uniformly well above the fiber saturation point in moisture content. During Stage I, the piece is free of defects and stress because although moisture is escaping from the surface, no shrinkage has yet taken place. Eventually, the surface moisture content drops below the fiber saturation point. The layer of wood near the surface, referred to as the **shell** (in contrast to the interior zone, known as the **core**) begins to shrink—or at least attempts to. But it cannot shrink as much as it wants because the fully

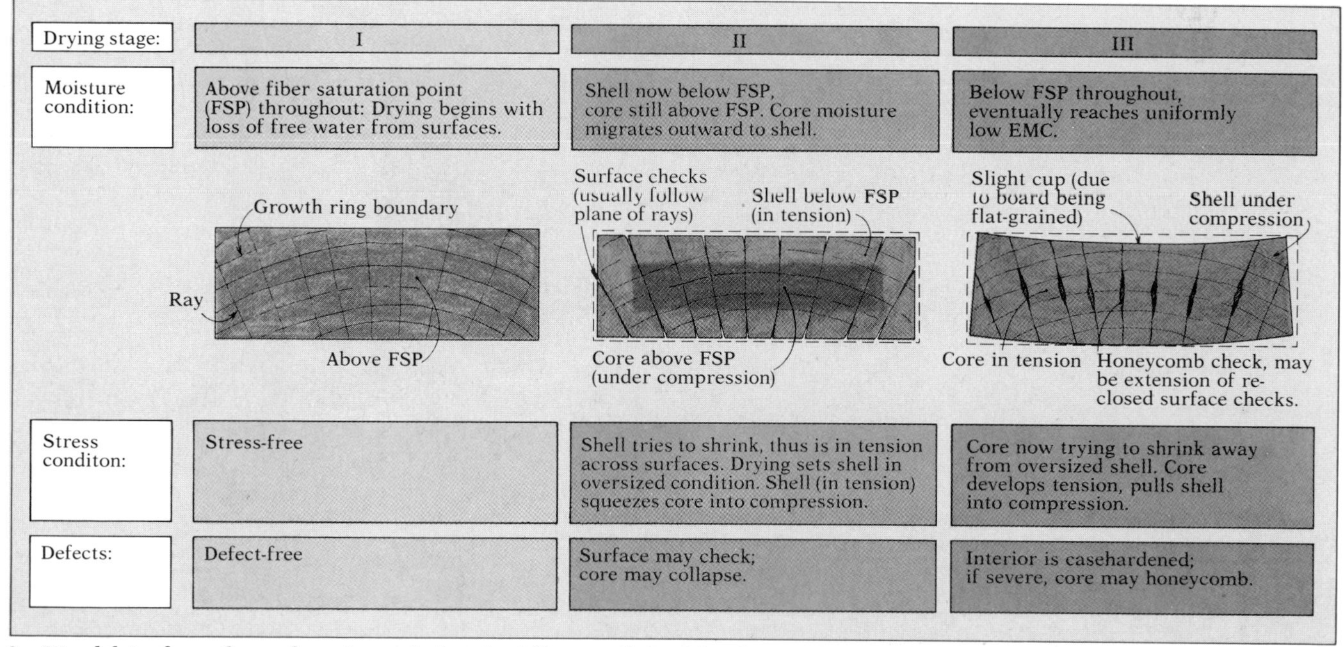

2—Wood dries from the surfaces inward, shrinks differentially and develops stress. The object of seasoning is to regulate the rate of drying, by slow air-drying or by controlled kiln-drying, to keep these stresses within tolerable levels and thus to avoid defects.

swollen core holds it in an oversized position. The shell therefore develops tension perpendicular to the grain around the outside of the board, characteristic of Stage II. If these stresses exceed the strength of the wood, surface checks develop to relieve a portion of the stress *(1)*. At the same time, the encircling tensile stress from the shell contracting around the core places the core in compression. This compression stress, aided by the capillary tension of free water being dried from the cells, may cause internal buckling of the wood cells in the core, called **collapse**, which may severely distort the board *(2)*.

But suppose the stresses are not severe enough to cause surface checks or collapse. The shell nevertheless starts to dry, to set in an oversized condition. As drying progresses, the shell surface begins to level out at a low moisture content. Subsequently, the core continues drying, eventually drops below fiber saturation point and attempts to shrink as Stage III is approached. As the core tries to attain a smaller dimension, it not only draws the shell inward but it gradually develops a reversal of stresses such that the core itself, being held outward by the shell, is now in tension, and the shell is in compression. Stage III has been reached, and the wood is said to be **casehardened**. With the shell in compression, any surface checks that developed in Stage II now close up. If the casehardening tensile stresses in the core are great enough, internal separation of the

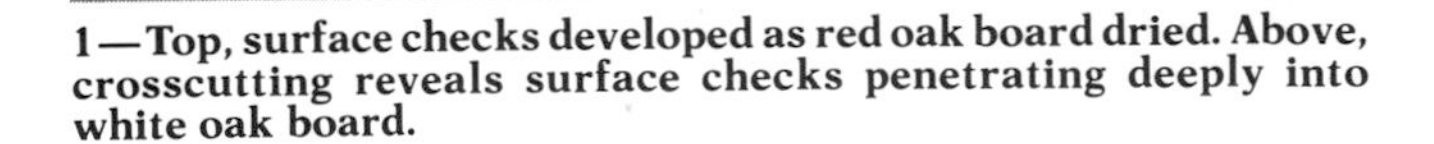

1—Top, surface checks developed as red oak board dried. Above, crosscutting reveals surface checks penetrating deeply into white oak board.

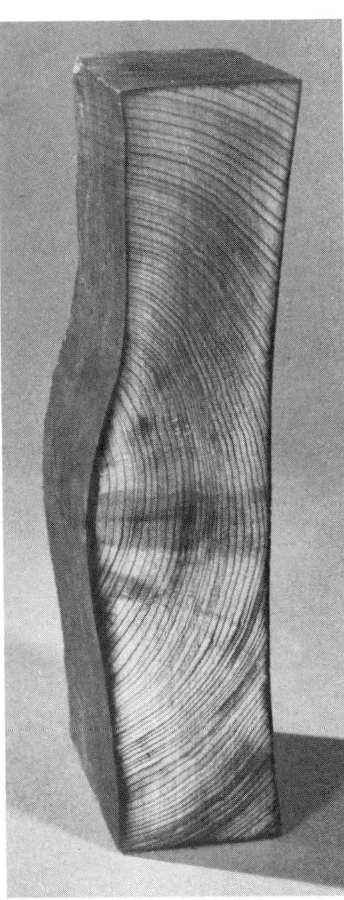

2—Extreme compression stress in drying may cause wood cells in the core to buckle or collapse, as in this imbuya board, which was sawn rectangular.

wood—called **honeycomb**—can occur. Honeycomb *(3)* is one of the worst defects that lumber can develop. These internal checks are often extensions of the surface checks developed in Stage II. Even when the piece is eventually at uniform moisture content throughout, it will remain casehardened. If such a plank is resawn, the two halves will cup *(4)*.

Anyone can duplicate these stages and develop a casehardened and even honeycombed board. Simply take a freshly cut flatsawn oak board, end-coat it well with a thermosetting sealer (resorcinol adhesive does nicely) and place it in an oven at 212°F or a bit warmer. In a day or two you'll probably have a casehardened board.

So here is the "Catch-22" that must be resolved. No moisture will move except from areas of higher moisture content to areas of lower moisture content, so to get moisture to move out of a board you have to set up a **moisture gradient**, that is, a condition of moisture difference within the wood. If the gradient isn't steep enough, moisture will scarcely move. But since variation in dryness causes variation in shrinkage, stresses develop. To eliminate these stresses, the gradient must be moderate, but then the moisture won't move fast enough. So, we must settle for a compromise between drying speed and drying defects.

Let's look first at how the drying compromise is handled in a commercial dry kiln, then at the problem of drying small quantities of lumber at home.

3—Above, honeycomb checks in a red oak board follow the planes of the large rays. Left, honeycombing in a maple square. Below, surface planing reveals the honeycombing in an oak board.

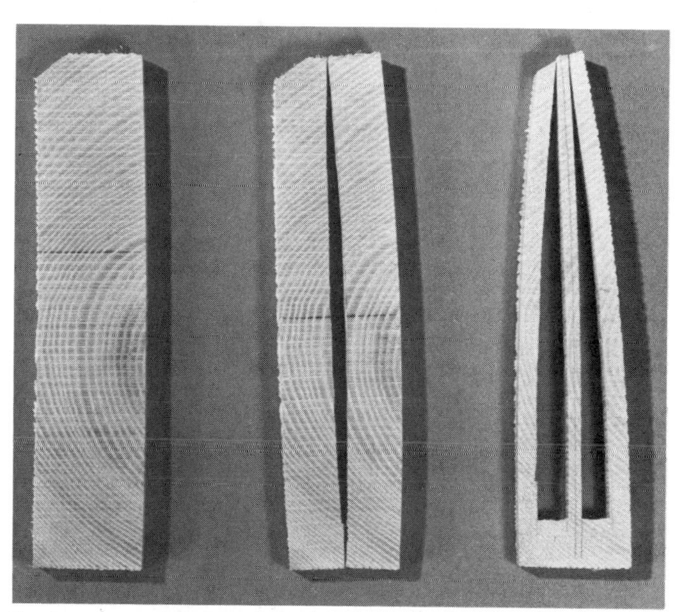

4—A wafer cut from a kiln-dried plank of white ash shows no symptoms of stress (left). Another section from the same plank, after resawing (center) reveals the casehardened condition (tension in core, compression in shell). Kiln operators cut fork-shaped sections that reveal casehardening when prongs curve inward (right).

The dry kiln

A typical dry kiln is a large, well-insulated room or chamber *(1)*. It has controlled air circulation, temperature and humidity. A neatly piled and stickered load of lumber is centered in the kiln. Air is driven by large fans from one side through the pile to the other side and then directed back to the fans. Because water moves through wood more readily at higher temperatures, heat is usually provided by banks of steam pipes in the path of the airflow. Humidity can be increased by water or steam sprays, or reduced by vents at the top of the kiln.

In drying lumber, the kiln is started at a fairly high humidity level and at only slightly elevated temperatures. Sample boards are monitored for moisture content. The kiln remains at the first step until the moisture content drops to a certain level. Then the temperature is raised slightly and the humidity lowered to the next step. A sequence of temperature/humidity values called a **kiln schedule** is followed which, from experience, is known to bring lumber from a designated species and thickness safely to the desired moisture content. By carefully controlling humidity, the kiln can utilize higher temperatures to increase the rate of moisture migration from the core to the surfaces of the board without a dangerously steep moisture gradient.

The kiln schedule is designed so surface checking or collapse will not develop. However, the cost of kiln operation demands that some casehardening be allowed, so long as it is not severe enough to develop honeycombing, for at the very end of the run, the casehardening can be removed. This is done after the lumber is brought to final dryness with the surface a little lower in moisture content than desired. Then, in the last step of the kiln schedule, the humidity is raised. This reintroduces moisture into the surface of the lumber to remove the residual moisture gradient and eliminate the casehardening. This final "equalizing and conditioning" leaves the lumber free of stress, with a uniform moisture content.

The kiln operator maintains a direct weight/moisture check on samples throughout the run and also makes a more thorough inspection with a meter on the finished load of lumber. To check for stress, he cuts cross sections from the boards and bandsaws a tuning-fork shape into the section. If the "tines" pinch in, as in Figure *4* on page 97, the lumber is still casehardened and must be conditioned further. If the tines remain straight, the job is done. The woodworker can likewise check for casehardening in lumber by cutting stress sections.

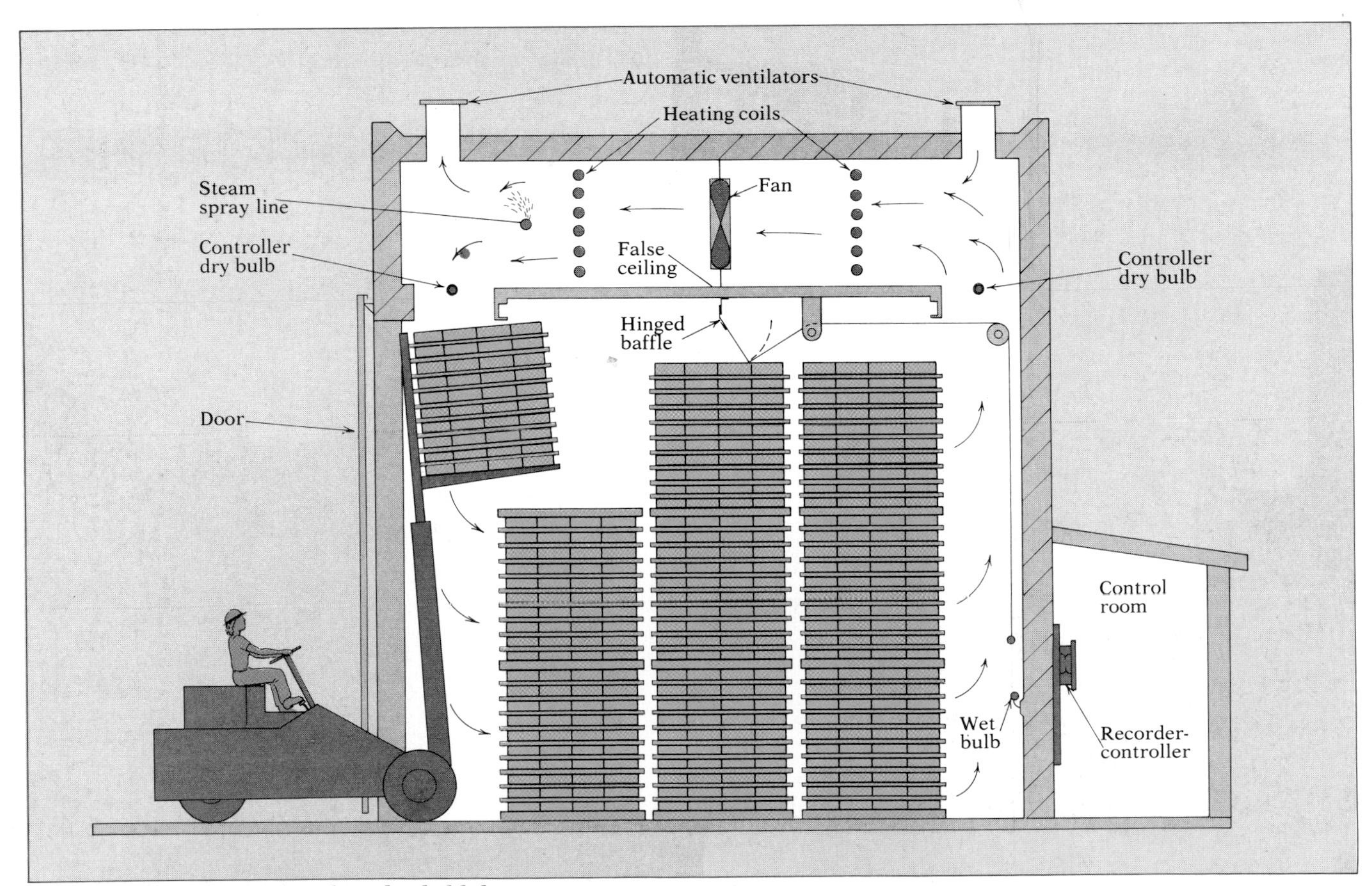

1—A typical commercial package-loaded kiln.

Drying your own wood

For the person interested in drying small quantities of wood, the same general guidelines apply, namely, proper cutting, preparation and stacking; control of drying rate; and monitoring the drying process. Let's review the application of these concepts to typical drying situations. We will consider the drying of short log segments or short thick stock, commonly used for woodcarvings or stout turnings, as well as regular lumber or boards. We will also assume that fairly small quantities such as several log chunks or up to a few hundred board feet are involved—as occurs when one suddenly falls heir to a storm-damaged tree or purchases enough lumber for several pieces of furniture.

First let's look at proper preparation of the material. Selection of pieces should favor those with normal structure and straight grain. If possible, avoid pieces with large, obvious defects. Lumber from trees with spiral grain will invariably twist upon drying. Irregularities such as crotch grain or burls are aesthetically interesting but chancy to dry, since their cell structure usually has unpredictable shrinkage. Knots are troublesome if they are large enough to involve serious grain distortion. Logs with sweep or from leaning trees having an eccentric cross-sectional shape probably contain reaction wood and will almost surely develop warp and stress due to abnormal shrinkage.

Whether preparing lumber or carving blocks, remember that normal shrinkage is about double tangentially as radially. My initial rule in splitting carving chunks from logs is to avoid pieces containing the pith. A half log or less that does not contain the pith can dry with a normal distortion of its cross-sectional shape, like slightly closing an oriental fan *(2)*.

Another advantage of not boxing in the pith is being able to see if any overgrown knots are present that may not have been apparent from the bark side. Every knot-causing branch developed from the pith, so it is important to examine pieces from the pith side to discover hidden branch stubs, especially if they have decay (Figure *1*, page 28). Additionally, the pith area is often abnormal juvenile wood that might best be eliminated.

In sawing lumber, minimize cup by favoring quartersawn boards, which have no tendency to cup, or flatsawn boards taken farthest from the pith. Boards sawn through the center of the log, containing the pith or passing close to it, will usually cup severely along the center (or split open if restrained) and might as well be ripped into two narrower boards before drying.

To prevent rapid end-drying, which will ruin carving chunks and the ends of lumber, the end-grain surface should be coated *(3)*. Any relatively impervious material (such as paraffin, aluminum paint or urethane varnish) in ample thickness will do nicely. End-coating can be applied to relatively wet surfaces by giving a primer coat of acrylic latex material first. It is important to end-coat as soon as possible after sawing, before even the tiniest checks can begin to develop. Once a check develops, the cell-structure failure will always be there, even if it later appears to have closed. Also, when normal drying stress develops, a small check can provide the stress concentration point for further failures, which otherwise might not have even begun in check-free wood. The purpose of end-coating is to force all moisture loss to take place from lateral surfaces.

In some species, radial drying may be significantly faster than tangential drying. Therefore, if the bark on larger carving blocks is tight (as with winter-cut wood), it may best be left on to slow the radial drying. If the bark has been removed from a heavy slab, it should be watched carefully during the early drying stages for

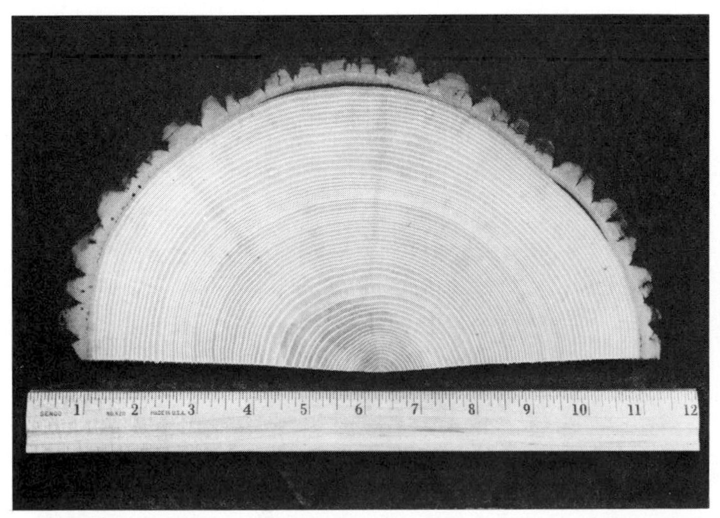

2—Half-log section, not containing the pith, was jointed flat before drying and has dried check-free—but not without distortion.

3—End-coated half-log sections of eastern white pine are stickered and open-piled for slow drying.

signs of surface checking. Another reason for prompt end-coating is to prevent ever-present airborne fungal spores from inoculating the surface. If the bark is loose, it should be removed; otherwise the layer of separation will become a fungal culture chamber.

Don't forget to mark a number and date on each piece. It is amazing how easily your memory can fail once you have several batches of wood in process.

Lumber must be correctly stacked so proper drying will result. Stacking must ensure maximum air circulation around virtually every surface of the material. With irregular carving blocks, merely piling them loosely may suffice, as long as flat surfaces do not lie against one another. No attempt should be made to restrain distortion of large chunks. With lumber, however, carefully designed, systematic stacking is best.

The usual stacking method for lumber is to arrange boards in regular layers or courses separated by narrow strips or **stickers** *(1)*. This permits free movement of air around the lumber, uniformity of exposure of the surfaces and restraint to minimize warp. The stickers should be dry and free of fungi and at least as long as the intended width of the stack. To ensure uniform restraint in a course, lumber and stickers should be as uniform in thickness as possible. In planning the stack, stickers should be placed at the very ends of each course (and at least every 16 inches along the length of the boards), since loose ends hanging out of the stack lack restraint, resulting in excessive warp or sag. Stickers should be lined up in straight vertical rows. It is best to have lumber uniform in length, but if random lengths are unavoidable, they should be arranged in a stack as long as the longest boards; within each course, stagger the position of alternate boards so their alternate ends are lined up with the end of the stack. This "boxed stack" system prevents excessive drying of overhanging ends. To prevent excessive drying degrade to the top and bottom courses or layers, extra outer courses of low-grade lumber or even plywood might be added to the stack.

In large stacks, the majority of boards are restrained by the weight of others above *(2)*. In small stacks, extra weight (old lumber, bricks, cinder blocks, etc.) should be

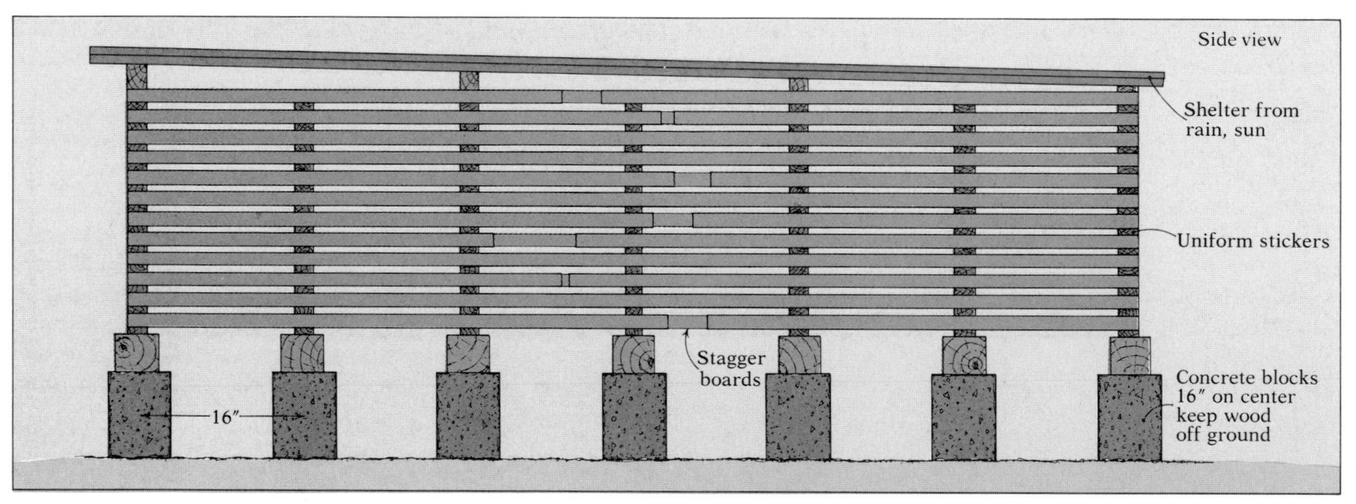

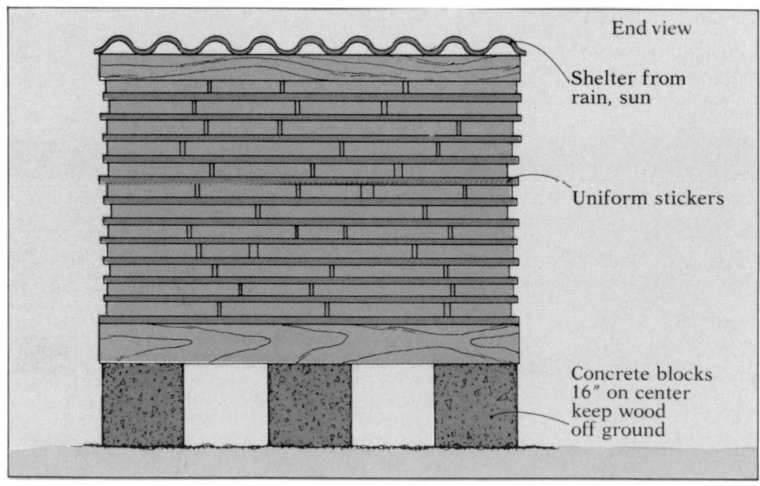

1—How to build a lumber pile.

placed atop. An alternate method of applying restraint is to assemble rectangular frames to surround the stack. The stack can be wedged against the frames *(3)*, and the wedges tapped further in to maintain restraint as the stack shrinks. Obviously the weighting or wedging should not be so extreme as to prevent shrinkage of the boards across their width.

Temperature and humidity in a dry kiln can be controlled, but to dry small quantities of wood in the home or shop you must make do with existing conditions. Choose locations or regulating conditions to allow only moderate drying at first, followed by more drastic conditions once the lumber has reached a lower moisture level. Remember, the drying compromise is especially delicate here. The gradient cannot become too extreme, since we do not want surface checks, and casehardening cannot be allowed, since we cannot condition at the end. Nor can we slow down drying too much, for surface drying must take place early enough to deter fungal activity.

One logical starting place is out-of-doors. Except for especially arid regions, the relative humidity is usually moderately high. Piles of blocks or. stacks of lumber should be kept well up off the ground to avoid dampness, and should be protected from direct rainfall and sun rays as well. Any unheated building that has good ventilation is ideal. Most garages serve well, and even unheated basements are suitable if plenty of air space around the stack is provided. In air-drying outdoors, some rather obvious seasonal variations will be encountered. In many eastern areas, slightly lower humidity and more prevalent winds favor drying in spring months. In winter, if temperatures drop to near or below freezing, drying may be brought to a standstill. You must therefore interpret conditions for each particular area. If wood is intended for finished items that will be used indoors, outdoor air-drying will not attain a low enough equilibrium moisture content. The material must be moved indoors to a heated location and again allowed to reach equilibrium before it is worked.

Surface checking should be closely watched. Minor shallow surface checking that will later dress out can be ignored. However, deeper checks should be considered

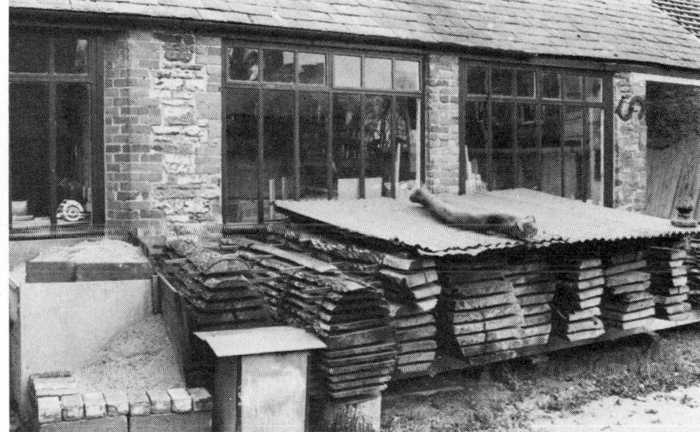

2—Most of the boards in the drying shed at left are restrained by the weight of the others. At right is a similar, simpler setup, where the wood is protected by a sheet of corrugated plastic. In both cases, the boards are stacked in the sequence they came off the saw.

3—Small quantities of lumber can be boxed inside wooden frames, with double wedges tapped in to maintain tension as the wood shrinks. If you are chain-saw milling logs yourself, keep the boards in the order they came off the tree.

unacceptable. The worst ones are those that open up but later reclose. Often they go unnoticed during subsequent machining, only to reveal themselves when staining and finishing of a completed piece is attempted. If any serious end checks develop, don't pretend they don't exist, or that they will ever get better or go away. For example, if a large carving block develops a serious check, this indicates fairly intensive stress; it is best to split the piece in half along the check, thus relieving the stresses, and be satisfied with smaller pieces.

If wood must be located indoors from the start, drying may be too rapid. Any signs of surface checks in the material suggest that some retardation may be necessary. This can be accomplished by covering the entire stack with a polyethylene film. Moisture from the lumber will soon elevate the humidity and retard the drying. However, this arrangement must be watched closely, since air circulation likewise will be stopped. Moisture condensation on the inside of the plastic covering or any mold on the wood surfaces may mean the stack has been turned into a fungi culture chamber and signals the need for speeding up the drying again. Common sense will suggest how often to check the wood and how to modify the storage location to speed up or slow down the drying. The seasonal humidity fluctuation encountered in heated buildings must also be allowed for in determining the equilibrium moisture level.

Drying progress can be monitored by weight. Weights should be taken often enough to be able to plot a fairly coherent graphical record of weight against time *(1)*.

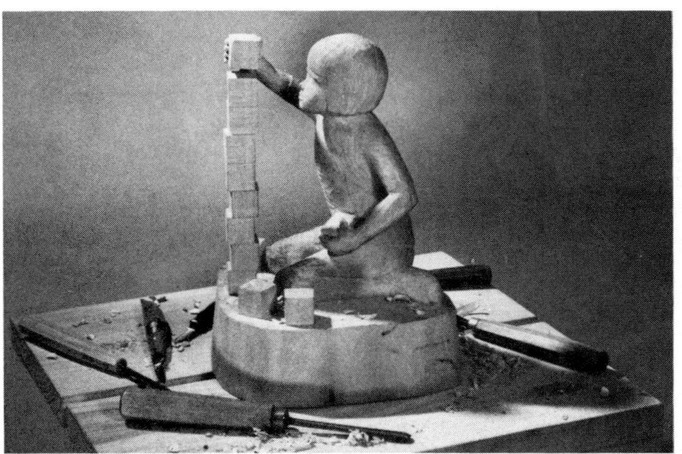

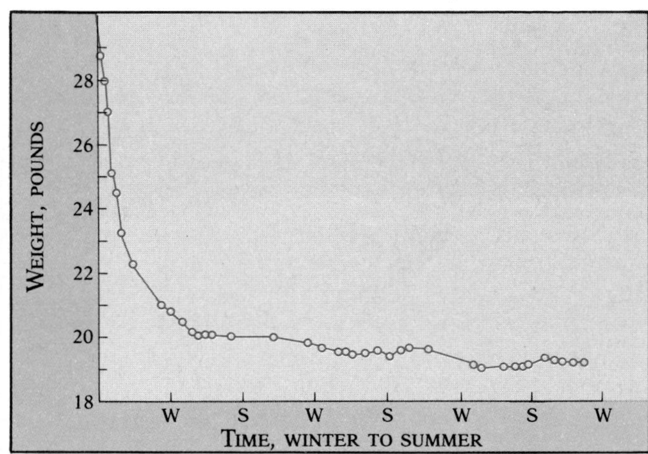

1—A drying block of wood can be weighed periodically, and its weight recorded on a graph against time. The graph levels out as equilibrium moisture content is reached. This basswood carving, shown as it neared completion, was made from timber seasoned and monitored as described here.

Weighing should be accurate to within one or two percent of the total weight of the piece. A large chunk in the 100-lb. to 150-lb. range can be weighed on a bathroom scale. Pieces in the 10-lb. to 25-lb. category can be weighed with a food or infant scale. Small stacks of boards can be monitored by simply weighing the entire stack if this is convenient. In larger stacks, sample boards can be pulled and weighed periodically. Electrical moisture meters are perhaps the simplest means of keeping track of the drying progress in boards.

The last stage of drying should be done in an environment that is similar to the one in which the finished item will be used. The weight of the pieces will eventually level out and reach a near constant equilibrium with only faint gains and losses of weight in response to seasonal fluctuations in humidity.

When material comes into equilibrium weight with the desired environment, it's ready. Don't pay attention to generalizations like "one year of drying for every inch of thickness." Such rules have no way of accounting for the tremendous variation in species characteristics or in atmospheric conditions. Basswood or pine decoy blanks 4 in. thick dry easily in less than a year, whereas a slab of rosewood 4 in. thick may take much longer to dry without defects. In general, low-density woods are easier to dry than high-density woods. Since the average cell-wall thickness is less, moisture movement is greater and this results in faster drying. In addition, the weaker cell structure is better able to deform in response to drying stresses, rather than resisting and checking.

Though exact drying times are impossible to predict, the drying times listed in *Table 9* give an idea of the comparative ranges of drying time for representative species of domestic woods. After some experience is gained for a particular species and thickness dried in a certain location, a fairly reliable estimate can be made as to the necessary drying time. Here, the initial date you mark on the piece will serve you well.

Whether drying log sections or boards, remember that the drying must be somewhat regulated; usually at the beginning, indoor drying proceeds too quickly and needs slowing down.

In drying your own lumber or carving wood, one common problem is hesitation. You can't wait! If you do, fungi or checks will get ahead of you. Try to think out all the details *before* you get your wood supply; don't wait until you get it home to decide how you are going to end-coat or where you are going to stack it.

Perhaps the greatest pitfall is greed. Most woodworkers never feel they have enough material and tend to overstock. With green wood, this can be disastrous. Don't try to handle too much. Don't even start to dry your own wood if you can't follow through. More material is ruined by neglect than by lack of know-how.

Table 9—Approximate time to air-dry 4/4 lumber to 20% MC.*

Softwoods	Days	Hardwoods	Days
Douglas fir	20-200	Ash, white	60-200
Hemlock, eastern	90-200	Aspen, bigtooth	50-150
western	60-200	Basswood, American	40-150
Larch, western	60-120	Beech, American	70-200
Pine, eastern white	60-200	Birch, yellow	70-200
southern yellow	30-150	Butternut	60-200
sugar	15-200	Cherry, black	70-200
western white	15-150	Elm, American	50-150
Redwood	60-365	Hickory	60-200
Spruce, red	30-120	Maple, red	30-120
		sugar	50-200
		Oak, northern red	70-200
		northern white	80-250
		southern red	100-300
		Sweetgum	60-300
		Sycamore	30-150
		Walnut, black	70-200
		Willow, black	30-150
		Yellow poplar	40-150

***Minimum days given refer to lumber dried during good drying weather, generally spring and summer. Lumber stacked too late in the period of good drying weather to reach 20% moisture content, or lumber stacked during the fall and winter, usually will not reach a moisture content of 20% until the next spring, which accounts for the maximum days given.**

Storing lumber

The previous discussion of relative humidity and the emphasis on wood as a hygroscopic material should serve nicely to guide the woodworker in sensible storage for wood once it is dried. Wrapping dry lumber in polyethylene film or even heavy kraft paper can be a tremendous help in getting through peak periods of dryness or humidity without excessive moisture exchange.

Woodworking and carving projects in progress are too often neglected in this regard. Rapid changes in relative humidity (RH) during the several days or weeks involved in the completion of a project must be taken seriously. Even as I have been writing this section, during the last of May and beginning of June, I have been startled by the change in humidity. It has been a late spring but I still expected an earlier rise in the indoor RH. Watching the dial hygrometer on my office wall, it seemed stuck in the 20% to 30% range all through May. I actually wondered if it were broken. Even through a few rainy but raw days around the first of June it only edged into the 40s, then stuck again. Then, in the second week of June, things began to happen. A spell of warm weather moved in with lots of rain and fog and within hours the hygrometer read 55% RH.

I had been working on some experimental wind turbine blades involving laminated spruce spars and molded maple plywood skins. I quickly unrolled some polyethylene, stapled up some 8-ft. bags and slid the half-finished blade components in. I'm glad I did. The hygrometer in the lab climbed to 76% RH and was headed higher. I can't imagine why more woodworkers and carvers don't give greater thought to wrapping their work between sessions.

Controlling humidity in the woodworking shop is too broad a subject to cover adequately in this book. In most cases, ordinary techniques of humidification or dehumidification can be used. Two improvisations mentioned below are an economical way to control humidity in a small chamber for small quantities of wood.

The first method is to determine the absolute humidity of the average ambient conditions, then to elevate the dry-bulb temperature sufficiently to reduce the relative humidity to the desired level. For example, suppose your shop is about 68°F and the humidity is 65%. The absolute humidity is about 5 grains per cu. ft. (See Figure 2, page 67.) Suppose you wish to condition some pieces of wood to 6% moisture content. Raising the temperature to 92°F will drop the relative humidity to 30%, just about right to give an equilibrium moisture content of 6%. A small closet or plywood chamber with enough light bulbs to maintain a temperature of 92°F will effect the desired moisture conditioning. A small thermoregulator to turn off the light when the desired temperature is reached would be a useful addition. But beware—the dry-bulb temperature will have to be adjusted accordingly as the humidity changes.

Another technique involves a tray of liquid that has a large enough surface area to influence the humidity in an enclosed chamber. If plain water is used, and the chamber area is small compared to the area of the water surface, the air soon will be virtually saturated (100% relative humidity) and an equilibrium reached in the atmosphere with equal numbers of water molecules leaving and re-entering the liquid surface. However, if chemical salt is dissolved into the water, fewer water molecules will leave the liquid surface. The equilibrium state

Table 10—Suggested salts for controlling relative humidity in closed containers.*

APPROX. EMC** (%)	RH*** (%)	CHEMICAL SALT	
4.5	20.0	$KC_2H_3O_2$	POTASSIUM ACETATE
6.0	32.0	$CaCl_2 \cdot 2H_2O$	CALCIUM CHLORIDE
6.5	33.5	$MgCl_2 \cdot 6H_2O$	MAGNESIUM CHLORIDE
8.0	42.0	$Zn(NO_3)_2 \cdot 6H_2O$	ZINC NITRATE
10.0	58.0	$NaBr$	SODIUM BROMIDE
12.0	66.0	$NaNO_2$	SODIUM NITRITE
14.5	76.0	$NaC_2H_3O_2 \cdot 3H_2O$	SODIUM ACETATE
16.0	80.5	$(NH_4)_2SO_4$	AMMONIUM SULFATE
17.5	84.0	KBr	POTASSIUM BROMIDE
20.0	90.0	$ZnSO_4 \cdot 7H_2O$	ZINC SULFATE
24.0	95.0	$NaSO_4$	SODIUM SULFATE

*** Since every potential reactive property or health hazard of chemical salts cannot be anticipated, the following precautions should be observed routinely in using any chemical salts:**
1. Store and use only at or near room temperature, avoiding excessive heat.
2. Handle carefully, avoiding skin contact, inhaling fumes or ingestion.
3. Use only as aqueous (water) solutions, and never intermix salts. Prevent contact with organic materials and metals. Mix only in glass or ceramic (nonmetallic) containers using glass stirrers.
**** for white spruce *(Picea* spp.)**
***** at 20°C (68°F)**

will therefore have fewer water molecules in the air, and the humidity will be lower. Since different salts have different solubilities, different humidities can be created by choosing particular salts. A list of salts is given in *Table 10*. The choice would depend on cost as well as desired humidity level.

This technique has been used successfully in the display of museum objects that must be kept at a prescribed humidity level. Trays of solution in the display case are hidden from view by visual baffles. This technique would doubtless have application for such items as musical instrument parts, in which precise conditioning before machining is desired. An aquarium with a glass lid makes an excellent chamber *(1)* because the entire bottom can be filled with saturated salt solution. The rack for wood pieces should allow air circulation. An excess of undissolved salt in the solution is advisable so that if the temperature fluctuates slightly there will be enough salt to keep the solution saturated. It is also important to keep the solution slightly agitated and the air moving. A standard fish-tank aerator installed as shown in Figure *1* will provide this agitation and air circulation.

Everything that has been discussed in this section is directly or indirectly related to the fact that wood seeks to establish an equilibrium moisture content according to the relative humidity of the surrounding atmosphere. If this point is kept in mind, the rest will follow logically.

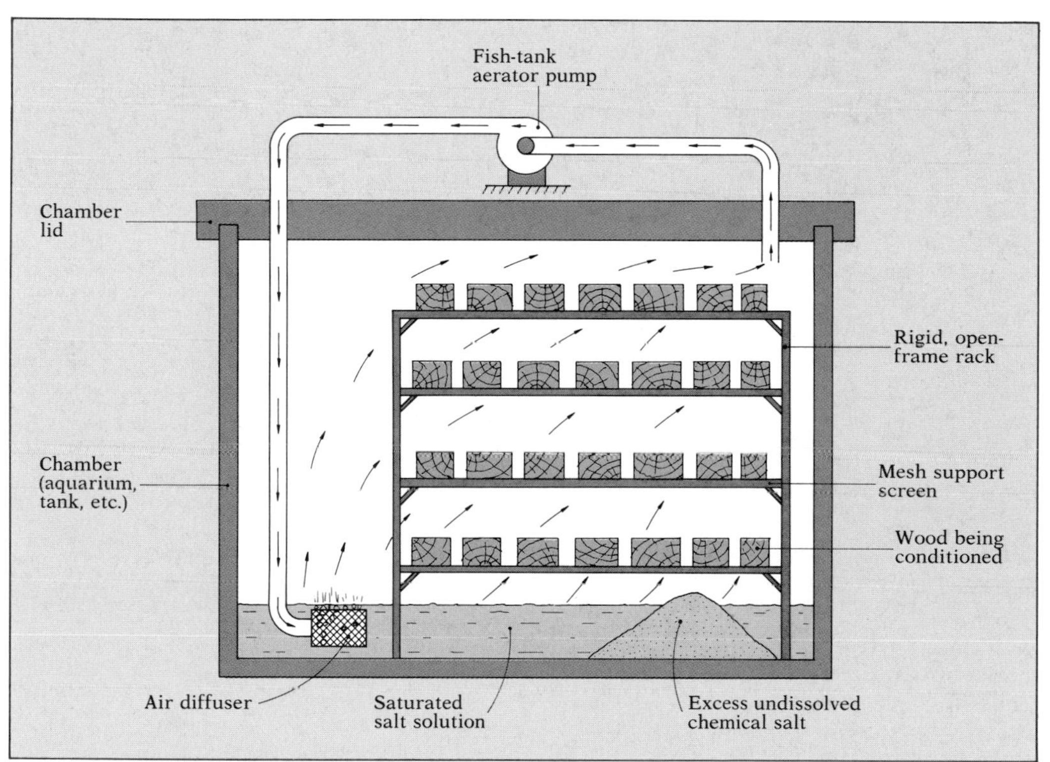

1—An ordinary aquarium can be converted to a conditioning chamber for small wood samples. Relative humidity is controlled by passing air through a saturated water solution of a selected chemical salt.

Strength of Wood

CHAPTER 6

A caller once asked bluntly, "How strong is oak?"

Puzzled by the question and wondering where to begin to answer I replied simply, "Oh, it's *very* strong."

"That's good to know, because I'm going to build a table and I want it to be good and strong, so I guess I'll use oak." He seemed content and went on to discuss something else.

My answer was horrible because it didn't really say anything, and when I hung up the phone I felt a little ashamed that I hadn't taken the time to try and unscramble the whole subject and discuss it thoroughly. But the whole subject of **strength** or **mechanical properties** of wood is far more complicated than meets the eye. A good answer to even an innocent question like "How strong is oak?" is nothing you can whisk up quickly or easily. For one thing, the mechanics of materials is in itself a complex field of science, even for "simple" materials that are **homogeneous** (uniform in composition) and **isotropic** (having equal properties in all directions), like steel. But on top of that, wood is an **anisotropic**, **heterogeneous** material, subject to species differences, biological variability and a wide array of natural irregularities and defects. Moreover, there are many kinds of strengths and many kinds of stresses and strains, which materials respond to in varying degrees.

As for the caller's question, I would rather have been asked, "What's involved in building a strong table?" Then I could have replied. First, you must have a good design, one that will anticipate the probable load to be imposed and efficiently carry and transfer the stresses among the components. Second, the table must be well-fabricated and assembled, especially with regard to joints and fasteners. Third, a wood must be chosen that is strong enough for the design, or preferably, the design would have been developed with the strength of the wood in mind. These three points are of course inseparable, but this section is mainly about the strength of wood, not design or fabrication.

Of all the properties of wood, strength is probably of most concern to woodworkers. It not only determines the mechanical performance of a finished piece, but it is also an important factor in drying, machining, bending, gluing and fastening. In considering strength it is important to recall the structure of wood, for its properties may be strikingly different in the longitudinal, radial and tangential directions. Familiarity with basic wood-moisture relationships is also critical.

A thorough discussion of strength properties could fill a thick textbook all by itself. Nevertheless, I will highlight some principles that I think will aid the woodworker and serve as a starting point for further study.

The study of strength is a numerical science, and in order to discuss the subject properly it is necessary to introduce mathematical formulas and symbols. To the uninitiated, the physics and engineering notation used may seem imposing, but even if all the details are not comprehended at the first reading, the basic principles given are meaningful. It will become apparent in this chapter that practical design or engineering with wood is not a numerically exact science, but rather more often is based upon judgment and intuition. Nevertheless, such judgment is best developed with some understanding of basic engineering principles.

In exploring the subject of strength, the terms **stress** and **strain** must first be defined.

Stress is **unit force**, that is, the amount of force or load acting on a unit of area. It is determined by dividing the load, *P*, by the area, *A (2)*. To illustrate, assume we have a piece of eastern white pine, 2 in. by 2 in. by 10 in. long, at 12% moisture content, straight-grained and free of defects. We place it in a press and apply a load of 2,000 lb. on the ends. The total force of 2,000 lb. acts on 4 sq. in., and develops a stress of:

$$\frac{P}{A} = \frac{2{,}000 \text{ lb.}}{4 \text{ in.}^2} = 500 \text{ lb./in.}^2$$

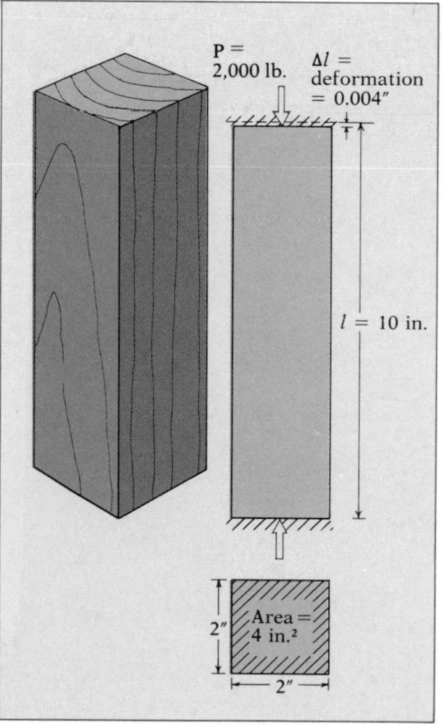

2—Stress, or force per unit area, is calculated by dividing the load (2,000 lb.) by the area (4 sq. in.), equaling 500 psi. The strain, or unit deformation, is calculated by dividing the deformation of the block (0.004 in.) by its original length (10 in.), equaling 0.0004 in./in.

1—Universal timber-testing machine, lower left, in Hoadley's laboratory is set up to measure the bending strength of a wooden beam under static loading. The machine's yoke can be arranged to apply stress in various modes, with the amount of force indicated by the large dial as well as by a running printout. Top, Hoadley adjusts anvil that will press a hardened steel ball into the end grain of a wooden block, to measure hardness.

(Pounds per square inch is usually abbreviated **psi.**)

Strain is **unit deformation**, that is, deformation per unit of original length. In the white pine block, suppose we precisely measure the amount by which the column was compressed while subjected to 500 psi stress and found it to be 0.004 in. The strain would be:

$$\frac{\text{deformation}}{\text{length}} = \frac{0.004 \text{ in.}}{10 \text{ in.}} = 0.0004 \ \frac{\text{in.}}{\text{in.}}$$

Strain is expressed in inches per inch, without canceling the units.

Strength is often defined as the ability to **resist** applied stress, and the strength of the material is synonymous with the **resistance** of the material. In this sense we are interested not simply in the total load or stress the material can resist, but also in how much deformation or strain results from a given level of stress. In considering the strength of wood, the relationship between stress and strain is of primary concern.

Going back to our block of eastern white pine, let's apply regular and equal increments of load or stress and with each increase measure the resulting deformation or strain. We can then accumulate a series of stress and strain measurements all the way from minimum stress to failure of the block and we can plot them in graph form *(1)*.

The data show that the maximum load carried by the block was 17,600 lb., the maximum stress, 4,400 psi. The test data reveal another important trait of the wood. Up to a stress level of 3,250 psi, stress and strain are proportional, that is, each increment of stress produces a proportional increment of strain. This characteristic proportionality of certain materials is known as **Hooke's Law**, after Robert Hooke, who discovered this behavior in 1678. Beyond the **proportional limit**, additional increments of stress result in increasingly larger increments of strain as the maximum stress, and ultimately failure, is approached. Figure *2* shows typical failure in compression parallel to the grain.

In wood, the proportional limit commonly occurs at between one-half and two-thirds of the maximum stress. The importance of Hooke's Law is that wood is elastic up to the proportional limit; that is, strains are recoverable upon removal of stress. As indicated by Figure *3*, a piece of wood subjected to a stress level x less than the proportional limit (σ') produces strain y. When the stress

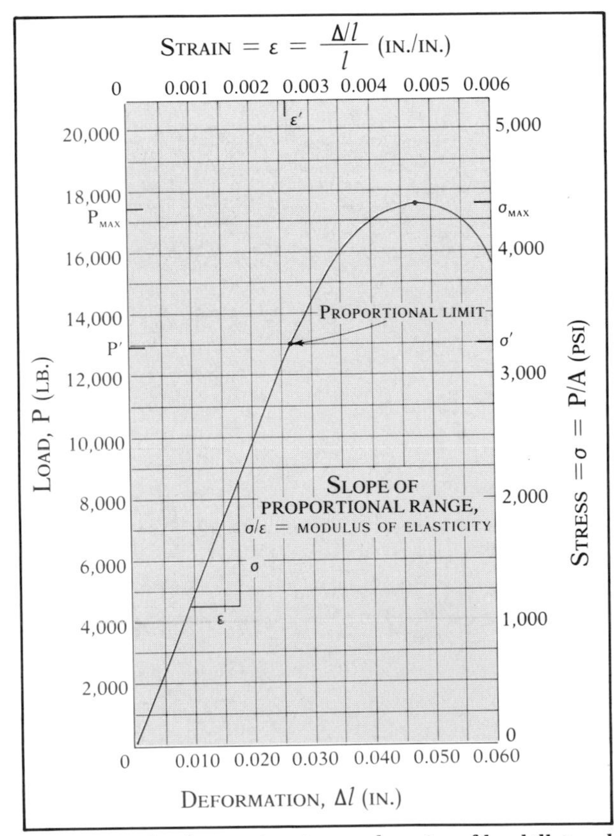

1—In a standard stress/strain graph, pairs of load (lb.) and deformation (in.) values taken during a destructive compression test are plotted to show maximum crushing strength, proportional limit and modulus of elasticity.

2—Failure in compression parallel to the grain.

is removed, the strain returns to zero. If the piece is stressed beyond the proportional limit, however, to stress level *X*, then removal of the stress recovers only part of the strain. Thus, strain equal to $0 - Y$ (called **permanent set**) will remain.

These fundamental mechanics are easily demonstrated with a hammer. Tap the surface of a board gently with a hammer, and no damage is apparent. The wood actually depresses under the force of the hammer but returns immediately as the hammer bounces back. But strike a hard blow—as when you miss in driving a nail—and the wood is permanently dented. A deeper dent occurs as the hammer strikes, but only partial elastic recovery takes place as the hammer rebounds. The remaining dent reflects the permanent set.

The wooden archery bow must be designed to keep deformation and bending stresses within the elastic limit, so that when the bowstring is drawn to arrow length and released, the bow springs back to original position. Drawing the bowstring back too far might stretch the wood fibers beyond proportional limit. Thus, when unstrung, the bow would not spring back to shape because it had taken on set.

As Hooke's Law states, the ratio of stress to strain for a given piece of wood within the elastic range is a constant. This ratio is called the **modulus* of elasticity** (also known as Young's Modulus and usually abbreviated as MOE or simply E), and equals the stress divided by the resulting strain. It can be calculated by choosing any set of values of stress and resulting strain, although the stress and strain values at the proportional limit are conventionally used. For our white pine block, we would calculate:

$$E = \frac{\text{stress}}{\text{strain}} = \frac{3{,}250 \text{ psi}}{0.0026 \text{ in./in.}} = 1{,}250{,}000 \text{ psi.}$$

The units representing strain (inches/inch) mathematically cancel out, and although the modulus of elasticity is simply expressed in psi, it is not an actual stress in the wood. We most often see E written as 1.25×10^6 psi, or simply abbreviated as 1.25E, as on a lumber grade stamp. It is useful in making comparisons. Wood with a value of 2.2E (that is, 2.2×10^6 psi) is twice as stiff as wood with a value of 1.1E.

* The word modulus means simply "measure."

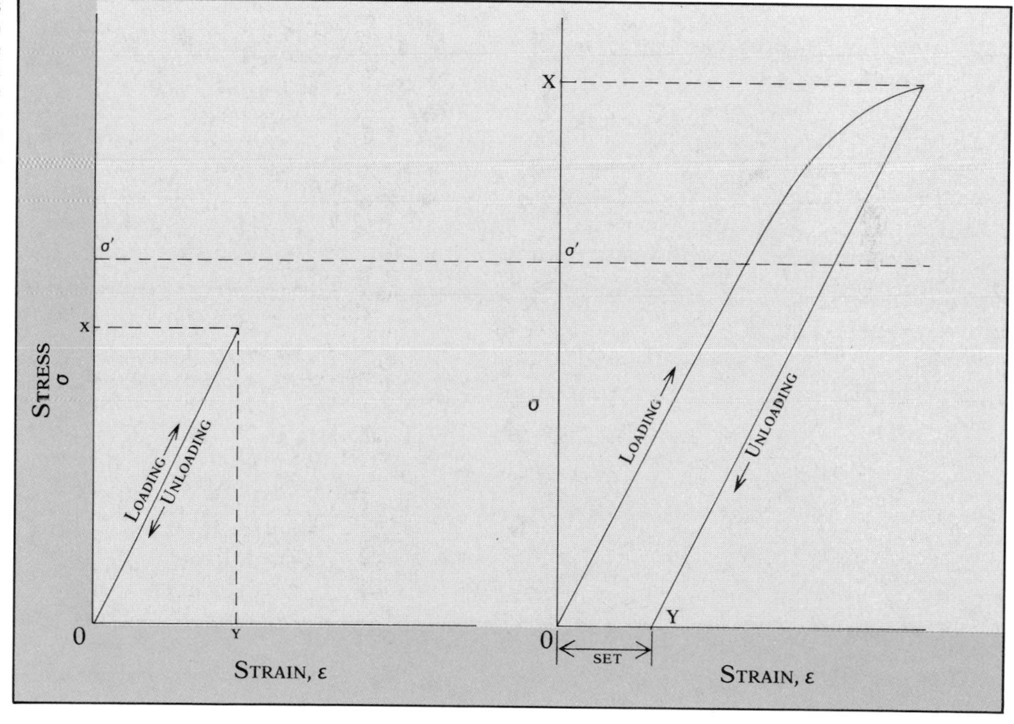

3—Wood stressed within its elastic range to some value (x) and unloaded recovers from the resulting strain. Wood stressed to some value (X) beyond the proportional limit and unloaded does not recover; some of the strain remains as permanent set (0-Y).

The relative **slope** of the stress-strain curve, as indicated by the modulus of elasticity, E, gives a measure of relative **stiffness**; the steeper the slope, the higher the E value and the stiffer the wood. Moreover, the higher the E value, the lower the deformation under a given load. Thus, under a given load, a floor framed with joists rated at 2.2E will sag only half as much as one using joists rated 1.1E.

We often think of the word strength in terms of failure, or maximum stress, and are satisfied to use maximum load values to rate one piece or species against another. At the same time, we depend upon the performance of the wood, not just at the failure level or slightly below it, but in fact well below the proportional limit, since we are depending on the elastic performance of the wood. Imagine a diving board or a baseball bat that remained slightly bent after use. Think of the consequences if the living-room floor joists didn't straighten up when the crowd went home after a party. We are usually well aware of instances where the proportional limit has been exceeded—those dents near the nail head left by errant blows of the hammer, chair rungs loose in their mortises and loose heads on hammer handles. Sometimes we intentionally make use of the plastic range (i.e., the non-elastic range beyond the elastic limit). A good example is in steam-bending wood, where the elastic limit is lowered by heating in order to produce a permanent set.

We can now characterize the compressive resistance ("strength") of our white pine block in terms of three important features of stress/strain behavior: maximum crushing strength; strength or fiber stress at proportional limit; and modulus of elasticity, E = stress/strain.

Our white pine block under compression is an example of only one of the three kinds of primary stress that can be applied to an object *(1)*. **Compression stress** shortens or compresses an object. **Tensile stress** elongates or expands the dimensions of an object. Tensile and compressive stresses are referred to as **axial stresses**, because they cause shortening or elongation along a common line of action. By convention, the Greek letter sigma (σ) is used to denote axial stress. (Sometimes σ_c designates compression, and σ_t, tension.) **Shear stress** causes portions of an object to move or slide in parallel but opposite directions. Shear is conventionally indicated by the Greek letter tau (τ).

Some strength properties are simply resistance to primary stresses, as in the example of the white pine block, where compression alone was involved. In other cases, the manner of loading causes combinations of stresses, wherein each type of stress must be sorted out and analyzed. An example of this is the bending strength of wood.

Since wood is anisotropic, each stress will be resisted differently according to growth-ring placement and grain direction. In order to determine relationships, an infinite number of tests of different wood-structure orientations could be done, but to minimize this array, the American Society for Testing and Materials (ASTM) has adopted standardized tests for the most important ones. The ASTM standards define the physical dimensions of the test pieces and specify conditions of moisture content, grain direction, growth-ring orientation and how free the piece is from defects. The tests are conducted in a standard testing machine (see Figure *1*, page 106) capable of holding pieces in appropriate fixtures, of applying force at prescribed rates of loading, of indicating the resistance and of measuring deflection. Not all strength behavior is routinely recorded for every test. *Table 11* shows strength properties for selected species.

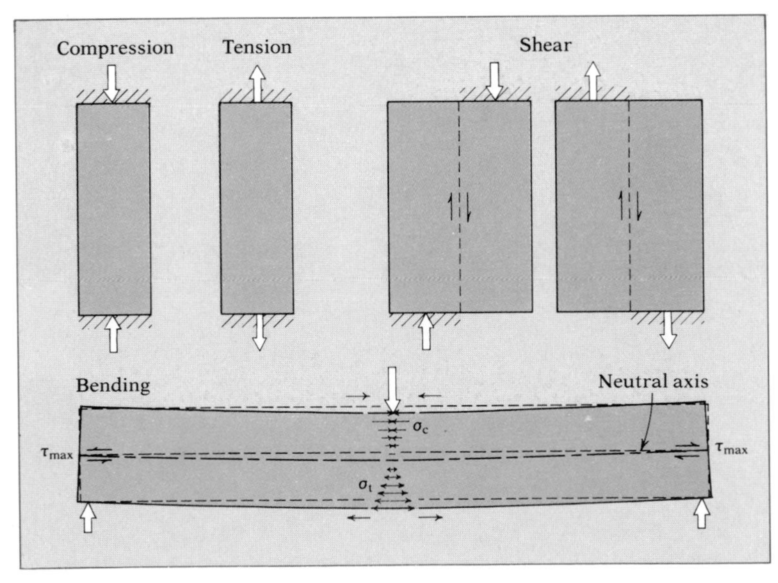

1—Compression, tension and shear are the primary forms of stress, and all three develop when a beam bends: compression (σ_c) along the upper portion, tension (σ_t) along the lower portion, and shear (τ_{max}) near the ends.

Table 11—Strength properties* at 12% MC of some commercially important woods grown in the United States.

*FSPL, fiber stress at proportional limit; MCS, maximum crushing strength; MTS, maximum tensile strength; MSS, maximum shear strength; MR, modulus of rupture; E, modulus of elasticity.	COMPRESSION			TENSION	SHEAR	STATIC BENDING		
	‖ TO GRAIN FSPL (σ') PSI	‖ TO GRAIN MCS (σ_{max}) PSI	⊥ TO GRAIN FSPL (σ') PSI	⊥ TO GRAIN MTS (σ_{max}) PSI	‖ TO GRAIN MSS (τ_{max}) PSI	FSPL (σ') PSI	MR (σ_{max}) PSI	E (σ/ε) 10^6 PSI
HARDWOODS								
ALDER, RED *(Alnus rubra)*	4,530	5,820	540	420	1,080	6,900	9,800	1.38
ASH, WHITE *(Fraxinus americana)*	5,790	7,410	1,410	940	1,950	8,900	15,400	1.77
BASSWOOD, AMERICAN *(Tilia americana)*	3,800	4,730	450	350	990	5,900	8,700	1.46
BEECH, AMERICAN *(Fagus grandifolia)*	4,880	7,300	1,250	1,010	2,010	8,700	14,900	1.72
BIRCH, PAPER *(Betula papyrifera)*	3,610	5,690	740	—	1,210	6,900	12,300	1.59
BIRCH, YELLOW *(B. alleghaniensis)*	6,130	8,170	1,190	920	1,880	10,100	16,600	2.01
BUTTERNUT *(Juglans cinerea)*	4,200	5,110	570	440	1,170	5,700	8,100	1.18
CHERRY, BLACK *(Prunus serotina)*	5,960	7,110	850	560	1,700	9,000	12,300	1.49
CHESTNUT, AMERICAN *(Castanea dentata)*	3,780	5,320	760	460	1,080	6,100	8,600	1.23
COTTONWOOD, EASTERN *(Populus deltoides)*	3,490	4,910	470	580	930	5,700	8,500	1.37
ELM, AMERICAN *(Ulmus americana)*	4,030	5,520	850	660	1,510	7,600	11,800	1.34
ELM, ROCK *(U. thomasii)*	4,700	7,050	1,520	-0-	1,920	8,000	14,800	1.54
HACKBERRY *(Celtis occidentalis)*	3,710	5,440	1,100	580	1,590	5,900	11,000	1.19
HICKORY, PECAN *(Carya illinoensis)*	5,180	7,850	2,130	—	2,080	9,100	13,700	1.73
HICKORY, SHAGBARK *(C. ovata)*	6,605	9,210	2,170	—	2,430	10,700	20,200	2.16
HONEYLOCUST *(Gleditsia triacanthos)*	5,250	7,500	2,280	900	2,250	8,800	14,700	1.63
LOCUST, BLACK *(Robinia pseudoacacia)*	6,800	10,180	2,260	640	2,480	12,800	19,400	2.05
MAGNOLIA, SOUTHERN *(Magnolia grandiflora)*	3,420	5,460	1,060	740	1,530	6,800	11,200	1.40
MAPLE, RED *(Acer rubrum)*	4,650	6,540	1,240	—	1,850	8,700	13,400	1.64
MAPLE, SUGAR *(A. saccharum)*	5,390	7,830	1,810	—	2,330	9,500	15,800	1.83
OAK, BLACK *(Quercus velutina)*	4,750	6,520	1,150	—	1,910	7,900	13,900	1.64
OAK, NORTHERN RED *(Q. rubra)*	4,580	6,760	1,250	800	1,780	8,500	14,300	1.82
OAK, POST *(Q. stellata)*	3,700	6,600	1,760	780	1,840	7,600	13,200	1.51
OAK, SOUTHERN RED *(Q. falcata)*	2,910	6,090	1,080	510	1,390	6,000	10,900	1.49
OAK, WHITE *(Q. alba)*	4,760	7,440	1,070	800	2,000	8,200	15,200	1.78
SWEETGUM *(Liquidambar styraciflua)*	3,670	6,320	660	760	1,600	6,600	12,500	1.64
SYCAMORE, AMERICAN *(Plantanus occidentalis)*	3,710	5,380	860	720	1,470	6,400	10,000	1.42
TUPELO, BLACK *(Nyssa sylvatica)*	3,470	5,520	1,150	500	1,340	7,300	9,600	1.20
WALNUT, BLACK *(Juglans nigra)*	5,780	7,580	1,250	690	1,370	10,500	14,600	1.68
YELLOW POPLAR *(Liriodendron tulipifera)*	3,730	5,540	560	540	1,190	6,200	10,100	1.58
SOFTWOODS								
BALDCYPRESS *(Taxodium distichum)*	4,740	6,360	900	270	1,000	7,200	10,600	1.44
CEDAR, ALASKA *(Chamaecyparis nootkatensis)*	5,210	6,310	770	360	1,130	7,100	11,100	1.42
CEDAR, INCENSE *(Libocedrus decurrens)*	4,760	5,200	730	270	880	5,900	8,000	1.04
CEDAR, WESTERN RED *(Thuja plicata)*	4,360	5,020	610	220	860	5,300	7,700	1.12
DOUGLAS FIR *(Pseudotsuga menziesii)*	5,850	7,430	870	340	1,160	7,800	12,200	1.95
FIR, BALSAM *(Abies balsamea)*	3,970	4,530	380	180	710	5,200	7,600	1.23
FIR, WHITE *(A. concolor)*	3,590	5,350	600	260	930	6,500	9,300	1.38
HEMLOCK, EASTERN *(Tsuga canadensis)*	4,020	5,410	800	—	1,060	6,100	8,900	1.20
HEMLOCK, WESTERN *(T. heterophylla)*	5,340	6,210	680	310	1,170	6,800	10,100	1.49
LARCH, WESTERN *(Larix occidentalis)*	5,620	8,110	980	430	1,410	8,300	13,900	1.96
PINE, EASTERN WHITE *(Pinus strobus)*	3,670	4,800	440	310	900	5,700	8,600	1.24
PINE, JACK *(P. banksiana)*	3,550	5,660	600	420	1,170	5,800	9,900	1.35
PINE, LODGEPOLE *(P. contorta)*	4,310	5,370	750	290	880	6,700	9,400	1.34
PINE, PONDEROSA *(P. ponderosa)*	4,060	5,270	740	400	1,160	6,300	9,200	1.26
PINE, RED *(P. resinosa)*	4,160	6,070	650	460	1,210	7,000	11,000	1.63
PINE, SOUTHERN YELLOW LONGLEAF *(P. palustris)*	6,150	8,220	950	470	1,500	9,300	14,300	1.93
PINE, SOUTHERN YELLOW SHORTLEAF *(P. echinata)*	5,090	7,270	750	470	1,390	7,700	13,100	1.76
PINE, SUGAR *(P. lambertiana)*	4,140	4,770	590	350	1,050	5,700	8,000	1.20
PINE, WESTERN WHITE *(P. monticola)*	4,480	5,620	540	—	850	6,200	9,500	1.51
REDWOOD, OLD-GROWTH *(Sequoia sempervirens)*	4,560	6,150	860	240	940	6,900	10,000	1.34
SPRUCE, ENGELMANN *(Picea engelmannii)*	3,580	4,770	540	350	1,030	5,500	8,700	1.28
SPRUCE, SITKA *(P. sitchensis)*	4,780	5,610	710	370	1,150	6,700	10,200	1.57
SPRUCE, WHITE *(P. glauca)*	3,700	5,470	540	360	1,080	6,500	9,800	1.34
TAMARACK *(Larix laricina)*	4,780	7,160	990	400	1,280	8,000	11,600	1.64

Compression parallel to the grain

When wood is stressed in a manner that shortens its fibers lengthwise, it is under compression parallel to the grain, as in a column supporting a porch, or our introductory test on eastern white pine.

In a standard ASTM test for compression parallel to grain *(1)*, the specimen is 8 in. long, fitted with a special deflection gauge that measures compression of the central 6 in. By taking load/deformation data beyond the proportional limit, fiber stress at the proportional limit (σ'), maximum crushing strength (σ_{max}) and modulus of elasticity (E) can be computed.

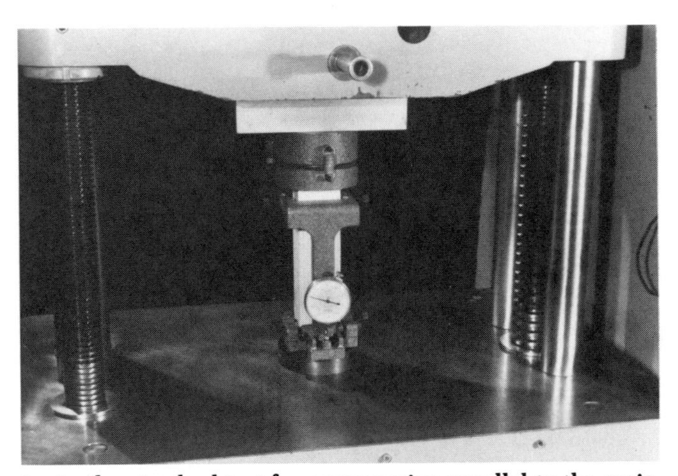

1—In the standard test for compression parallel to the grain, the testing machine applies a load to the ends of the column. A deflection gauge fastened to the specimen measures shortening of the column.

For its weight, wood is surprisingly strong in compression parallel to the grain. I once saw a hickory chair whose four legs were each 1¼ in. in diameter. The chair looked quite graceful, not at all overdesigned. But let's make a few quick calculations. The four legs have a total cross-sectional area of:

$$A = 4\,(\pi)\,r^2 = 4\,(\pi)(0.625)^2 = 4.91 \text{ in.}^2$$

From *Table 11*, (page 111), we see that hickory has an average proportional limit strength of 6,605 psi, so the four legs could support a total load of about 32,430 lb. Thus, compression parallel to the grain is hardly a structurally limiting factor. And this is the common experience. Even calculations of supporting members of buildings usually show a wide margin of overdesign in compression parallel to the grain. I cannot remember ever seeing a structural failure, or even hearing of one, due purely to compression stress parallel to the grain. If compression were the only factor, a 250-lb. person could be supported by four hickory dowels, each ⅛ in. in diameter. We see this in the formula:

$$A = 4\,(\pi)(\tfrac{1}{16})^2 = 0.0491 \text{ in.}^2$$

$$\sigma = \frac{P}{A} = \frac{250 \text{ lb.}}{0.049 \text{ in.}^2} = 5{,}102 \text{ psi.}$$

Clearly, this load is within the proportional-limit compression strength of hickory.

But we can hardly conceive of sitting on a seat supported by four ⅛-in. dowels. Our intuition alone would warn us of the danger, but we can also calculate the fault in the design. First of all, the dowels will buckle, since columns having a l/d ratio (l = length, d = least dimension) of greater than 11 will buckle at less than full load. (If our dowel legs were 17 in. long, the l/d ratio would be 17/0.125 = 136.) As in a building, the supporting frame must be stabilized by lateral connection to prevent buckling. Even if our dowel legs were stabilized by a multitude of small rungs, any sideways thrust against the chair would cause bending. So the legs of a chair must be thick enough to be more than just supporting posts. However, the bottoms of the 1¼-in. chair legs could be tapered down like sharpened pencils to ⅛-in. diameter tips and be more than adequate to sustain a 250-lb. person sitting with all four chair legs squarely on the floor.

Compression perpendicular to the grain

When heavy objects rest upon the surface of a wooden table or on a wooden floor, they apply loads (equal to their weights) that stress the wood in compression perpendicular to the grain. To determine the strength of wood in compression perpendicular to the grain, a 6-in. long, 2-in. by 2-in. cross section is supported horizontally and loaded over its central 2 in. *(2).* Load/deflection data are recorded until the proportional limit is reached. Beyond this limit, however, as the piece is compacted more and more, the resistance increases and no meaningful maximum load is reached *(3).* Therefore the only strength value routinely determined is the fiber stress at proportional limit (FSPL). Typical values for fiber stress at proportional limit in compression perpendicular to the grain are 440 psi for eastern white pine, 2,170 psi for hickory. In general, values are drastically lower for compression perpendicular to the grain than for compression parallel to the grain.

Published values for strength properties commonly list a single value for perpendicular-to-grain compression strength that is the average of both radial and tangential properties. In some species there may be insignificant differences between the two. In others, however, the anatomical structure may result in noteworthy radial and tangential differences. For example, in ring-porous hardwoods such as ash or catalpa and in uneven-grained softwoods such as southern yellow pine or Douglas fir, a piece stressed in the radial direction will be no stronger than the weakest layer of earlywood. This fact is used to advantage in pounding loose strips of ash for basket-making. Batting a baseball against the tangential face of an ash bat (that is, stressing the wood radially) deadens the impact by crushing the earlywood layers *(4).* For this reason, the trademark is always imprinted on the tangential face of the bat, and the batter is instructed at an early age not to hit the ball on the trademark (which of course leaves many a ball player with the notion that printing the trademark has weakened the bat). Such woods will support greater loads when loaded tangentially (i.e., against the radial face), because the stress is then shared equally by the layers of stronger latewood. Interestingly, compression strength is usually least when applied perpendicular to the grain, at a 45° angle to the growth rings.

Compression perpendicular to the grain is very often a limiting strength. For example, suppose our hickory chair had legs tapered to ⅛ in. Although the 5,102 psi de-

2—In the standard test for compression perpendicular to the grain, the probe of a free-standing deflectometer automatically records the strain as the test proceeds.

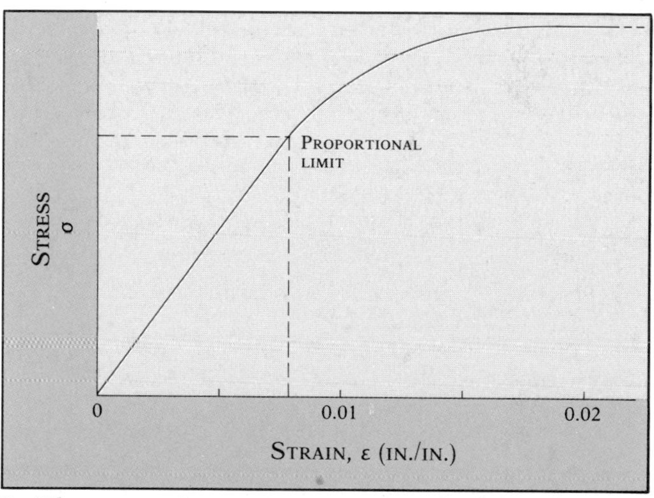

3—When measuring compression perpendicular to the grain, there is no meaningful maximum load, because after reaching proportional limit the piece is compacted more and more and resistance increases.

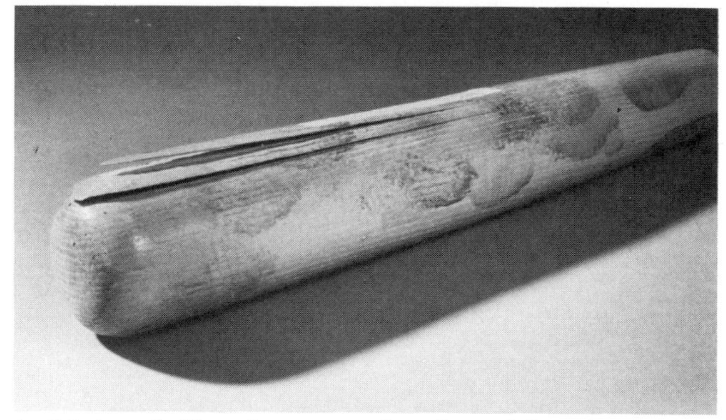

4—Hitting a ball on the tangential surface of this white ash bat has caused 'chipping,' or separation of the growth layers.

veloped at the tips could be carried by the chair legs, it would easily crush down into the surface of an eastern white pine floor board whose FSPL (fiber stress at proportional limit) is listed at only 440 lb. In fact, a 120-lb. woman placing her full weight on a shoe with a ½-in. by ½-in. heel will develop a stress as high as:

$$\frac{120 \text{ lb.}}{\frac{1}{2} \text{ in.} \times \frac{1}{2} \text{ in.}} = 480 \text{ psi.}$$

Eastern white pine obviously is not a logical choice for flooring, but the 2,170-psi perpendicular-to-grain strength of hickory would comfortably resist heel denting.

In internal-external joints such as a mortise and tenon, where racking is involved, strength in compression perpendicular to the grain can be especially critical *(1)*. The side-grain strength (compression perpendicular) of the tenon is no match for the end-grain strength (compression parallel) of the mortise when the same species is used for both.

A particularly critical situation develops when wood is restrained from swelling, as in the end of a hammer handle inserted into the hammer head, or even in a mortise-and-tenon joint. As mentioned in Chapter 4, the amount of swelling perpendicular to the grain that takes place as a result of natural humidity changes may be substantial. If swelling is restrained, the effect is that of compressing swollen wood by that amount *(2)*. However, the elastic limit strain is less than 1%, as can be seen in Figure 3. Therefore, compression set may develop. Upon re-drying, the piece "unloads" itself by shrinking to a smaller than original diameter, thus loosening the joint.

Moisture cycling is probably a greater cause of mortise-and-tenon loosening than mechanical stressing. We have placed brand-new chairs through moderately severe moisture cycles (90%-30%-90%-30% relative humidity) and had them loosen up so much you could hear them rattle without ever sitting upon them. The "1% compression limit rule" is a rough approximation for compression perpendicular to the grain, and is an important guideline for a woodworker to consider.

As an example, imagine a hickory handle conditioned to 7% moisture content in a heated shop and then fitted and tightly wedged into the eye of a hammer head *(4)*. The hammer is then moved to a garage where the handle

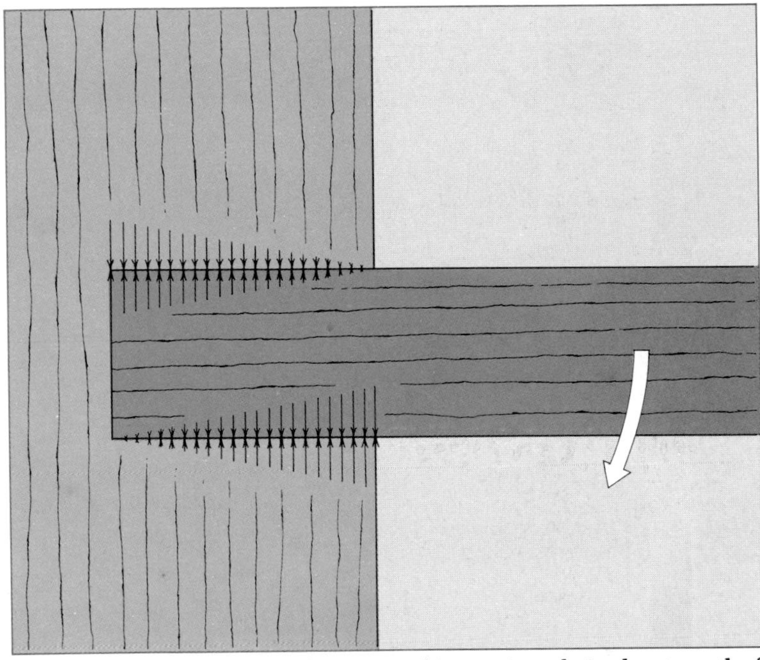

1—In a mortise and tenon subject to racking, as in a chair, the strength of the tenon in compression perpendicular to the grain is critical.

2—An experiment demonstrates the development of compression shrinkage. Three sections were cut from a board at 7% moisture content, then mounted in a steel frame and attached at the lower end by a woodscrew. The third section was free to swell, but the first and second were restrained by the upper part of the frame, to which the upper end of the first section was also fastened by a woodscrew. The humidity was slowly raised until the moisture content of the wood was 18% or more. The third section swelled; the first and second, confined by the frame, developed internal stress beyond their elastic limits. Though no damage was apparent, they took on permanent compression set. The moisture content was then restored to 7%, and the results can be seen in the photo. The third section shrank back to original size; the second became shorter due to compression shrinkage; the first, restrained from shrinking, developed internal tension sufficient to break it apart.

reaches an equilibrium moisture content of 14%. Recalling the shrinkage formula:

$$\Delta D = D_i S(\frac{\Delta MC}{fsp})$$

and noting from *Table 7* on page 74 that S_t = 10.5% for shagbark hickory,

$$\Delta D = D_i(0.105)(\frac{0.07-0.14}{0.28}) = -0.026\ D_i\ .$$

The handle will try to swell tangentially by 2.6% of its dimension. Perpendicular-to-grain compression tests on hickory have shown a proportional limit strain of about 1%. Thus we might expect more than 1.5% of set to occur. If the hammer is then brought back to its original 7% equilibrium moisture content (EMC) environment, we might expect it to shrink tangentially to about 1.5% less than its original dimension at that same moisture content.

Suppose that the hammer is left out in the rain, where the handle might absorb moisture to virtual fiber saturation point. An attempted swelling of 7.9%

$$\Delta D = D_i(0.105)(\frac{0.07-0.28}{0.28}) = -0.079D_i$$

would be restrained, so about 6.9% compression set could be expected.

To check out these predictions I ran some controlled tests in which shagbark hickory discs were restrained in stainless-steel rings. The discs were carefully machined to slip snugly into the rings at 7% moisture content. They were allowed to absorb moisture until fiber saturation was attained. Then they were redried to original weight and remeasured. They showed an average loss in diameter of 7.5% tangentially and 5.4% radially. Similar experiments with beech *(5)* and other species also showed that results could be approximately predicted. Little wonder tool handles loosen when moved from place to place.

Incidentally, I now keep my adze, axes and sledgehammers in the garage, with plastic trash bags wrapped around the heads. This has ended my handle-loosening problems, since the average EMC in the garage is more uniform than cyclic indoor conditions.

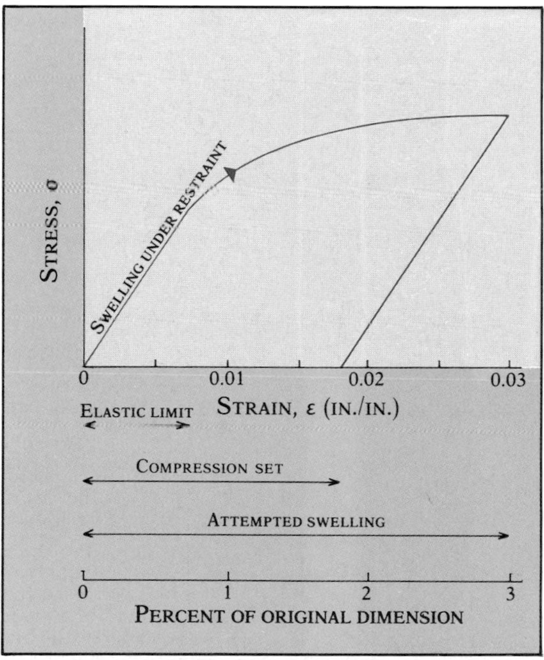

3—In response to normal humidity variation, wood may swell as much as 3% perpendicular to the grain. Restrained swelling is similar to compression and may cause wood to load itself beyond the proportional limit, thereby causing compression set.

5—Pairs of American beech discs illustrate the effect of cyclically varying moisture content on wood under restraint. The top pair is shown as originally turned at 7% moisture content. The middle pair was moistened to fiber saturation. The bottom pair was moistened to fiber saturation, then reconditioned to original weight. The wood (in compression set) has shrunk away from the restraining ring.

4—Severe compression shrinkage in hickory handle of commercially manufactured hammer was caused by moisture cycling. When the hammer was originally sectioned, the handle was tightly wedged in the eye. After storage in a damp place, the hammer was moved to drier conditions, and the joint loosened. Moisture cycling, not pounding, is the usual cause of loose tool handles.

Tension perpendicular to the grain

Perpendicular-to-grain tensile failures occur when we split firewood, when nails are driven into unyielding wood, when cupped boards are forced flat, when a karate demonstration is performed and when we plane a long, curling shaving from the edge of a board. But we generally are aware of the weakness of wood in this regard and have pretty much learned to design and use wood to avoid such failures. Perhaps the most common and misunderstood tensile failures are due to self-induced tension from uneven dimensional change. As mentioned in the previous chapter, end-checking, surface-checking and internal honeycomb checks are classic examples, as are the radial cracks developed in pieces embracing the pith.

The standard ASTM test for tension perpendicular to the grain *(1)* makes no provision for measuring strain. Only stress at maximum load is determined, and this is the only extensively published strength property in tension perpendicular to the grain. Typical values for maximum tensile strength are 310 psi for eastern white pine, and 1,010 psi for American beech. Such values are important when comparing resistance to splitting among various woods.

Fortunately, some testing has been done on specimens that are fitted with gauges to measure strain *(2)*. As indicated in Figure 3, the elastic behavior is similar to that of compression perpendicular to the grain, with approximately the same proportional-limit strain. Above the proportional limit we find a striking difference, in that tensile failure occurs totally and abruptly at strains of less than 2%, at a maximum stress about double the proportional-limit stress. It is unfortunate that more extensive data is not available for more species, because this information would give us valuable insight into strain limits. If the moisture content of one portion of a piece of wood drops 8% to 10% lower than an adjacent portion, the drier portion will attempt to shrink by about 2% of its tangential dimension. Our strength information suggests that it will probably crack wide open.

The strain limit in tension perpendicular to the grain can be critical when boards are restrained and undergo shrinkage due to moisture loss. The all-too-common example of this problem occurs in items constructed of air-dry lumber at 14% or 15% moisture content that eventually reaches equilibrium at 6% or 7% indoors. The resulting shrinkage in a flatsawn board averages at least 2%. That kind of example is usually clearly understood and easy to evaluate. The most baffling problem

1—In the standard test for tension perpendicular to the grain (resistance to splitting), a piece of wood is pulled apart until it splits. Only maximum stress can be determined.

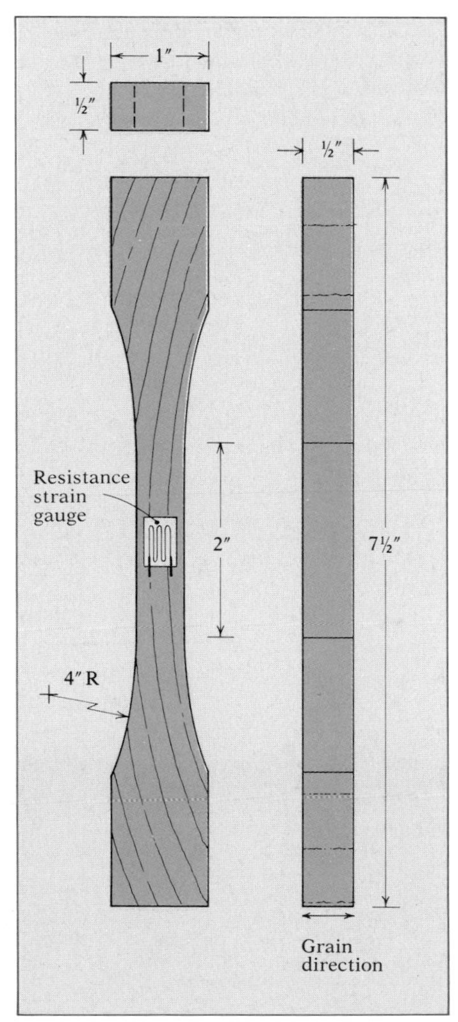

2—In Young's test for strength of wood in tension perpendicular to the grain, a gauge attached to a sample of standard form and dimensions measures strain as the wood, gripped by its ends, is put under tension.

crops up when things are made of wood at an appropriate moisture content, say 7%, and crack wide open the next year. The typical example involves a design in which wide boards or panels are fastened at their edges to a fixed frame. The side panels of a carcase built with the longitudinal-grained drawer runners fastened at each end to the panel edges is one case in point. If exposed to high humidity (such as furniture stored in the cellar or a cool basement game room) it cannot swell, and up to 2% or 3% compression set may be developed. No problem may be apparent, but when it returns to the "correct" moisture content, the piece cannot shrink to its earlier shape, again due to edge restraint by the drawer runners. This is when things begin to pop. The result is a check as soon as the attempted shrinkage exceeds the tensile-strain limit. The unfortunate misinterpretation of this problem is often something like, "Oh, we never have any trouble with high humidity, it's low humidity that causes all our problems." The unfortunate conclusion is that wood should be kept at high humidity, or that dryness, per se, is detrimental to wood.

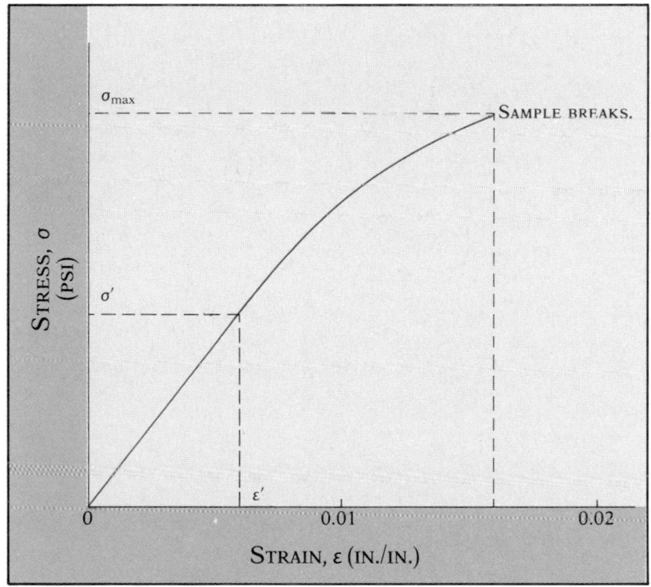

3—A typical stress/strain graph of tension perpendicular to the grain. The proportional limit is reached at less than 1% strain, and failure occurs at less than 2% strain.

In dowel or mortise-and-tenon joints, good gluing can help in restraining some compression-set shrinkage that would ordinarily open a joint. However, as the tensile strain limit is approached, the failure usually occurs at or near the glueline *(4)*. Failure adjacent to the glueline is probably due to a number of things, especially the machining damage to the surfaces of the mortise and tenon as well as the perpendicular grain direction conflict involved. I have never seen a tenon fail internally, such as with a honeycomb check, because the glue wouldn't let it shrink. It is always the glueline or a layer of cells adjacent that fails first.

I once did some research with round mortise-and-tenon joints. I made half with regular tenons, and in the other half I took a chisel and split the tenons radially as far down as the length of their insertion into the mortise. After moisture cycling severe enough to break gluelines in the regular joints, the split tenons had opened up internally, and the gluelines were largely intact *(5)*. This approach of creating a plane of failure (and therefore strain relief) where it won't do any harm is worth a lot of further research. I suspect that various wedged tenons turn out to be successful because of this mechanism. I have seen through-wedged tenons in plank bench tops where the tenon looked tight along the outer edge but there was an obvious opening beside the wedge.

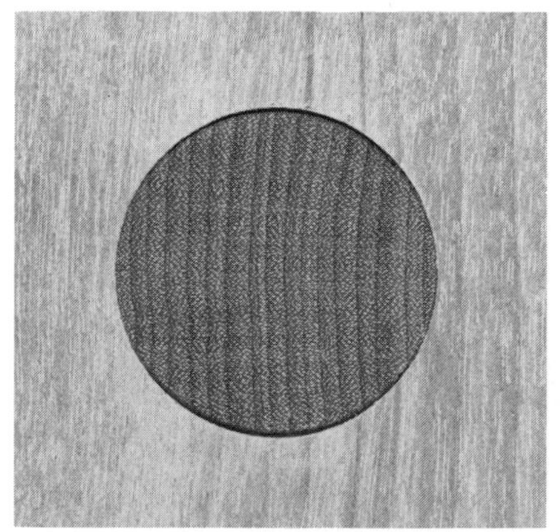

4—After moisture cycling, a dowel joint without glue, left, takes on compression set and shrinks away from the mortise top and bottom. With glue, right, only one side of the joint opens, usually tearing some tissue from the wall of the mortise. But the remaining glueline is no match for racking stress and quickly fails in tension.

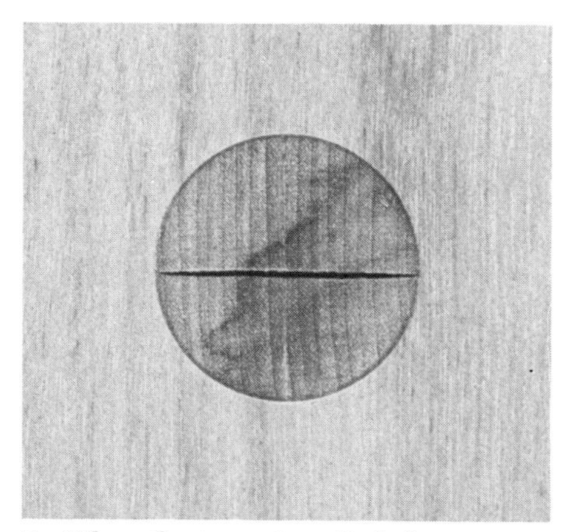

5—When the tenon is split before assembly, thereby creating a plane of failure, the crack will open under moisture cycling but the glueline is likely to remain intact. This is probably what happens when tenons are wedged.

Tension parallel to the grain

Wood is strongest in tension parallel to the grain, but this fact does us little good. A typical test specimen *(1)* suggests why. In order to make wood fail in tension parallel to the grain, it must be increased in dimension on each end where it is being held. To put it another way, if you just try to grip or fasten a piece of wood at its ends and pull it apart, the fastenings will split or shear off the ends. When a structural member is designed to be large enough so its attachment fastenings can safely transmit the required loads to other members, the cross section of the piece turns out to be far greater than needed along its entire length to carry tensile loads. Except for uses such as in aircraft, where weight is a critical factor, it is usually cheaper to leave the material full size than to machine the excess away.

Ultimate tensile strength for eastern white pine (specific gravity 0.36) has been measured at about 13,000 psi; for hickory (specific gravity 0.81) 30,000 psi. Compared to structural steel, having a specific gravity of 7.8 and ultimate tensile strength in the range of 60,000 psi, it can easily be argued that on a weight-for-weight basis, wood is stronger than steel (in tension parallel to the grain, of course).

Tension parallel to the grain has been of somewhat academic importance to date. In fact, because of the difficulty of conducting tests and the limited use for such data, this property has been largely ignored. However, with our developing technology in composite and assembled products, it may be possible to design structural members that can take advantage of the superior strength of wood in tension parallel to the grain.

Shear perpendicular to the grain

Wood is extremely resistant to shearing perpendicular to the grain, because of the alignment and structure of its longitudinal cells. Thin veneers can be cut perpendicular to the grain with scissors or with a paper cutter, but when any appreciable thickness is involved, no meaningful shear plane develops and the wood simply crushes or tears. As a result, no standard test has been developed to measure shear perpendicular to the grain.

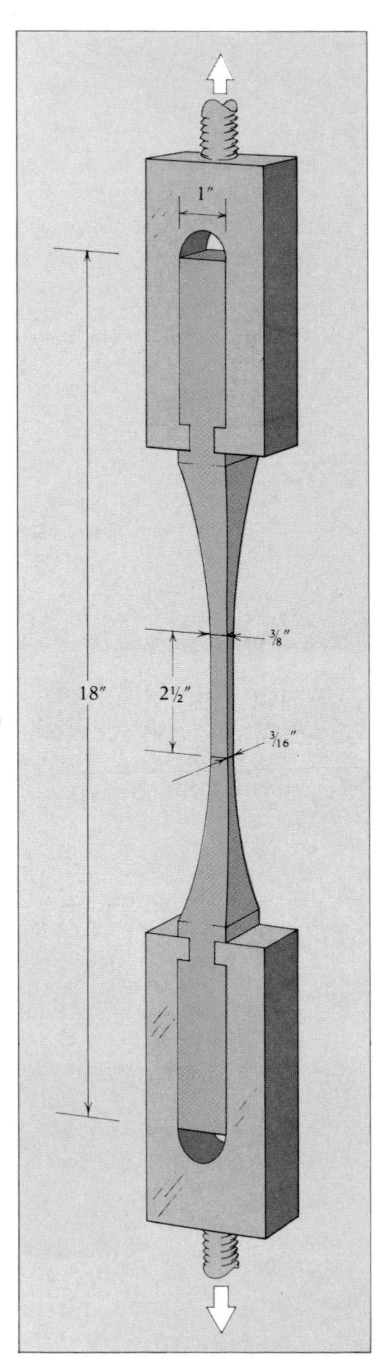

1—ASTM test for tension parallel to the grain.

Shear parallel to the grain

Along the grain, wood separates in shear more readily. A common example of shear failure is where fasteners or interlocking joints shear out to the nearest end-grain surface *(2)*.

The ASTM test for shear is performed on a notched block *(3)* held in a special shear tool *(4)*, because it is important to prevent rotation of the block as load is developed. The shear strength per square inch is determined by dividing the maximum load, P, by the shear area, A. Typical values are 900 psi for eastern white pine and 2,330 psi for sugar maple.

As with tension perpendicular to the grain, shear strength parallel to the grain may be drastically affected by anatomical features such as large rays or earlywood-latewood variation. As will be discussed later, shear may become critical in bending, especially in short

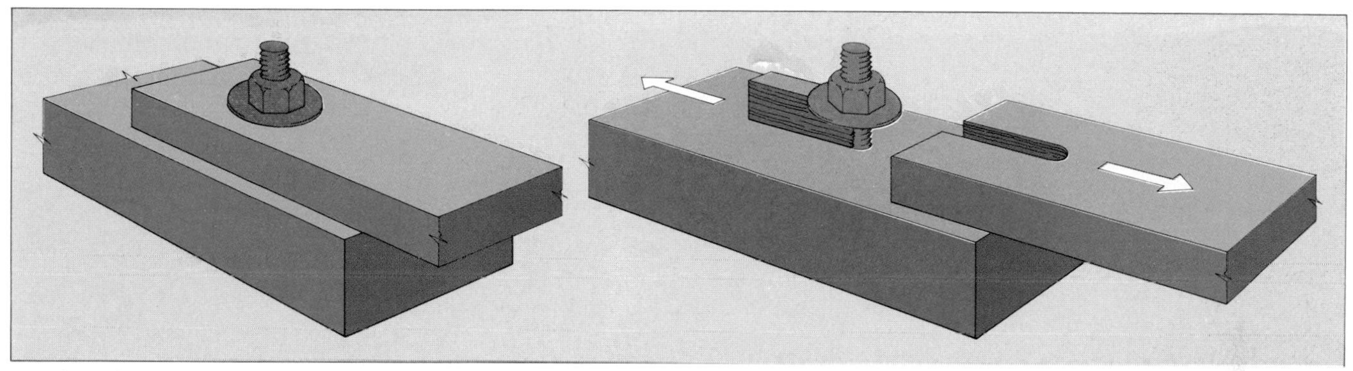

2—Shear failure occurs where fasteners shear out to the nearest end-grain surface.

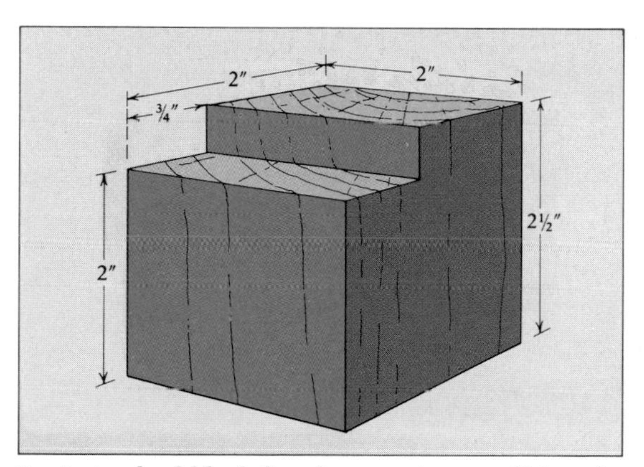

3—A standard block for shear-testing parallel to the grain.

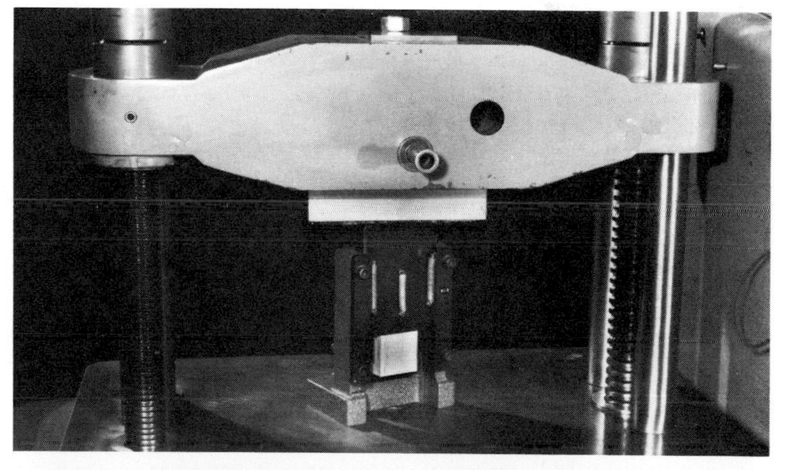

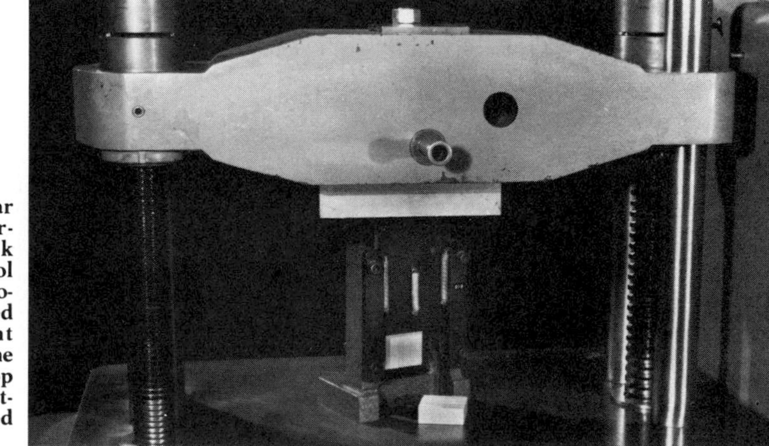

4—The standard test for shear parallel to the grain is performed on a notched block (Figure 3) held in a special tool that keeps the sample from rotating while a blade (returned by springs) pushes straight down, shearing the block at the inside edge of the notch. Top right, the sample ready for testing; right, the sample has failed in shear parallel to the grain.

beams and composite products.

When the shear plane is developed parallel to the grain, but produced by stresses perpendicular to the grain, **rolling shear** may result *(1)*. This behavior is characteristic of fibrous materials and results in groups or bundles of fibers separating and rotating. This type of failure commonly develops in the core ply of plywood and it is initiated by knife checks, which often occur in thick veneers.

A special shear test is the mainstay for evaluating wood adhesives. In the standard block shear test for adhesives, the double-notched shape of the specimen *(2)* prevents rotation and directs the stress along the plane of the glueline. Sugar maple is the standard designated species for the test, and the quality of bonding is judged on the combination of stress developed and the percentage of wood failure (versus glue failure) along the shear plane *(3)*.

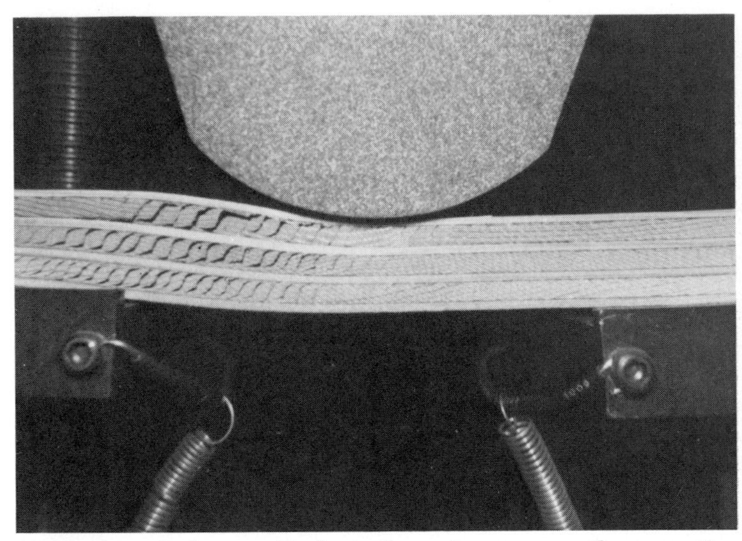

1—Stress applied perpendicular to the grain may cause the separation and rotation of small bundles of fibers, resulting in a shear plane parallel to the grain, above. This is called rolling shear. It commonly occurs in the core of plywood, as in the block-test samples at right, where rolling shear was initiated by knife checks created when the plies were manufactured.

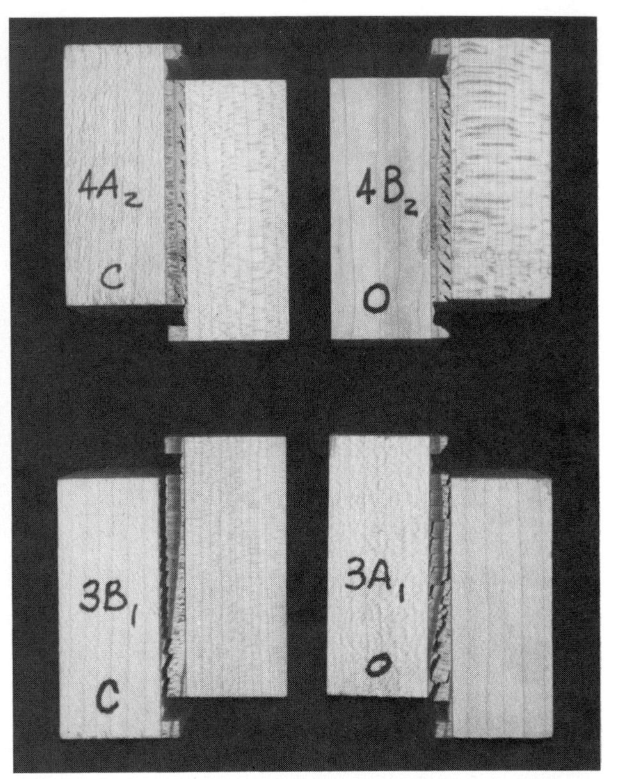

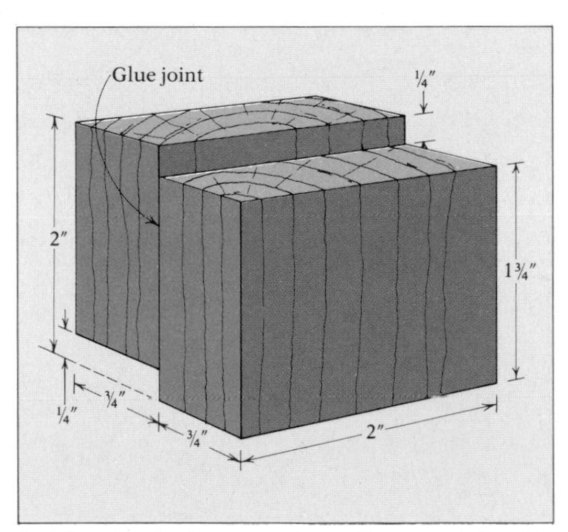

2—The standard test specimen for measuring the shear strength of a glueline. A special fixture holds the specimen in the testing machine, to direct the force along the plane of the glueline and prevent rotation of the test block.

3—Failed glue joints may show that the glue broke, the wood broke, or both. Glue failure before wood failure may be acceptable if the specimen resists stress equal to the normal shear strength of solid wood parallel to the grain.

Bending theory

No chapter on the strength of wood would be complete without a discussion of **bending** or **flexural** behavior. Nature evolved the tree to be a combination column and beam, and it is not surprising that wood is so beautifully efficient when loaded in compression parallel to the grain or in bending. We take greatest advantage of the strength of wood when we use it in the form of beams. Understanding beam mechanics is fundamental to structural design, and it is somewhat complicated. Let's try a brief survey, step by step.

A **beam** is defined as an elongated member, loaded perpendicular to its long axis. Many examples of beams come quickly to mind: A **simple beam** is one supported at the ends and loaded in between, such as the joists in a floor, the seat of a child's swing and the treads of open stairs. Even a seesaw is a simple beam (in reverse, supported in the middle and loaded on the ends). There are more complicated types, such as **continuous beams** (supported and loaded in several places), **fixed-end beams** (like the rungs of a chair), and **cantilever beams** (fixed on one end—a tree limb for example). We'll stick with simple beams.

Let's consider a center-loaded, simply supported beam *(4)*. The bending resulting from the load tends to shorten or compress the upper surface and to stretch the lower surface. The stresses developed are axial stresses: compression along the upper surface, tension along the bottom. Both stresses are maximum at the very surface and diminish to zero at the central horizontal plane of the beam, which is called the **neutral axis**. The surface axial stresses are greatest at midspan and decrease to zero at each end of the beam. These axial tension and compression stresses are called **bending stresses**, or **fiber stresses**. In materials such as steel, which have equal resistance in compression and tension, the bending stresses at the upper and lower surfaces are equal.

A classic engineering formula, the **flexure formula**, is used to determine the magnitude of bending stresses based on the known load or loads (P) and the length *(l)*, the width (b) and the depth (d) of the beam. The strength of wood is greater in tension than in compression and so the flexure formula does not strictly apply. It is nevertheless assumed to be valid. The apparent axial stresses are determined by subjecting wood beams to a bending load, collecting the appropriate load data and applying the flexure formula. In the standard ASTM static bending test, a 2-in. by 2-in. by 30-in. beam of wood is sup-

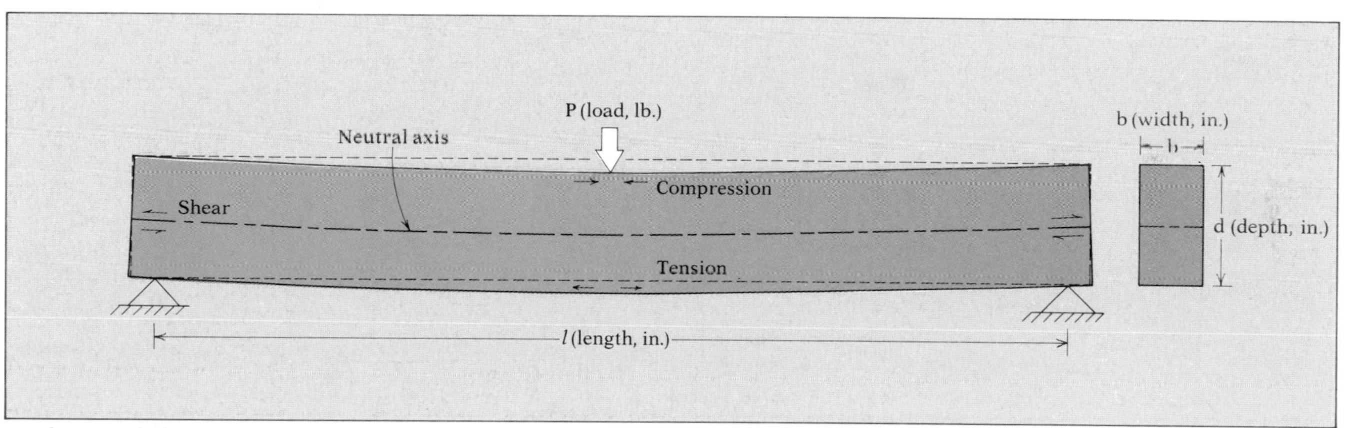

4—This simple beam is supported at the ends and loaded in the middle. The bending that results shortens the upper surface and stretches the lower surface, while the neutral axis stays the same length. The associated axial stresses are compression along the upper surface and tension along the lower.

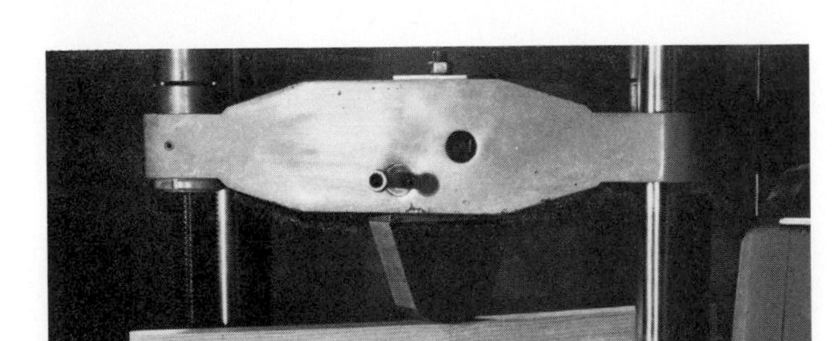

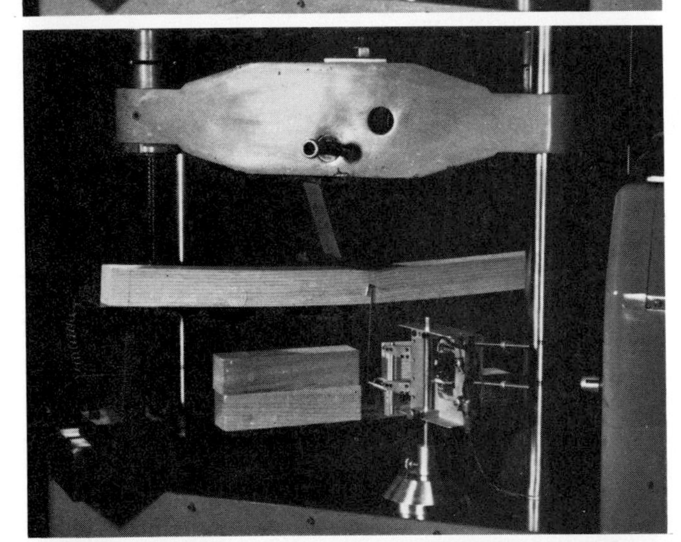

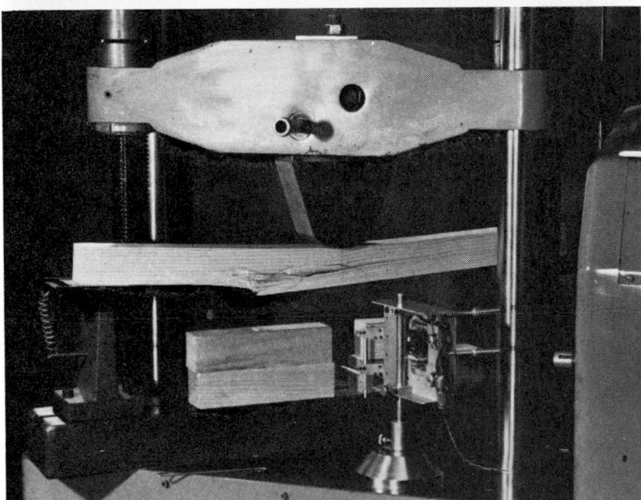

1—ASTM static bending test: As the bearing head descends at 0.1 inch per minute, simultaneous deflection measurements are continuously sensed by the deflectometer and printed out as a graph.

ported at both ends over a 28-in. span and center-loaded *(1)*. The test is called a static test because it is conducted at a fairly slow rate of speed (0.1 in. per minute). (In an impact bending test, loading is almost instantaneous.) With suitable instruments, simultaneous measurements of load and deflection of the neutral axis are taken continuously throughout the test.

A typical load/deflection graph *(2)* was recorded in testing a beam of eastern white pine. Prior to testing, the beam, measured at midspan, had a width (b) of 2.01 in. and a depth (d) of 2 in. The graph shows a maximum load (P_{max}) of 1,590 lb., a proportional limit load (P′) of 990 lb., and a deflection at proportional limit (y′) of 0.28 in. The relative slope of the curve within the elastic range indicates the stiffness of the beam. (A special formula, given below, is used to convert the load/deflection ratio to stress/strain or modulus of elasticity.) In a typical bending failure *(3)*, the beam's upper part fails first in compression, forming buckled layers of cells, while the lower portion shows typical splintering tension failure.

When adapted to a simply supported, center-loaded beam of rectangular cross section, the flexure formula becomes:

$$\sigma = 1.5 \frac{Pl}{bd^2},$$

where σ = bending stress at midspan, upper and lower surfaces of beam, psi; P = load on beam, lb.; l = span of beam, in.; b = width of beam, in.; d = depth of beam, in. To calculate the maximum bending stress, referred to as the **modulus of rupture**, or MR (also abbreviated as MOR or simply R), we use the maximum load, P_{max}:

$$\sigma_{max} = \text{MR} = 1.5 \frac{P_{max} l}{bd^2} = 1.5 \frac{(1{,}590 \text{ lb.})(28 \text{ in.})}{(2.01 \text{ in.})(2.00 \text{ in.})^2}$$

$$= 8{,}306 \text{ psi.}$$

The bending stress at proportional limit load, called **fiber stress at proportional limit** or FSPL, is calculated by using the proportional limit load, P′:

$$\sigma' = \text{FSPL} = 1.5 \frac{P'l}{bd^2} = 1.5 \frac{(990 \text{ lb.})(28 \text{ in.})}{(2.01 \text{ in.})(2.00 \text{ in.})^2}$$

$$= 5{,}172 \text{ psi.}$$

MR and FSPL are the values of stress along the upper and lower surfaces of the beam at midspan, which are the points of maximum stress in the beam.

Another formula is used to calculate the modulus of elasticity, E, as follows:

$$E = \tfrac{1}{4} \frac{P'l^3}{y'bd^3} = \tfrac{1}{4} \frac{(990 \text{ lb.})(28 \text{ in.})^3}{(0.28 \text{ in.})(2.01 \text{ in.})(2.00 \text{ in.})^3}$$

$$= 1{,}206{,}716 \text{ psi,}$$

where y′ = the center deflection of the beam at proportional limit load, P′.

E measures the apparent ratio of stress to strain of fibers along the upper and lower surfaces of the beam, which of course would be related to how much deflection or "sag" would result in a beam under load.

Although P′ and y′ (the values of load and deflection at proportional limit) are used by convention, any pair of values could be used, because the ratio of load to deflection (P to y) is constant over the entire elastic range. (The importance of this fact will be emphasized in discussing non-destructive structural grading.)

The average values of bending strengths, MR, FSPL and E are given in *Table 11* on page 111 for various woods.

In structural engineering with wood, modifications of the above formulas are used for different placement or distribution of loads. The formulas can be used in various ways by rearranging them algebraically. For example, for a given structural design, the load-carrying capacity, P, or expected deflections, y, can be determined. Given a requirement for load-carrying capacity or deflection limit, the necessary design specifications (dimensions, species, grade) can be established. In practice, especially for standard framing systems using stock lumber sizes of established grades, structural design is accomplished in cookbook fashion, using tables of values that have been worked out for relating practically all combinations of spans, dimensions, loads and deflection. It would be an awesome task, even for an engineer, to analyze from scratch the stress on every stick in a structure.

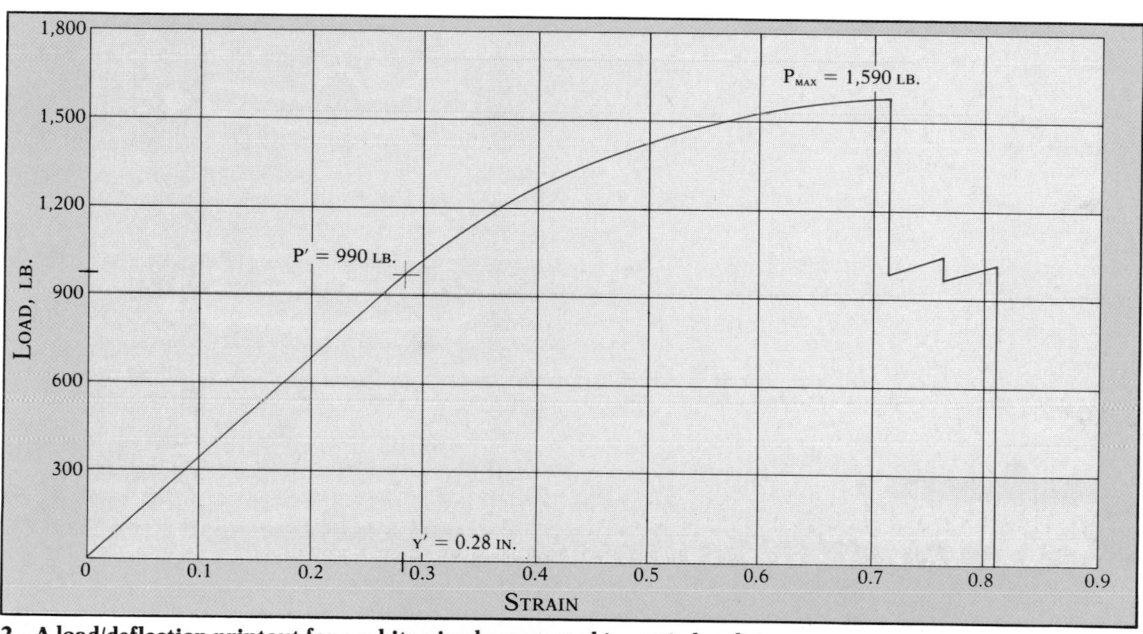

2—A load/deflection printout for a white pine beam tested in static bending.

3—Typical splintering tension failure on a standard bending specimen of sugar maple. The nail is located at the neutral axis, which runs lengthwise through the beam.

The stiffness of beams

A tremendous amount of insight into the strength of wood members subject to bending can be gained by rearranging the two formulas used to indicate strength and stiffness of beams.

The two general formulas for bending stress and modulus of elasticity are:

$$\sigma = k\frac{Pl}{bd^2} \quad \text{and} \quad E = K\frac{Pl^3}{ybd^3}.$$

(k and K are constants according to the placement and distribution of loads. For the center-loaded beam we have been considering $k = 1.5$, $K = 0.25$.)

The woodworker, however, is not involved in determining the strength of wood, but rather in assessing the performance of objects made of wood for which strength information is already available. The two most relevant questions facing the woodworker are how much the beam will carry and what factors influence its carrying capacity, and how much the beam will deflect and what factors influence its deflection.

To answer the first question it is convenient to rewrite the first equation:

$$P = \frac{1}{k}\frac{\sigma bd^2}{l} = \frac{1}{k}\frac{F_b bd^2}{l}.$$

(For σ we have substituted F_b, the commonly used symbol for **allowable bending stress**, as designated in stress grade tables. On grade stamps on lumber, it is sometimes written simply as **f**.)

This allows us to calculate the load P, which the beam can carry.

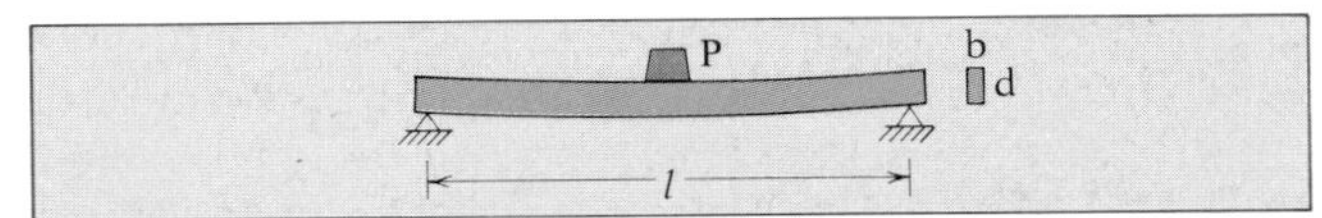

It also allows us to see that load-bearing capacity will vary:

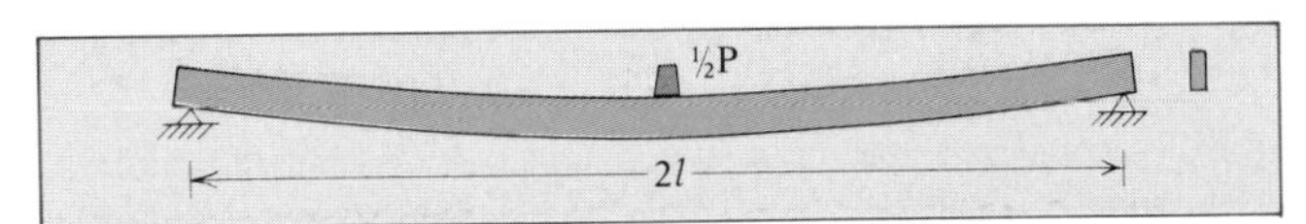

—indirectly as the length (double the length, you cut the load-bearing capacity in half);

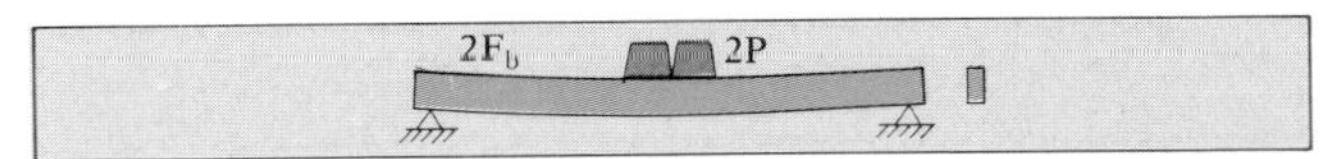

—directly as the allowable fiber stress, F_b or f (choose a lumber grade that is twice as strong, the beam will carry twice as much);

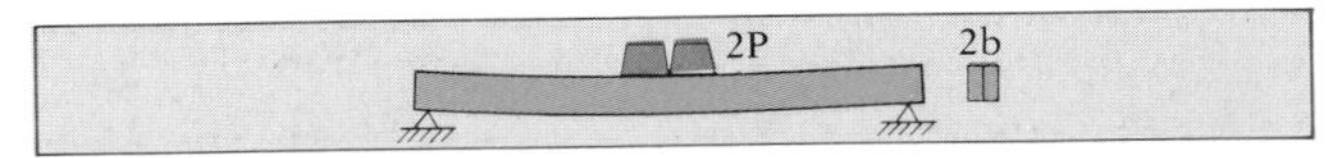

—directly as the width (make the beam twice as wide, it will carry twice as much);

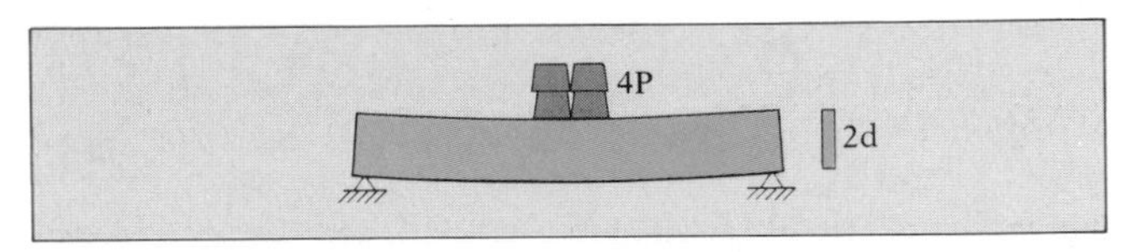

—directly as the *square* of the depth (double the depth and it will carry *four times* as much).

Similarly, we rewrite the second equation:

$$y = K\frac{P}{E}\frac{l^3}{bd^3}.$$

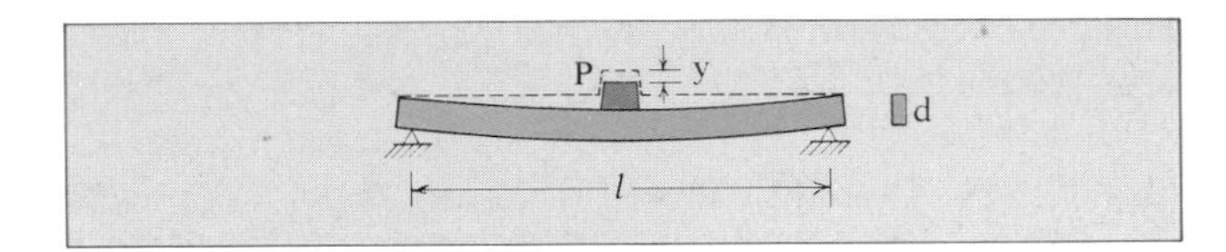

The deflection, y, of a beam varies:

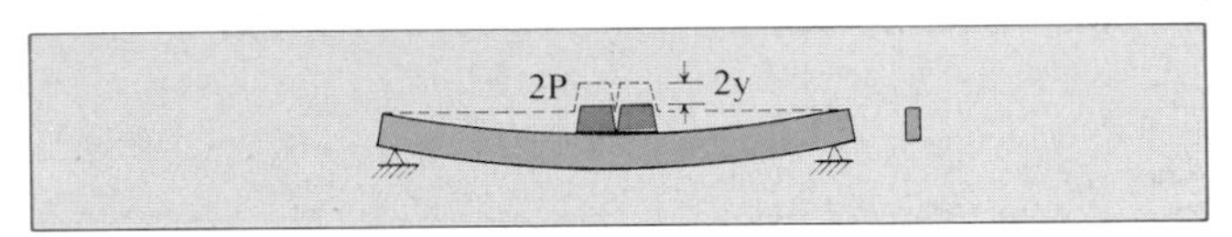

—directly as the load (double the load on a beam, it will deflect twice as much);

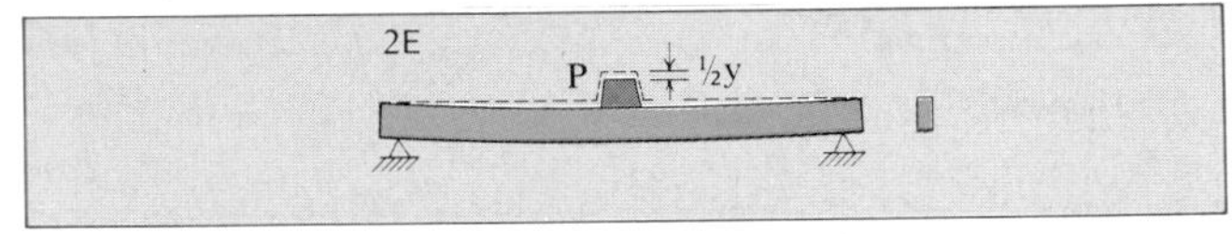

—indirectly as the modulus of elasticity, E (choose a grade with double the E, the deflection will only be half as much);

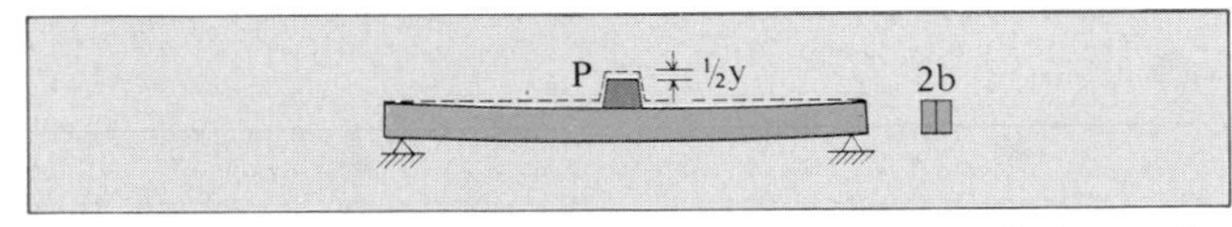

—indirectly as the width (use two beams side by side, the deflection will only be half as much).

However, the deflection also varies:

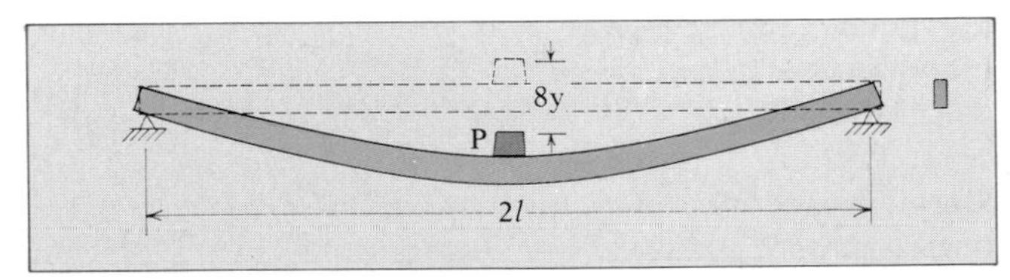

—directly as the *cube* of the span (double the span, the beam will deflect *eight times* as much);

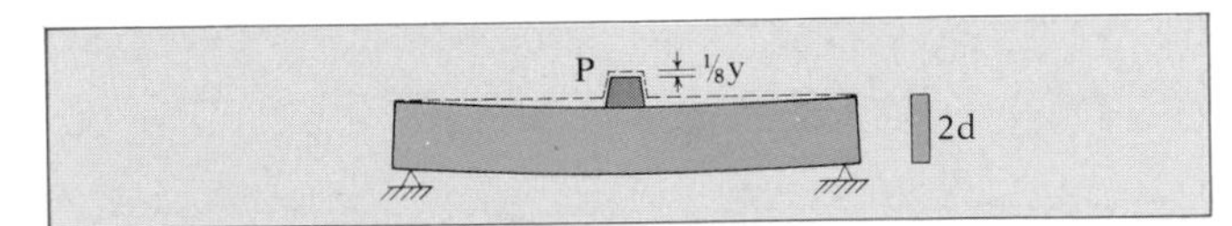

—indirectly as the *cube* of the depth (double the depth, you reduce the deflection to *one-eighth*).

A few points are worth emphasizing. Carrying capacity and stiffness are each affected directly according to the numerical value of the strength property involved. Likewise, varying the width has its expected proportional effect, so two beams are "twice as strong" as one.

In many cases, little can be done to manipulate the span. If you want a 16-ft. wide living room, that is how long the ceiling joists will have to be. The length of bed rails is fairly well fixed by our average height. A canoe paddle has to be long enough to dip well into the water. The rails of a ladder must be far enough apart for ease of climbing and stability of the ladder. But where the span can be shortened, the effect on strength, deflection in particular, can be amazing. A good rule of thumb: Cut the span by one-fifth and you double (approximately) the stiffness (0.8 x 0.8 x 0.8 = 0.512); increase the span by one-quarter, and you double the deflection ($[1.25]^3$ = 1.95)*. Another hypothetical case: As you plan to finish off your basement into a game room, workshop or utility room, suppose it is possible to position a partition wall directly under the middle of your living-room floor system. Cutting the span of the joists in half will make your living-room floor seem eight times as rigid.

But it is the depth of a beam that can usually be manipulated to best advantage in influencing strength. Another excellent case in point of deflection mechanics is the structural advantage of using a board as a joist (loaded edgewise) rather than as a plank (loaded flatwise). Consider the standard 2x10, which has actual dressed dimensions of 1½ in. by 9¼ in. The original carrying capacity is given by

$$P = \frac{1}{k}\,\frac{F_b bd^2}{l}$$

(where the board is loaded as a plank, so b = 9.25 in., d = 1.5 in.).

Now if we rotate the board to its edgewise position to be loaded as a joist, the relative value of b is reduced to

$$\frac{1.5}{9.25}\,b = 0.162b, \text{ compared to the original b,}$$

but the depth has been increased to

$$\frac{9.25}{1.5}\,d = 6.167d, \text{ compared to the original d.}$$

Carrying capacity now equals

$$\frac{1}{k}\,\frac{(0.162b)(6.167d)^2}{l} = \frac{6.16}{k}\,\frac{F_b bd^2}{l} = 6.16\ P.$$

*$0.8l \times 0.8l \times 0.8l = 0.512l^3$ instead of l^3, giving 0.512 y.
$(1.25l)^3 = 1.95l^3$ instead of l^3, giving 1.95 y.

The carrying capacity is more than six times as great for a 2x10 joist than for a 2x10 plank.

Now let's consider deflection.

$$y = K\,\frac{P}{E}\,\frac{l^3}{bd^3} \text{ becomes}$$

$$K\,\frac{Pl^3}{E(0.162b)(6.167d)^3} = \frac{K}{38.00}\,\frac{Pl^3}{Ebd^3} = \frac{y}{38.00}.$$

The 2x10 joist will deflect only $\frac{1}{38}$ as much as the 2x10 plank.

The fact that depth has a more drastic influence on deflection than on carrying capacity is extremely important in framing flooring systems, because deflection rather than carrying capacity is the key performance criterion. Common design guidelines are 40 lb. per sq. ft. load and 1 in 360 deflection limit (i.e., 1-in. deflection in 360 in. of span). We have no way of sensing (or caring) what critical stress levels are as long as they are never reached, and usually when the 1 in 360 deflection limit is met, the stress developed is well below that which would cause failure. But deflection becomes increasingly apparent to our senses. I have never heard of a floor actually failing, although I have heard many a complaint about floors being too "bouncy" or "springy."

Under the provision of 40 lb./ft.^{2} load and 1 in 360 deflection limits, the "allowable span" tables tell us that a 2x8 with stress rating of E = 1,000,000 psi is adequate for a 12-ft. span. This, however, is marginal and although it satisfies the load and deflection requirements, it may not be as rigid as the homeowner wants. You may wish to overdesign for added satisfaction. I once knew a builder who proudly revealed that he had doubled all the flooring joists (obviously at twice the cost) in a condominium to give better quality and increased satisfaction to the occupants. While commendable from the consumer's point of view, it was a questionable design choice from the cost standpoint. Instead of doubling 2x8s (at double the cost), 2x10s might have been used. Since 2x10s of the same species and grade were selling at the same price per board foot, this would have required a material cost increase of only 25%. By increasing the depth from 7.25 in. to 9.25 in., or an increase from d to 1.276d, we find:

$$K\,\frac{Pl^3}{Eb(1.276d)^3} = \frac{y}{2.077}$$

So by using 2x10s in place of 2x8s, performance could have been doubled for 25% more cost.

There are countless applications where the principles of beam mechanics can be applied generally, without getting into precise calculations, to improve the mechanical performance of a wood member. For example, in building stairs we realize the importance of firmly fastening the risers to the treads, so the risers will serve as very deep "joists" and stiffen the treads.

A canoe paddle is a modified simple beam that develops maximum bending stress along the shaft at the point of the lower handhold *(1)*. In making a canoe paddle, you want the maximum strength for the least weight. In shaping the critical area of the paddle, then, you would want to make the shaft elliptical in cross section, with the long axis of the ellipse oriented in the loading direction, that is, perpendicular to the plane of the paddle blade. The elliptical shape will of course be stronger (that is, it will develop less axial stress for a given loading, or will enable a greater loading to develop the same level of maximum stress) than a round one of equal cross-sectional area (and therefore equal weight).

I sometimes think of a simple beam as two cantilever beams attached to one another. Put another way, a cantilever beam can be thought of as half a simple beam *(2)*. The same mechanical principles that govern the relationship of simple beam depth to midspan axial stresses can be applied to attachment stresses on a cantilever beam.

This suggests further application to certain joints. For example, when you lean back in a chair, the seat apron is cantilevered onto the back posts *(3)*. Here we see the importance of the depth of the apron member (or the height of the connecting mortise) in determining the magnitude of stresses in the joint. Similarly, where such joints are doweled, the distance between the dowels, as controlled by the depth of the apron, also determines stress levels in the dowels.

There is a practical limit to how much a beam can be improved by increasing the depth in comparison to the width, for just as the stability of columns is limited by their relative length, joist-type beams may become unstable or difficult to fasten. In floor systems, diagonal crossbracing must be installed to ensure stability as well as to help distribute concentrated loads.

Shear strength may also be critical in bending. A beam may be thought of as consisting of horizontal layers that attempt to slide past one another as the beam bends under load *(4)*. The internal stress that causes sliding, called **horizontal shear**, is greatest near the ends of the beam along the neutral axis, and it is zero through the

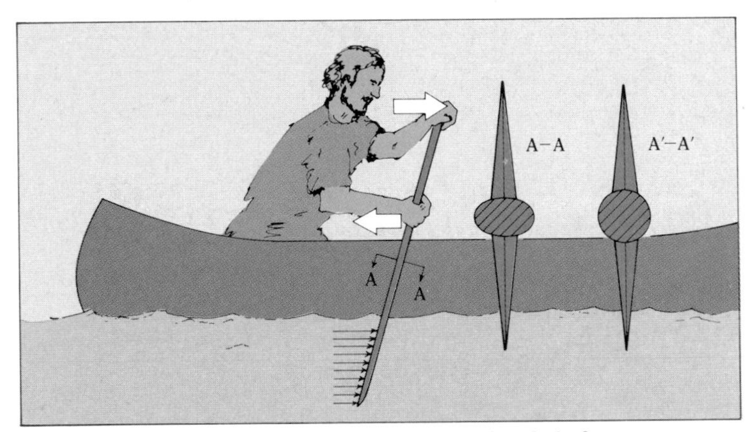

1—A canoe paddle is a simple beam. The elliptical shaft is stronger in bending than a round one of the same cross-sectional area.

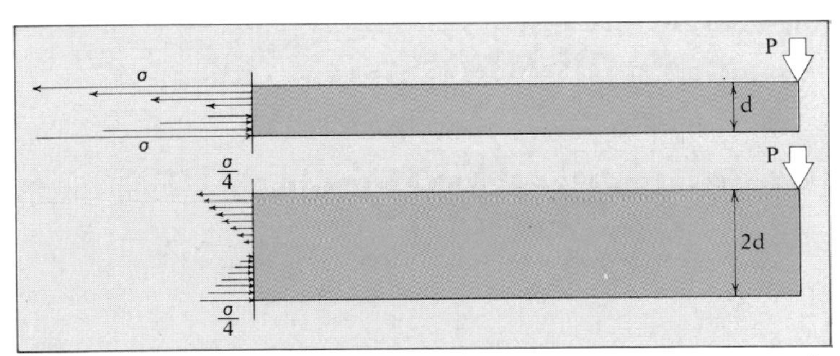

2—In a cantilever beam, the maximum board stresses vary inversely with the square of the depth of the beam.

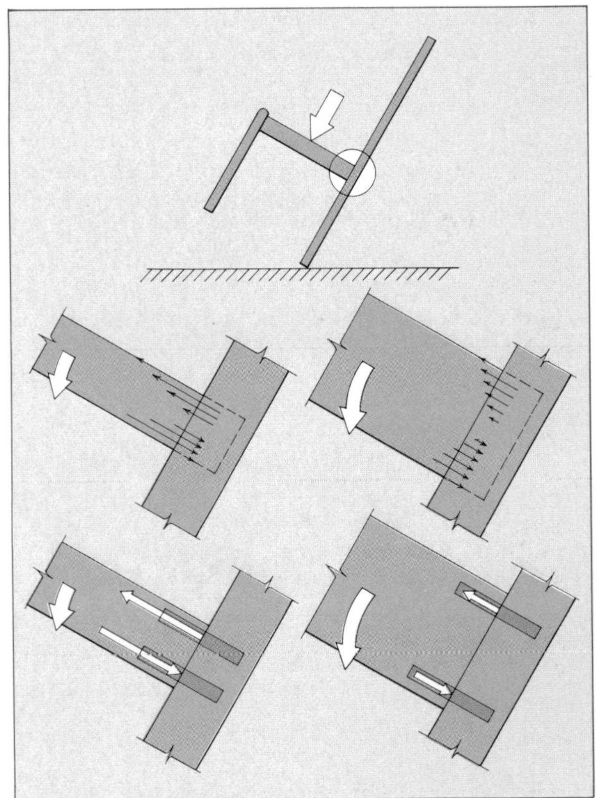

3—When a person leans back in a chair, the seat apron is cantilevered onto the back posts. The depth of the apron member determines the magnitude of stresses in the joint.

midspan of a center-loaded beam. In beams that are fairly long relative to their depth, axial bending stresses become critical long before shear stress becomes very great. However, in beams that have a relatively small length-to-depth ratio (i.e., beams that are short), shear stresses may reach critical levels before axial stresses, resulting in horizontal shear failure. Although they are not too common, instances of such failures come to mind. When you attempt to use a wooden pole as a lever to raise a heavy object, the tip of the pole may broom over. When bulldozers push against tree stumps, the stumps sometimes broom over rather than uproot or break off.

Horizontal shear failures in bending are also likely in woods with natural planes of weakness, such as the earlywood pore zones in white ash, when the weak layer of structure coincides with the plane of maximum horizontal shear. This is an additional reason for not hitting a baseball on the trademark side of the bat.

Horizontal shear can be especially critical in composite products having relatively weak internal layers. The classic case in point is thick plywood, in which the perpendicular grain direction of interior plies represents weaker layers, often further weakened by knife checks in the veneer, so rolling shear develops *(5)*. Such failures often result when plywood is used in place of solid wood to construct a short cantilevered overhang, as in a tabletop. Heavy loading of the overhang may cause failure in horizontal rolling shear, even though the surface plies have not failed from axial bending stress. This separation of the plies is often incorrectly regarded as delamination and usually blamed on adhesive failure, but close inspection will usually identify the real problem.

Various types of particle board may also be prone to horizontal shear failure due to their low internal bond strength. This is especially true in boards having a stratified density.

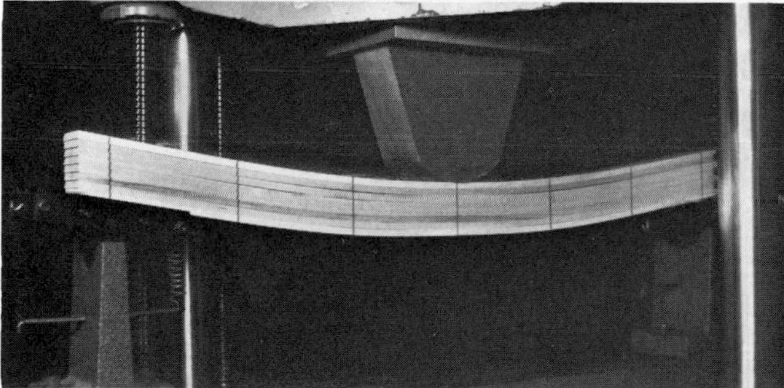

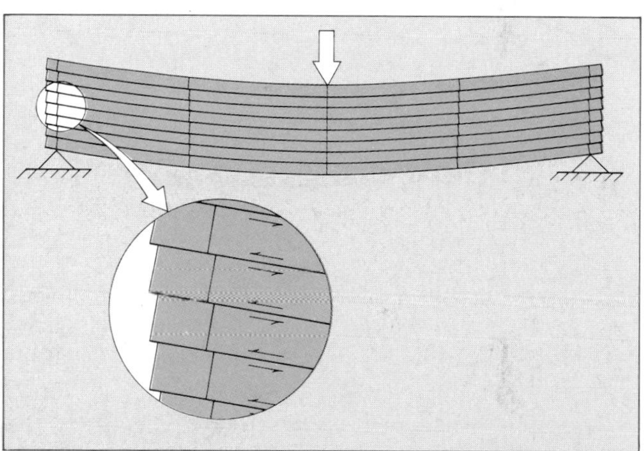

4—Horizontal shear in a beam is illustrated in the photos at left and in the drawing above by subjecting to bending a stack of free-sliding slats arranged as a beam. The stiffness that would be gained by gluing the layers together suggests the role of shear resistance in a solid beam.

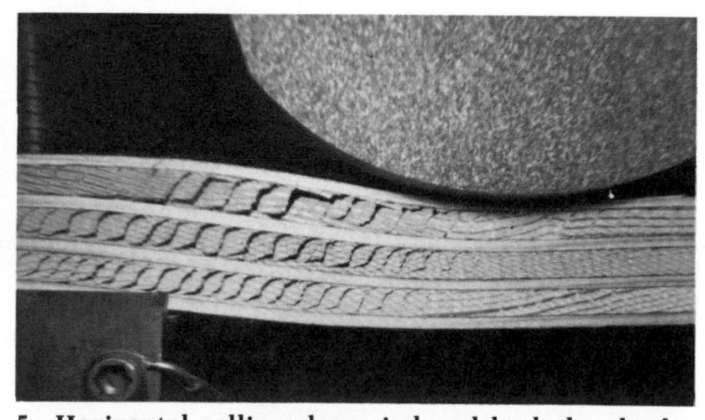

5—Horizontal rolling shear, induced by lathe checks, develops when a 12-in. plywood beam is loaded in the center. This is analogous to putting weight on the edge of an overhanging plywood tabletop.

Factors affecting strength properties

Much of the previous discussion has been aimed at the basics of strength properties, and for simplicity we made some assumptions and generalizations that would seldom apply in reality. For example, calculations of stress in chair parts were made with average static strength test results, without considering variability or defects in the wood, moisture content, time under load or safety factors. In the real world, however, lumber can have defects, and we must have dependable, "allowable" stress values that take into account the possible effects of defects or uses that affect strength. It is important for the woodworker to understand the extent to which various factors affect strength, so that intelligent judgment can be made in the many cases where one can't go by the book. Some factors reflect environmental conditions in which the wood is used, others reflect natural characteristics or abnormalities inherent in trees, while others are defects inflicted upon the wood.

Moisture content—The change in tear resistance and flexibility of paper in response to wetting and drying dramatizes the influence of moisture content on the strength of wood substance. Although less striking, the increase in the strength of wood as moisture content decreases below the fiber saturation point is nevertheless obvious. Many woodworkers and carvers have discovered the relative ease with which green wood can be worked with hand tools. In cases where a partially shaped piece can be dried without defect, the rough work is often easiest to accomplish in the green state. Nails easily driven through many hardwoods while green would bend over if the wood were dry.

The rate of strength improvement is not directly related to loss of bound water, as is the shrinkage rate, but strength does increase as the wood gets drier *(1)*. The amount of added strength varies, depending on the property. For example, maximum crushing strength in compression parallel to the grain and fiber stress at proportional limit in compression perpendicular to the grain are approximately tripled in drying from green to oven-dry. Modulus of rupture in bending is more than doubled in the process, but stiffness is increased by only about half.

Time under load—The test results listed in *Table 11* on page 111 were obtained from static strength tests in which specimens were under load only for a matter of minutes. If loaded more rapidly they would show higher strength, while if loaded more slowly they would show lower strength. For example, wood loaded to failure in bending within one second (rather than the standard 5 to 10-minute duration of the static bending test) would show about 25% higher strength. If held under sustained load for 10 years, it would show only 60% of the static bending strength. To put it another way, if a beam must support a load for 10 years, it can carry only 60% as much load as it carries in the static bending test. This reveals that in addition to the immediate elastic response which is apparent upon loading, there is additional time-dependent deformation called **creep**. At low to moderate loads, creep is imperceptible. Over extremely long periods of time, or when loads approach maximum, creep may result in objectionable amounts of bending or even failure, such as sagging timbers in old buildings. The study of time-dependent stress and strain behavior is called **rheology**.

Since most structures are built for an anticipated life of many years, long-term loading is an important consideration in determining allowable stress ratings for lumber. Wood performs well under repeated short-term or cyclic loads without fatigue failure or becoming brittle, unlike certain metals and concrete.

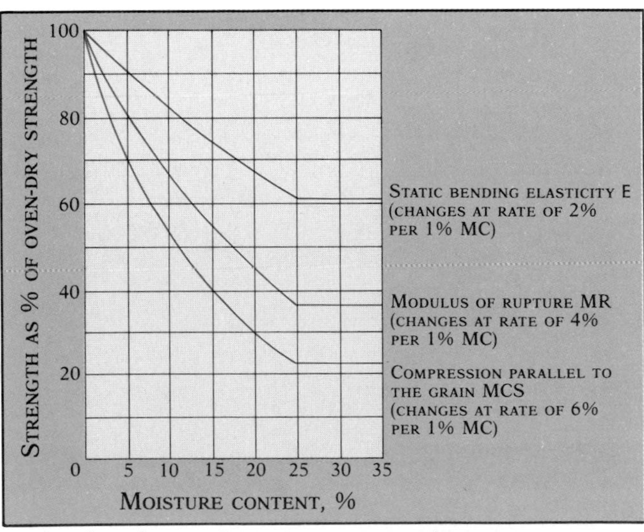

1—Wood weakens as its moisture content increases.

2—Striations in the bark indicate spiral grain in this hophornbeam stem.

Temperature—Strength in wood responds immediately to changes in temperature, increasing as the temperature drops, decreasing as the temperature rises. Through the range of temperatures found in nature, air-dry wood changes in strength by an average of 2% to 5% for every 10°F change in temperature. Such changes are reversible. That is, when returned to room temperature, the wood returns to its original strength. High temperatures may result in some permanent loss of strength, the extent depending on the temperature reached, the duration of heating, the heat source and the moisture content of the wood. For example, wood of some species at low moisture content that is heated momentarily in dry air to near ignition temperatures and immediately recooled shows no strength loss. However, sustained heating of moist wood in hot water at temperatures near 212°F can cause some permanent loss of certain strength properties. Nobody can predict exactly how much strength will be lost as a result of heating.

Cross grain—In most wooden items—boards, beams, posts, turnings and so forth—it is intended or assumed that the grain direction is parallel to the long axis of the piece. Wood with such ideal grain orientation is said to be **straight-grained**. Deviation of the grain from the direction of the longitudinal axis is termed **cross grain**.

Cross grain occurring as deviation from the edge of a flatsawn board is called **spiral grain**, the result of helical grain direction in the tree *(2)*. Spiral grain may also result from misaligned sawing. **Diagonal grain** is cross grain resulting from deviation of the plane of the growth ring from the longitudinal axis, resulting from failure to saw boards parallel to the bark of the log.

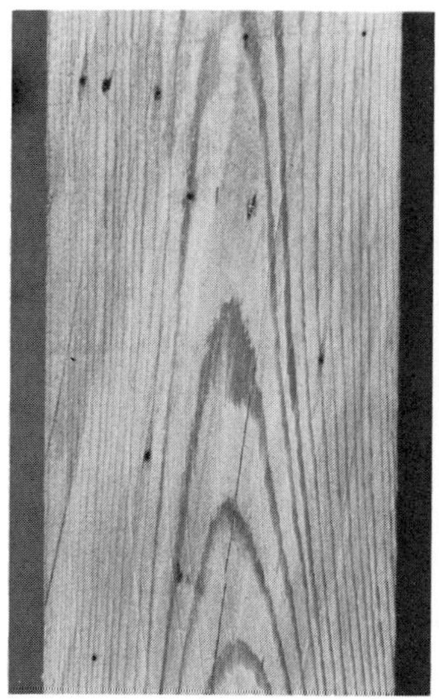

3—Flatsawn lodgepole pine has centrally arranged growth rings, but surface checks betray spiral grain. Slope of grain is 1 in 5.

4—Round-edge red maple board first appears normal, but pith flecks indicate spiral grain (slope: 1 in 6).

Cross grain is measured quantitively as **slope of grain**, taken as the ratio of unit deviation across the grain to the corresponding distance along the grain *(3, 4)*. The slope of the grain is expressed "1:12" or "1 in 12." When the grain deviates from two adjacent faces, the **combined slope of grain** must be calculated by geometric methods *(5)*.

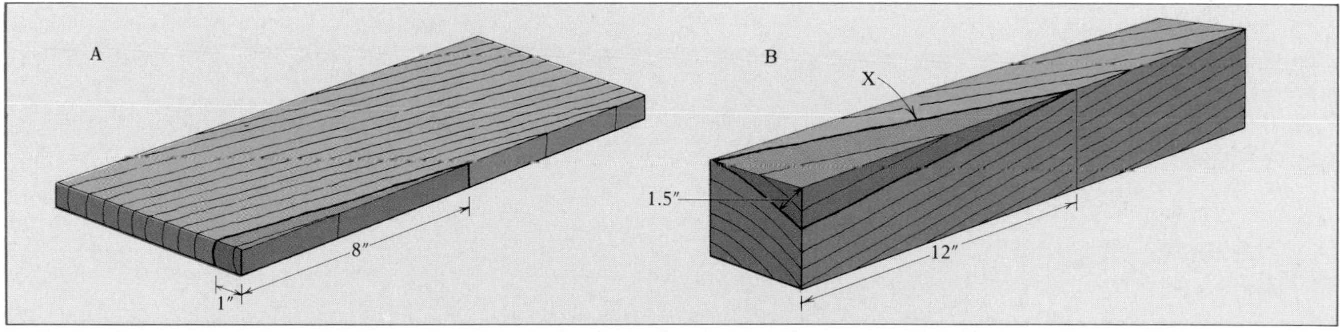

5—In simple cross grain, as in the diagonal grain in A, the slope of grain is easily measured by observing the growth rings on the surface (here, 1 in 8). More commonly, however, as in B, the growth-ring appearance on the surface does not coincide with the grain direction (X). Slope, here 1.5 in 12 (1 in 8), must be calculated from end grain and edges.

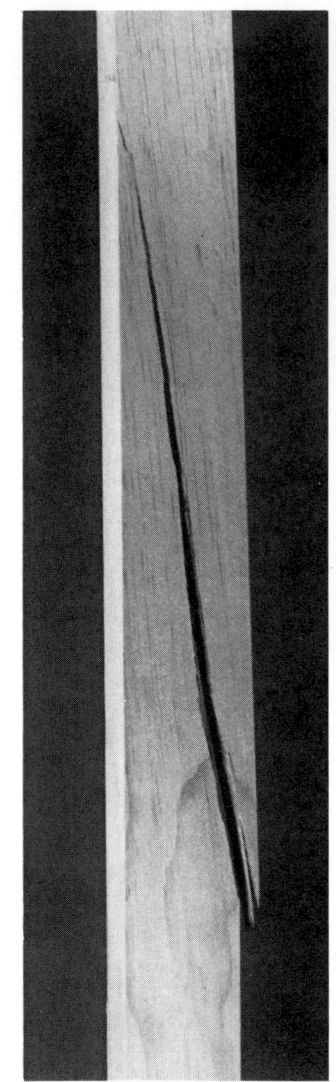

1—Above, cross grain in a chair leg. Left, bending failure due to cross grain.

Visual determination of grain direction can be most deceiving. Except on true radial surfaces, the temptation to use growth rings as an indicator of grain direction should be avoided. Except on tangential surfaces, rays also should not be relied upon. Linear anatomical features visible to the eye, such as softwood resin canals or hardwood vessels, are quite helpful. It may also help to mark the wood with ink or dye, using a felt-tipped pen. The grain direction then will be indicated by hairline extensions of the ink along longitudinal elements. There are also special scribing tools that follow the grain when drawn along the surface of a board. It is good practice to test your ability to analyze slope of grain on short scraps of wood, then split them apart to check your evaluation.

Straightness of grain is important because of the effect it has on strength properties. Limitations on permissible grain deviation are therefore often specified in structural lumber, particularly in ladder stock. It is assumed that stresses induced by loads on the wood, when developed parallel to the axis of a piece, will coincide with the grain direction of the wood and be resisted by the superior parallel-to-grain strength of wood in ten-

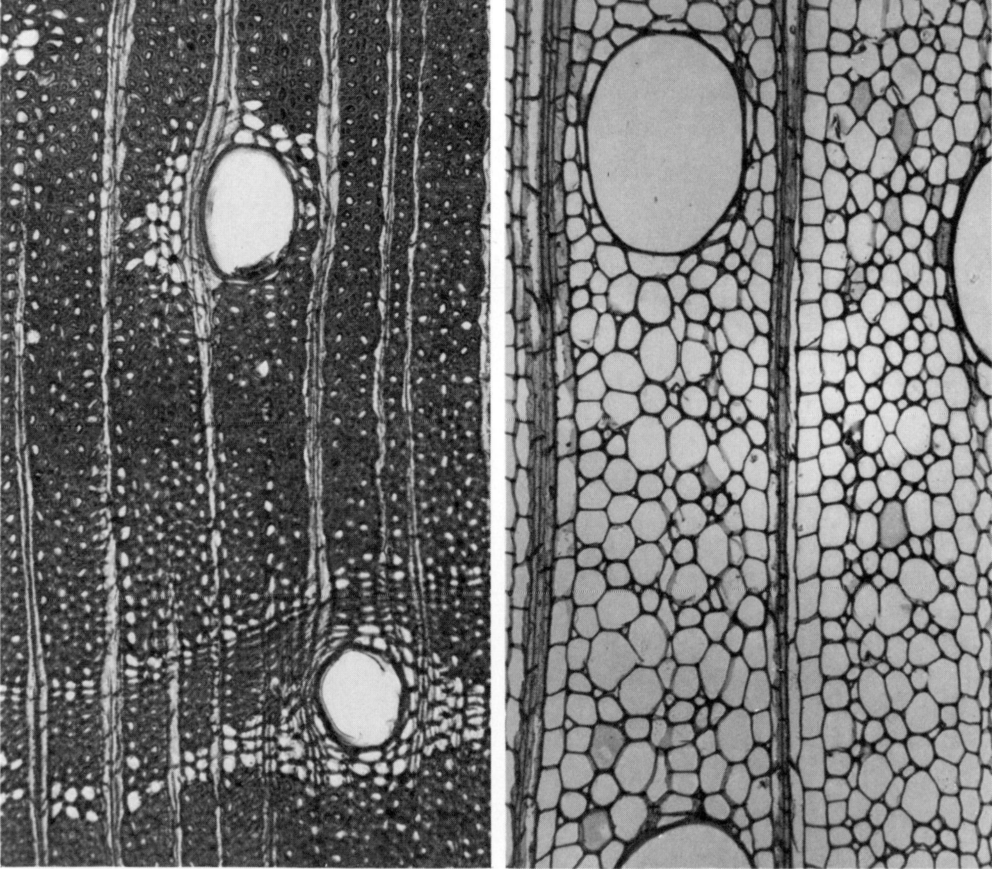

2—The denser the cell structure, the stronger the wood. Contrast rosewood, left, with balsa, right.

sion or compression. The consequence of cross grain depends on the extent to which a component of weaker perpendicular-to-grain strength is brought into play *(1)*.

Compression strength appears least affected by this characteristic. Slopes of grain no worse than 1:10 have a negligible effect, and a slope of 1:5 shows only about a 7% reduction in strength. Bending and tensile strengths are more drastically affected. Modulus of rupture, for example, is reduced by about 20% by a slope of 1:10, and by 45% by a slope of 1:5.

Variability—In Chapter 1, and particularly in *Table 3* on page 8, the twelvefold range of density among species was discussed. This variation among species can be seen in the relative proportions of different cell-type dimensions and cell-wall thicknesses *(2)*. The average specific gravity of a species is about the best possible single predictor of strength.

Within a species there is considerable variation in density and strength, and the strongest piece of wood (among straight-grained, defect-free pieces) of a species typically has at least double the strength of the weakest. For example, if the average maximum crushing strength in compression parallel to the grain for a given species is 4,500 psi, individual pieces of that species could be expected to have strengths ranging from about 3,000 psi to about 6,000 psi. Most of this strength range is associated with a density variation and can therefore be sensed by the weight of a piece. However, some is due to subtle differences in cell structure and cannot be predicted.

In certain species some of this variability is predictable based on growth rate. In conifers, especially uneven-grained species such as Douglas fir, the width of the denser, stronger latewood is least affected by changes in growth rate. As a tree grows faster, however, the wider rings have a greater proportion of earlywood; as the growth rate slows down, earlywood is narrower *(3)*. The wider rings of fast-grown softwoods therefore have a greater percentage of earlywood and the wood will be weaker on the average. Growth rate, usually measured in rings per inch, is extremely valuable in visually stress-grading structural softwood lumber.

In ring-porous hardwoods like oak and ash, the width of the large-pored earlywood doesn't vary much, so the rate of growth is reflected in the amount of denser latewood. Therefore, fast growth produces denser material *(4)*. Among diffuse-porous hardwoods, growth rate has

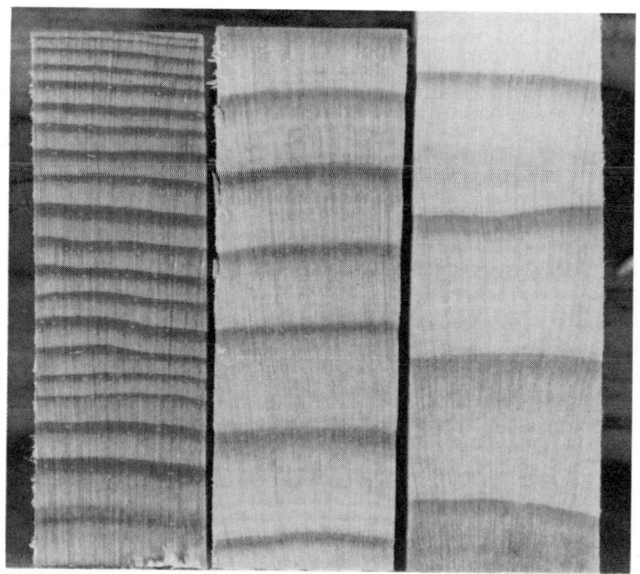

3—Variable rate of growth in red pine is shown by spacing of growth rings. The more rings per inch, the slower the growth, the narrower the earlywood band, and the stronger the wood.

4—In ring-porous hardwoods, the faster the growth, the stronger the wood. Compare slow-grown red oak, left and center, with fast growth, right.

no predictable relationship to strength. Beyond these few indications, variability in strength of clear wood is difficult to detect.

There is no significant difference in the strength of sapwood and heartwood *per se*, within a given species. In some cases, heartwood may be preferred because of its decay resistance and its natural ability to retain strength longer. On the other hand, sapwood takes preservative better and can therefore be treated to retain its strength even longer.

Localized defects—The word defect is a risky term to use without qualification. In this section it refers to irregularities or features affecting strength. A knot, for example, may be a visual asset, but it is clearly detrimental to strength.

Knots are unquestionably the most commonly encountered defects, since all trees have some form of branching system. Knots reduce strength in two ways. The knot itself has abnormal cell structure that runs at an angle to the surrounding grain direction, and an encased knot is not connected to the surrounding tissue. Furthermore, the area around the knot usually contains cross grain that results in severe strength reduction. The degree of weakening caused by knots is quite variable and unpredictable—ranging from negligible in the case of small round pin knots to total in the case of large, loose spike knots that actually cause a board to fall apart. To be on the safe side, one should assume that the knot has the same effect as physically cutting out the entire knot and the surrounding distorted tissue from the piece. In judging the strength of a piece of knotty lumber, imagine all the knots cut away, then ask yourself whether what is left will be strong enough.

Checks, shakes and splits are voids and discontinuities in the wood tissue. They should always be suspected of extending farther than their obvious visual appearance. Seasoning checks may have partially closed, for example. Although reclosed checks or hidden checks may be visually acceptable, their effect on strength will always be there. Failures in tension perpendicular to the grain or shear parallel to the grain can occur by the extension of pre-existing checks at much lower stress than would cause failure in undamaged wood.

Defects must be carefully evaluated according to the manner of loading. For example, a ring check, totally enclosed within a boxed heart timber, would affect strength if the timber was used as a beam, but would not if it was used as a short column. In bending, the placement of a defect can be profoundly important. A large round knot along the neutral axis of a joist would have little effect, since stresses are minimal in this area. But if the same knot were along the edge of the beam, the rule of assumed physical removal would have to be applied, which would reduce the apparent depth of the joist. Our previous discussion (p. 124) pointed out that reducing the depth of a beam also reduces its load-bearing capacity as the square of the depth. Exactly where you notch or bore a joist is therefore critical. A hole bored through the neutral axis of a joist to pass a wire or pipe will have negligible effect compared to the same-sized notch at the beam surface.

Decay, even in the early stage, has an effect on the impact strength of wood. As decay progresses, every other property is in turn affected, and eventually total loss of strength may result. It is safest to assume that where any sign of decay is visually apparent, the wood is already weakened. Although stain fungi do not reduce strength, fungal stain indicates that conditions exist or did exist that favor decay and therefore the wood should be held suspect. Once dried and kept dry, stained wood can be used without fear of loss of strength.

Miscellaneous small defects of wood could form an impressively long list, including insect holes, pitch pockets, bark pockets, pith, wane, bird peck, nail holes, etc. By carefully examining a piece of wood for physical damage, tissue failure or associated grain distortion, common sense will suggest how badly the strength has been affected.

Compression failures

During severe wind storms or under heavy snowload the bole of a tree may be bent excessively. This also sometimes occurs during blowdown or in felling a tree. When the bole of a tree is bent like one giant beam, the wood in the concave side can fail in compression without tension failure developing in the convex side. The compression damage to the tree goes unnoticed. In the wood, the result is irregular planes of buckled longitudinal cells, called **compression failures**. These localized planes of damage generally run crosswise to the grain. In rough lumber they are virtually impossible to detect, but on planed or sanded longitudinal surfaces they are usually visible, appearing as wrinkles across the grain *(1)*. Compression failures are extremely detrimental to members stressed in bending, for the buckled cell structure fails readily in tension, resulting in sudden beam failure at lower-than-normal stress levels. Critical structural members, such as ladder rails, scaffolding, aircraft parts or boat spars, must be carefully inspected to exclude material with compression failures. Wood containing compression failures is also unsuitable for steam-bending work.

Brashness—Normal wood is characteristically tough. When it breaks in bending, the cell structure progressively separates as splintering fractures. Audible cracking and creaking accompany the gradual failure of the wood. There is considerable deformation, well beyond the elastic limit, before total failure finally takes place.

Under certain conditions, wood fails abruptly, with minimal deflection and at lower than expected loads. These failures are characterized by cross breaks that lack the usual evidence of fibrous or splintering separation. Such brittle and weak behavior in wood is called **brashness** *(2)*. The fibrous breaking of normal wood might be compared to the breaking of a stringy stalk of celery, while the brash failure of wood would be analogous to the abrupt snapping of a carrot.

As described above, wood damaged by compression failures will exhibit brashness when stressed in bending. Reaction wood in conifers (compression wood) also shows brashness, even though its density is greater than normal wood. Extreme growth rates may also produce brashness. In ring-porous hardwoods, extremely slow growth results in low-density wood because of the large percentage of earlywood vessels. This wood is characteristically brash and is especially troublesome in steam-bending. Among conifers, moderately slow but healthy growth produces dense, strong wood. Extremely slow growth, as in stunted or overmature trees (wood with 35 or more rings per inch) may also produce brash wood. Extremely fast growth in conifers, reflecting the low density of the dominant earlywood, also tends to be brash. The wide-ringed growth of pronounced juvenile wood is especially prone to brashness. Prolonged exposure to high temperature and fungal decay are other causes of brashness in wood.

1—Compression failures appear on longitudinal faces of surfaced lumber as wrinkles across the grain.

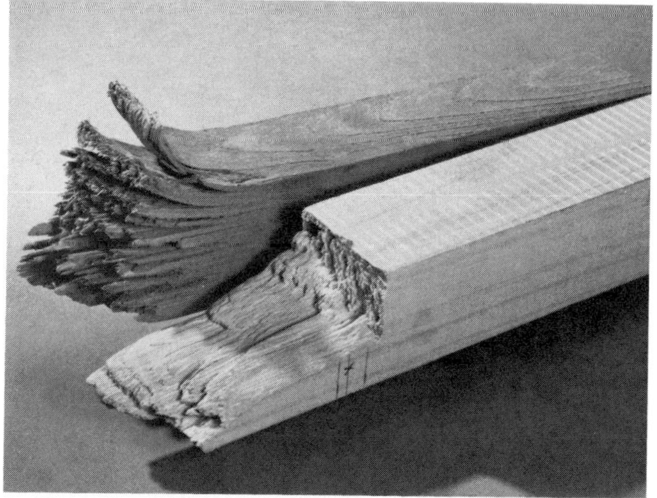

2—Normal gradual splintering failure in catalpa beam, top, contrasts with abruptly cracked (brash) white pine beam that contains reaction wood, above.

Structural grades

After reviewing the many factors that can affect the strength of wood, it is obvious that few (if any) pieces of structural lumber have the same properties as straight-grained, defect-free pieces, and that the ordinary conditions of use are unlike short-duration testing under controlled conditions. The difference is indeed striking.

Allowable stress ratings have been established for safe and reliable structural use of wood. This is done by stress grading, which groups lumber into categories or grades within which all pieces can be relied upon to have prescribed minimum strength properties. Starting with the strength properties of small clear specimens, adjustments are then made for normal variability, duration of load, moisture content and a factor of safety. The traditional system of visual grading sorts the pieces into groupings based on visible characteristics, such as rate of growth, latewood percentage (indicating density), knots, slope of grain, reaction wood, shake, wane and other defects. The groups are usually designated by unique grade names, as well as allowable stress values for as many as six properties: extreme fiber stress in bending (F_b), modulus of elasticity (E), tension parallel to grain (F_t), compression parallel to grain (F_c), compression perpendicular to grain (F_{cp}), and horizontal shear (F_v). As an example of the contrast between average strength values for small, clear, straight-grained specimens and the allowable stresses assigned to graded lumber of the same woods, values for southern yellow pine are summarized in *Table 12.* The test values are the averages for six major species of southern yellow pine, adjusted to 12% moisture content (MC). The allowable stress values are for structural joists and planks, kiln-dried to 15% MC.

First let's compare fiber stress in bending (F_b). For small clear specimens, the test averages range from 16,300 psi down to 12,600 psi for modulus of rupture (MR). However, the allowable design stress in structural material is reduced to the range of 2,250 psi down to 800 psi among the various grades. This indicates the extent to which defects and other allowances affect the

Table 12—Comparison of highest and lowest average clear wood strength properties among the principal species of southern pines (longleaf, shortleaf, loblolly and slash) with equivalent allowable design stress values of various grades of structural lumber.

Average clear wood strength properties of wood tested at 12% moisture content (values in psi).

TEST	STATIC BENDING	COMPRESSION PARALLEL TO GRAIN	SHEAR PARALLEL TO GRAIN	COMPRESSION PERPENDICULAR TO GRAIN	STATIC BENDING
PROPERTY	MODULUS OF RUPTURE	MAXIMUM CRUSHING STRENGTH	MAXIMUM SHEAR STRENGTH	FIBER STRESS AT PROPORTIONAL LIMIT	MODULUS OF ELASTICITY
HIGHEST AVERAGE	16,300	8,260	1,680	1,010	1,980,000
LOWEST AVERAGE	12,600	6,940	1,390	750	1,750,000

Allowable stress ratings of joists and planks, surfaced at 15% moisture content and used at a maximum of 15% moisture content (values in psi).

DESIGN STRESS	EXTREME FIBER IN BENDING	COMPRESSION PARALLEL TO GRAIN	HORIZONTAL SHEAR	COMPRESSION PERPENDICULAR TO GRAIN	MODULUS OF ELASTICITY
DESIGNATION	F_b	F_c	F_v	F_{cp}	E
COMMERCIAL GRADE					
DENSE SELECT STRUCTURAL	2,250	1,900	95	475	2,000,000
SELECT STRUCTURAL	1,950	1,650	95	405	1,900,000
No. 1 DENSE	1,900	1,700	95	475	2,000,000
No. 1	1,650	1,450	95	405	1,900,000
No. 2 DENSE	1,550	1,450	95	475	1,800,000
No. 2 MEDIUM GRAIN	1,350	1,350	95	405	1,700,000
No. 2	1,150	1,050	80	345	1,500,000
No. 3 DENSE	925	875	95	475	1,600,000
No. 3	800	750	80	345	1,500,000

strength of clear wood and must be accounted for in designing structures against the chances of failure.

But for modulus of elasticity (E), relatively minor reductions are made, even in the lowest grades. In fact, it is apparent that visual sorting is capable of successfully selecting the stiffest pieces, since the best allowable stress value is 2.0 x 10^6 psi, which is higher than the strongest species average of 1.98 x 10^6 psi. This brings out an interesting point about modulus of elasticity. It senses the performance of the entire piece, because under load all areas of the piece, weak and strong, make a representative contribution to resisting strain. However, in the case of breaking strength, defects may act as the weak link in the chain and severely reduce load-carrying capacity. The percentage reduction in E therefore is less over the range of grades than is the reduction in bending strength.

This suggests an important strategy in design of framing where deflection is a controlling factor. Many times it is advantageous to use a low grade of high modulus of elasticity species, such as Douglas fir or southern yellow pine, rather than a high grade of a low E species, such as spruce. For example, in one case a house design calls for 14-ft. spans, for which 2x10s of grade No. 2 and better eastern spruce with an allowable E of 1.0 would marginally satisfy the deflection requirement. But consider No. 3 2x12 Douglas fir. Its allowable fiber stress value is lower, but the deeper beam reduces the stress enough to remain safe. No. 3 Douglas fir is rated E = 1.5. Let's look again at the formula for beam deflection:

$$y = K = \frac{Pl^3}{Ebd^3}.$$

When we raise

E from 1.0 to 1.5 and

d from 1.0d to $\frac{11.25}{9.25} = 1.216d$,

$$K \frac{Pl^3}{(1.5E)b(1.216d)^3} = \frac{K}{2.69}\frac{Pl^3}{Ebd^3} = \frac{y}{2.69} = 0.370y.$$

We see that deflection is reduced by 60%, i.e., the joists are more than twice as rigid.

The increase in lumber board footage is only 20%. But the cost of Douglas fir No. 3 is currently only 59% of the cost of No. 2 and better spruce. Therefore, the 20% additional Douglas fir still costs only 71% as much as the spruce. In summary, the floor system framed with No. 3 Douglas fir 2x12s would cost 29% less but would be twice as rigid, and would be safe enough.

In any of the above examples involving calculations of strength behavior in floor systems, for the sake of simplicity only the joists were considered. In reality, the subflooring, finish flooring and cross-bridging are integral parts of the total floor system and will be involved in the total effect. Effects, as demonstrated for different joist depths and species properties, must be taken as approximate, but comparisons are nevertheless valid.

Machine stress grading—In recent years, the practice of machine stress grading has been developed. Rather than sorting pieces visually, each piece is flexed in a machine with enough load (but well below the proportional limit [P']) to establish the slope of the elastic curve. In this way, the modulus of elasticity (E) is determined directly. This results in an actual modulus of elasticity for the piece, not an estimated one. Based on thousands of previous tests carried to failure, closer predictions of modulus of rupture (MR) can be made from these E values, so that more precise F_b (extreme fiber stress in bending) values can be assigned.

Although such a non-destructive and accurate method of stress rating should have been welcomed, it has caused an interesting reaction. It has created a case of what I call "the baker's dozen syndrome." Some of us can still remember when the baker gave you 13 cupcakes when you bought 12. We can also remember the day when we bought a dozen and got 12. In shock we thought, "Hey, where's my other cupcake. I paid for a dozen and I got only 12!" Likewise, builders and homeowners had been getting extra F_b and E. But now the more precise is recognized as something less. In particular, boards that don't *look* strong are a special cause for concern. So regrettably, machine stress rating, although seemingly a major advancement in lumber quality control, has been slow to become established.

Other Properties

So far we have been concerned with the biological nature of wood and with its mechanical properties. I'd like to digress briefly into its thermal properties, into the little-known property of fluorescence, and into what I call wood's psychological properties.

Thermal conductivity

We are all familiar with the relative thermal conductivity of wood. Or perhaps we should say the relative nonconductivity of wood, because wood is more noteworthy as an insulator than as a conductor of heat. The image of wood being "warm" is quite understandable on the basis of thermal conductivity. The sensation of hot or cold depends not only on the temperature of the object touched but on the rate at which it conducts heat away from or into our skin. For example, if you touch cool pieces of wood and metal, heat is conducted out of your skin and into the metal hundreds of times faster than into the wood. Although they are at the same temperature, the wood feels warmer. Consequently, wood has long been preferred for gunstocks, tool handles and seating. Comparative values of conductivity are listed in *Table 13*. The coefficient of thermal conductivity, K, indicates the relative rate of heat flow through the various materials listed.

Since the **K value** indicates relative conductivity, its reciprocal, $1/K = R$, measures relative insulating value. The higher the **R value,** the better the insulator.

As given in *Table 13,* the values for oven-dry wood are meaningful only to the extent that they show the range represented by different species. This is because wood is seldom used in its oven-dry condition. The thermal conductivity for a certain species of wood depends on both specific gravity and moisture content, and can be calculated by the formula:

$$K = S\,(1.39 + 0.028\ MC) + 0.165,$$

where K = the coefficient of thermal conductivity in Btu/ft.^{2}/°F/hr./in.; S = specific gravity; and MC = moisture content in %. Thus, both increasing specific gravity and increasing moisture content will result in higher conductivity (i.e., lower insulating value).

For most species of softwood structural lumber at commonly encountered moisture contents, the K value will generally be about 1 or slightly less, the R value about 1 or slightly above. For example, in spruce structural lumber having a specific gravity of 0.40 and an average moisture content of 10%,

$$K = 0.40\,(1.39 + 0.028\,[10]) + 0.165$$
$$= 0.833.$$

1—A generous cord of 4-ft. long mixed hardwood firewood, stacked 8 ft. wide and more than 4 ft. high.

Given the critical condition of our energy resources, heat loss in buildings is of utmost concern. *Table 13* makes it clear that wood is a better thermal insulator than most other structural materials. It is more than seven times as effective as concrete, 300 times as effective as steel and 1,400 times as effective as aluminum of comparable thickness. At the same time, materials produced specifically for insulation, such as glass fibers, rock wool and foam products, are three to four times better than solid wood. In many cases where strength, beauty and insulating value are desirable, wood is an appropriate compromise and a logical choice.

Noting from the table that water has a K value of 4 and ice has a K value of 15, it is obvious that keeping wood and other materials dry is vital to maintaining their insulating potential. In cold climates, vapor barriers must be provided as close as possible to the interior walls of buildings, to prevent condensation from accumulating within the walls, lowering the R value as well as encouraging decay.

In most cases where thermal conductivity is critical, the direction of heat movement is perpendicular to the grain, so the values mentioned above have been for this direction. Conductivity parallel to the grain is two to three times greater than across the grain.

Table 13—Approximate thermal properties for various materials.

Material	K*	R**
Air	0.16	6.25
Water	4	.25
Ice	15	.07
Glass	5	.2
Brick	4.5	.22
Concrete	7.5	.13
Marble	17	.06
Steel	310	.003
Aluminum	1,400	.0007
Building insulation (foam, rock wool, fiberglass)	0.2 to 0.3	3.3 to 5
Wood, oven-dry, perpendicular to grain	0.4 to 1.2	.8 to 2.5

***K = Coefficient of thermal conductivity, expressed as the number of Btu's (British thermal units) that will flow through a material per hour, per inch of thickness, per square foot of surface area, per °F difference in temperature from the warmer to the cooler side.**
****R = 1/K, referred to as the resistivity of a material, represents the relative insulation value for a material.**

Effect of temperature on wood

Like any other material, wood expands when heated and contracts when cooled. This change is referred to as linear thermal expansion or contraction. The unit amount by which a material expands (per unit of original length per degree rise in temperature) is called its **coefficient of thermal linear expansion**. In the grain direction, wood changes dimension by only about 2 millionths of its length per Fahrenheit degree change in temperature. This means that if an 8-ft. Douglas fir framing stud were to change from 90°F to −10°F, it would become only 0.018 in. shorter, slightly more than 1/64 in. Over the same temperature range, steel rods of the same length shrink by more than three times this amount, and aluminum more than seven times. Perpendicular to the grain, the coefficient of thermal expansion is as much as ten times that of wood parallel to the grain direction. However, when extreme changes in temperature occur in the environment, the associated humidity changes will probably produce shrinkage or swelling of the wood that far overshadows the insignificantly small thermal expansion or contraction. For routine uses of wood, changes in dimension caused by changes in temperature can usually be neglected entirely.

The effect of temperature on the strength of wood is extremely important. As will be considered later in discussing strength and steam-bending of wood, elevated temperature may cause both temporary and permanent strength reductions, the net effect depending on such factors as the species and moisture content, the heat source, and the level and duration of the heat.

Burning of wood

When the temperature of wood is raised well above levels reached in nature or above the boiling point of water, chemical degradation begins to take place. At higher temperatures or with prolonged heating, darkening or charring becomes apparent. In the presence of ample oxygen, wood ignites when the temperature reaches 500° to 550°F. If oxygen is excluded, gases are driven off without combustion. This process is called **destructive distillation** or **pyrolysis**, and the eventual product is charcoal. When oxygen is present, the gases and charcoal burn together. However, all the gases do not burn until temperatures reach 1,100°F. In turn, the burning gases may develop temperatures up to 2,000°F.

The molecules forming the wood's major constituents were originally synthesized by taking energy from the sun to drive the necessary chemical reactions. As these components break down in combustion, this energy is released as heat, up to 9,000 Btu's per pound of dry wood. Although a thorough discussion of wood as a fuel is beyond the scope of this book, readers who harvest trees for woodworking will probably want to convert the residues to fuel.

The most popular unit of measure of fuel wood is the **standard cord**, defined as a pile of 4-ft. long wood, stacked 4 ft. high and 8 ft. wide. Although it occupies 128 cu. ft. of space, an average cord contains about 80 cu. ft. of solid wood. Depending upon species and moisture content, a cord weighs from one to two tons.

Firewood is commonly cut to short lengths and sold in units called **face cords**. A face cord is a stack 4 ft. high and 8 ft. wide whose depth is equal to the length of the pieces in the stack. A 16-in. face cord is a 4-ft. by 8-ft. stack of pieces 16 in. long. It would occupy a third of the space of a standard cord, but would probably contain slightly more wood than a third of a standard cord, because wood stacks more tightly when cut to shorter length. A 4-ft. face cord equals a standard cord.

I have yet to discover a wood that I cannot use satisfactorily for fuel (provided it has been adequately dried). Among the various species, however, I have undeniable favorites—some for their flame color, some for their aroma, some for their ability to produce cooking coals, some for their slow burning rate. But heating value is closely related to dry weight, so specific gravity is the most reliable predictor of thermal value. For example, eastern white pine has a fuel potential of 12,000,000 Btu's per cord, while hickory has a fuel value of 24,000,000 Btu's per cord. In determining whether fuel wood is an economically attractive alternative, calculations have shown that a cord of white pine has the heating value of about 100 gallons of No. 2 fuel oil, a cord of hickory, 200 gallons. But these calculations are based on two important considerations: the stove efficiency and the moisture content of the wood. Modern airtight or "high-efficiency" wood stoves have a conversion efficiency of 50% to 60%, an assumption included in the above calculations. By contrast, a standard box stove has only about 25% efficiency and an open fireplace less than 10%. The calculations further assume that the wood has been air-dried, or "seasoned," to 20% MC or less. If the wood is too wet, a great deal of heat is lost in heating and evaporating the excess moisture. An additional consequence of burning wet wood is that the moisture driven off condenses in the flue, along with other materials, forming a layer of creosote, a flammable substance that may later support a chimney fire.

In cities and suburban areas, burning wood for heat is an expensive luxury, justified only for the aesthetic and psychological pleasure it brings. In most rural areas, the cost of conventional fuels compared to available cordwood makes wood a logical choice. It is estimated that in Vermont, the amount of timber that decays and disintegrates by neglect represents far more fuel than would be needed to satisfy the entire energy budget of the state. Although the reaction rate is drastically different, the breakdown of wood by burning and by biological deterioration are similar in that the principal products are thermal energy, carbon dioxide and water. Thankfully, burning wood produces only negligible amounts of sulfuric pollutants, in contrast to many other common fuels.

The burning of wood is also used as a craft technique, to some extent. Carvers of birds use fine wood-burning pencils to create delicate veining of feathers. Sculptors and other craftsmen use blowtorches to decorate surfaces by charring, which achieves an attractive effect because of differences in the relative density of earlywood and latewood.

Fluorescence

A rather unfamiliar property of wood—fluorescence under ultraviolet radiation (black light)—is a fascinating visual phenomenon, well worth investigating. Certain woods viewed under black light emit a mysterious glow or fluorescence, which is almost sure to inspire ideas in woodcraft.

Fluorescence is the absorption of invisible light energy by a material capable of transforming it and emitting it at wavelengths visible to the eye. The human eye can see light over the spectrum of wavelengths varying from about 8,000 Angstrom units (red) down to about 3,800 Angstrom units (violet). (Angstrom units, which equal about 4 billionths of an inch, are used to measure wavelengths of light.) Above 8,000 is invisible infrared light; below 3,800 is invisible ultraviolet light. The black light commonly used for visual effects is in the 3,800 to 3,200 range and is referred to as "long-wave" or "near" ultraviolet light. It won't harm the eyes. Light in the 3,200 to 2,900 range includes rays that cause tanning and sunburn. Below 2,900 is shortwave or far ultraviolet. This light is used to kill bacteria and is very dangerous to the eyes. Sterilization units or other sources of shortwave ultraviolet should never be used to view fluorescent materials.

The long-wave or black-light lamps available commercially emit light averaging about 3,700. However, the light may range from as low as 3,200 up to about 4,500, well into the visible range. We therefore often see a purple glow, although most of the light emitted is invisible. Chemicals in a fluorescent material absorb this invisible light but transform the energy, so that the light emitted from the material is within the visible range and is seen as a particular color. The term fluorescence was given to this phenomenon in 1852 by Sir George Stokes, who first observed the response in the mineral fluorite.

Many domestic species of wood exhibit some fluorescence. *Table 14* lists the principal species, although there are doubtless others. Fluorescence is scarcely recognizable in some species and strikingly brilliant in others. The colors listed are typical for each species, but variation of both hue and brilliance occur among individual samples. Countless other species from around the world, too numerous to list, also show fluorescent properties. Some 78 genera, 45 of which are in the family *Leguminosae*, yield fluorescent woods. However, our native species are as attractive as those found anywhere in the world.

Yellow is the predominant color, and also the most brilliant, as in black locust, honeylocust, Kentucky coffeetree and acacia. Barberry, a lemon-yellow wood under normal light, is also among the most brilliant yellows, but because it is a shrub it is difficult to locate pieces large enough for anything but small items, such as jewelry.

Perhaps the most interesting is staghorn sumac. The sapwood has a pale lavender-blue fluorescence *(1)*. In the heartwood, each growth ring repeats a yellow, yellow-green, lavender-blue sequence. Many species show less spectacular but nevertheless interesting colors. Yucca and holly have a soft bluish to grey fluorescence. Purpleheart emits a dim coppery glow. Badi exhibits a mellow pumpkin orange.

In addition to normal sapwood and heartwood fluorescence, certain anatomical features such as resin canals, oil cells, vessel contents, bark, fungal stains and pigment streaks show selective fluorescence. In aspen, for example, a brilliant yellow fluorescence usually occurs at the margins of areas stained by fungi. Fluorescence is of obvious value in identification, and many otherwise confusing species are easily separated on the basis of fluorescence.

The unusual effects of fluorescing woods suggest interesting applications for the craftsman, especially in carving and marquetry *(2)*. Inspiring sculpture and religious statuary can have moving effects when viewed in darkness or in subdued light with hidden ultraviolet

Table 14—Partial list of North American woods exhibiting noteworthy fluorescence under ultraviolet light.

Scientific Name	Common Name	Color of Fluorescence
Acacia greggii	Catclaw acacia	Deep yellow
Annona glabra	Pond-apple	Dull yellow
Asimina triloba	Pawpaw	Faint yellow-green
Berberis thunbergi	Japanese barberry	Bright yellow
Cercidium floridum	Blue paloverde	Yellow green
Cercidium microphyllum	Yellow paloverde	Yellow
Cercis canadensis	Eastern redbud	Bright yellow
Cladrastis lutea	Yellow-wood	Pale yellow-light blue
Cotinus obovatus	American smoketree	Deep yellow
Gleditsia aquatica	Waterlocust	Pale yellow
Gleditsia triacanthos	Honeylocust	Bright yellow
Gymnocladus dioicus	Kentucky coffeetree	Deep yellow, bright
Ilex verticillata	Common winterberry	Light blue
Magnolia virginiana	Sweetbay	Faint pale yellow
Mangifera indica	Mango	Pale orange
Piscidia piscipula	Florida fishpoison-tree	Dull yellow
Rhus copallina	Shining sumac	Bright yellow
Rhus glabra	Smooth sumac	Bright yellow
Rhus integrifolia	Lemonade sumac	Bright yellow
Rhus ovata	Sugar sumac	Yellow
Rhus typhina	Staghorn sumac	Bright yellow, greenish yellow to pale blue
Robinia neomexicana	New Mexico locust	Bright yellow
Robinia pseudoacacia	Black locust	Bright yellow
Robinia viscosa	Clammy locust	Bright yellow
Torreya taxifolia	Florida torreya	Dull yellow
Yucca brevifolia	Joshua-tree	Yellowish grey
Yucca elata	Soaptree yucca	Silvery grey
Zanthoxylum clava-herculis	Hercules-club	Sapwood pale yellow to light blue, heartwood bright orange

lamps. With the increased use of black light for entertainment areas such as game rooms and cocktail lounges, fluorescent figures and decorative carvings are quite popular. Desk-set bases or light-switch covers can be given unusual effects. Fluorescence can add extra excitement to wooden jewelry and personal accessories. An African mask or Polynesian "tiki" carved in a fluorescent species makes an unusual pendant or pin.

Fluorescent woods can be used in combination by laminating or inlaying. Menacing fluorescing teeth can

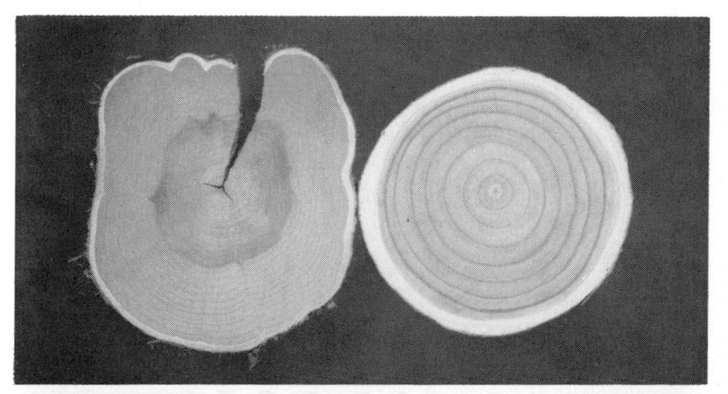

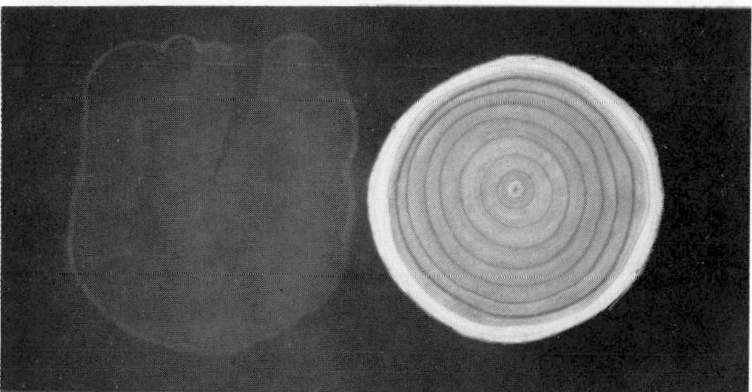

1—Different woods exhibit various degrees of fluorescence. At top are cross-sectional slices (4 in. in diameter) of eastern redcedar, left, and staghorn sumac, right, under normal light. Under ultraviolet light, above, the sumac exhibits noticeable fluorescence, the redcedar does not.

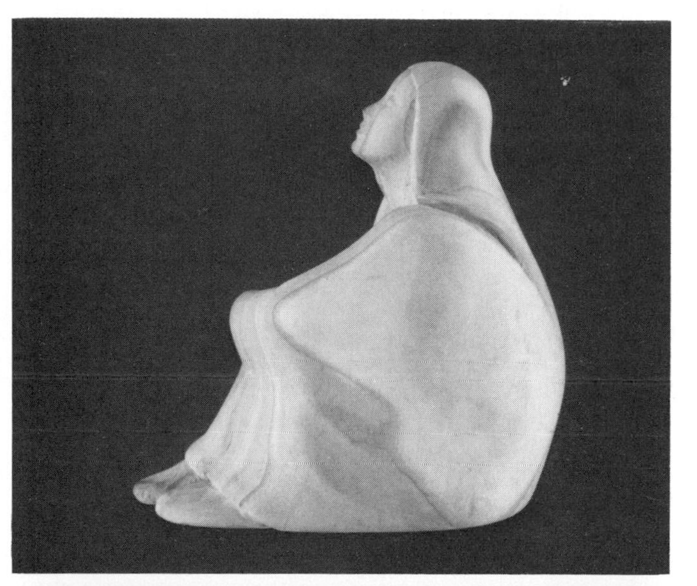

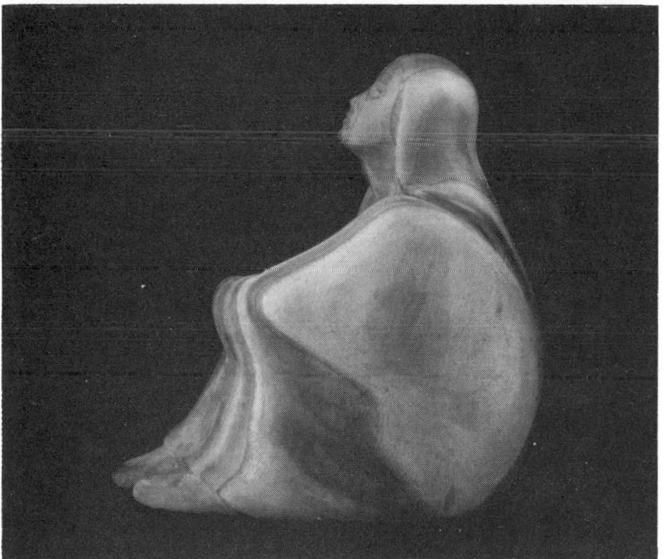

2—Seated figure, 6 in. high, carved by the author from staghorn sumac. Under normal light, top, the large area of sapwood along the upper arm and body is a pale greyish-cream color; heartwood is drab yellow green to olive green. Under ultraviolet light, above, the sapwood fluoresces to pale lavender-blue, and the heartwood repeats a luminescent lavender/yellow-green/bright yellow sequence in each growth ring.

be set in the mouth of a carved dragon. Spooky yellow eyes that "light up" can be inlaid into a carved owl. Laminated woods can be carved or turned into unusual lamp bases—especially for black lights. And don't throw away carving chips and planer shavings—children delight in gluing chips and sawdust to cardboard to create fluorescing designs and pictures.

The fluorescent response of wood is subject to surface chemical degradation, however, apparently associated with the familiar darkening or aging effect. The brilliance of fluorescent wood is most rapidly faded by exposure to daylight, especially direct sunlight. The original brilliance of a carving will remain for years if it is kept in a dark place. In normal indoor daylight conditions, a year's exposure will fade a piece to about half its original brilliance. This dulling effect is at the very surface, and a light recarving or sanding to expose unaged wood will renew the original fluorescence.

Most finishes reduce the brilliance of fluorescence, but in the long run may help maintain it by minimizing aging. Clear paste wax seems to have the least dulling effect.

Anyone who has not yet witnessed the strange luminescence of woods under ultraviolet light has a delightful surprise in store.

Psychological properties

The word properties usually suggests scientific information about physical, mechanical, chemical or anatomical characteristics. We commonly express such properties as moisture content, density, strength and thermal conductivity in standardized units of length, volume, mass or degrees. This book has so far dealt closely with these scientific properties because they allow us to work out so many of the routine technical problems encountered in woodworking. But this book would be incomplete if it did not acknowledge what I call the psychological attributes of wood.

Wood has values or powers that cannot be quantified in scientific terms. These aspects of wood, without being well understood or even explainable, may well be among the most important and powerful. However difficult they are to describe or define, we can at least demonstrate their certain existence.

Part of what we are dealing with here might be termed *sensual properties*, where there is a close relationship to our senses of touch, sight, hearing, smell and taste. These properties include color, odor, resonance and so forth, and are in part explainable and describable in scientific terms. For example, the somewhat emotional reference to the warmth of wood is largely the physical reality that wood feels warm to the touch (especially when compared to ceramics and metals), which we understand scientifically in terms of dry wood's low coefficient of thermal conductivity. But there remains an aura of warmth suggestive of "friendliness."

In considering psychological properties we need not strive for (nor expect) much agreement among people. Unlike physical and even sensual properties, which can be described by the universal language of science and thereby be conveyed to others, psychological properties are subjective. A psychological feature may exist for one person and be totally meaningless to the rest of humanity. Similarly, an individual piece of wood may have meaning that does not attend other pieces of the same species, or even the same tree. For example, deep sentimental value may be attached to an object made from the wood of a familiar or famous tree. Souvenirs carved of wood that is native to a particular place are treasured because of the memories they evoke. If you have ever made a thing of wood taken from a particular tree and presented it to someone who had special attachment to that tree or its locale, you know how precious such a thing can be. A similar attachment to something made of synthetic materials is improbable. Of my entire collection of a hundred or more pieces of wood earmarked for carving, I know the exact place and the very tree from which each came. It is extremely important to the meaning and satisfaction I derive from carving to be familiar with the "roots" of each piece of wood.

I suggest that the psychological appeal of wooden objects develops through the interaction of two vital elements working together: nature and mankind. That wood is a direct and unchanged product of nature undeniably attracts us. In contrast, we bemoan the aesthetic loss in particle board and hardboard, or in products like rayon carpets or molded egg cartons, which are made of wood but are so transformed from the tree as to be unrecognizable. I am sure that anyone deeply involved with wood must react with a mixture of amusement and dismay to the pictures of wood that are frequently pasted on the sides of otherwise beautiful—and expensive—automobiles.

Subconsciously if not consciously, many people today resist being pushed gradually into a synthetic environment, further and further removed from nature, and they seek to retain every possible remnant of the natural world. The importance of wood in interior decoration and the continued importance of leather, wool and stone testify to the aesthetic and psychological value of natural materials.

Closely related to the value of natural wood in an object is the element of human involvement. Wood was originally used because it was the most appropriate, available and logical material to satisfy functional needs, but along with production skill there developed a high level of artistic creativity and aesthetic appreciation. The fact that our large-scale commodity production has lost much of this sensitivity is one of the reasons why real wood, especially when hand-crafted, remains so meaningful.

It is not easy to analyze and understand the value of working in wood by hand, nor of the products made thereby. Some people want information about wood in order to become better technicians, the better to satisfy some of their own commodity needs themselves—the "do-it-yourself-and-save-a-buck" syndrome. But I believe that just as many people begin woodworking with an interest in the material itself, rather than any need or desire to make something out of it. Of course, many people begin as technicians and upon "discovering" wood, turn into devoted artisans.

There may also be some confusion about the so-called craft revival. For example, one writer commenting on a recent popular exhibition of early Americana opined, "America is turning back to the old way of doing things." However, the apparent nostalgia for the old ways is more likely just a sensitive response to the natural materials and human skill so characteristically expressed in old things. For we can just as surely realize the pleasure and integrity of natural material and human craftsmanship in modern designs and fresh ideas, which are no less an important part of what I see as the craft revival. And on the other hand, we cannot seriously contemplate substituting hand-crafts for our whole system of commodity production.

These are all nebulous ideas. I'll try to illustrate my point by recounting an "experiment" I have repeated with several groups of wood technology students at the University of Massachusetts. In our department seminar room we have a long walnut conference table. It is handsomely proportioned, with slightly bowed edges so the top is narrower at each end than in the middle. The table is expertly built, with a top of figured sliced walnut veneers, carefully matched. The ends and edges are neatly banded with deep strips of solid walnut, giving a plank effect to the whole.

With a group of students seated around the table, I call attention to its features from the overall appearance and design, the figure on the top and the well-fitted construction down to its general heft and resonance. I ask for a consensus on the table's merits, and the students enthusiastically approve. Then I ask, "If this is such a beautiful table, aesthetically pleasing and well crafted, does it really matter that it isn't real wood? You all *do* realize that it's actually plastic, don't you?"

The utter shock on every face reveals that indeed it does matter. The bewilderment deepens as I extol the sophistication of modern technology, which can precisely imitate the physical properties of the material, and of photographic methods capable of imitating the fine cellular detail of the wood.

When I feel that the point has been driven home, I reveal my prank with the assurance that the table really is genuine walnut. After a short period of emotional confusion, the students forgive me when they realize the purpose of my deception. The group then agrees that "all other things being equal" (as textbooks like to put it), there is indeed a difference between a real wooden table and a synthetic lookalike. That difference, whatever it is, is the mystique of wood.

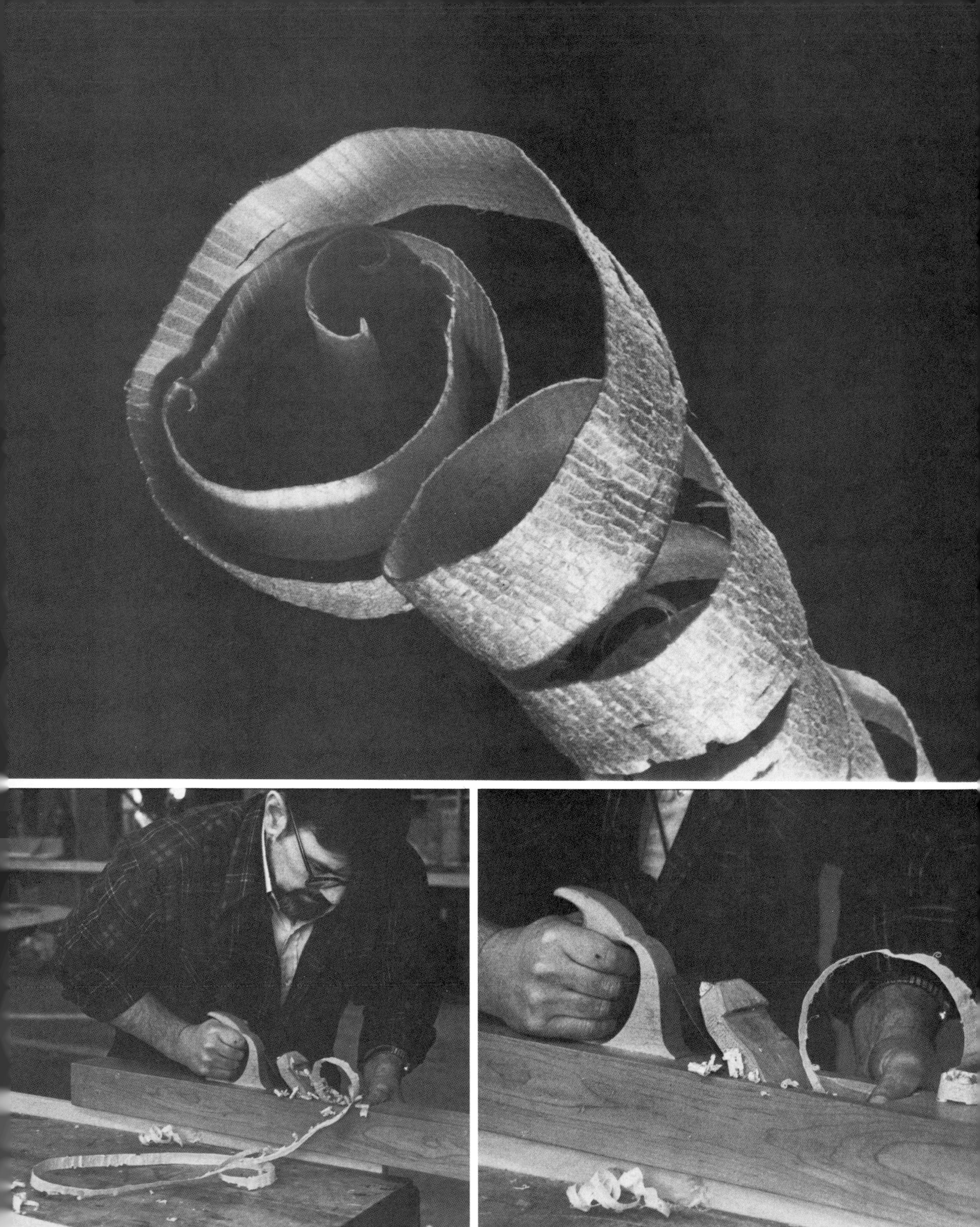

Machining Wood

CHAPTER 8

My dictionary defines "technology" as "applied science; systematic knowledge of the industrial arts." The implication for the woodworker is that while much practical knowledge can be gained by studying the basic science of wood (as we have been doing up to now), much also can be gained by investigating modern industrial practices. Far more is learned through solving the daily problems of production woodworking than any individual craftsman could hope to accumulate in a lifetime of custom handcrafting.

The key to learning from contemporary technology is this: Ignore inapplicable production-scale methods, and concentrate on the pertinent underlying principles. The interpretation and adaptation of these principles to the problems at hand is the constant challenge. Thus I try to separate out that part of technology which has to do with wood itself from that which is involved in each specific application. The wood doesn't know or care whether it is being made into part of a handcrafted masterpiece or into a mass-produced commodity. If the conditions are similar, the wood will behave the same way. A piece of birch will form the same chemical bond with urea-formaldehyde adhesive in your cellar as it will in a four-acre furniture factory. A length of ash will plasticize the same way under exposure to saturated steam in your garage as will a thousand lengths under similar treatment on the assembly line.

In the chapters that follow I will not attempt to present "systematic knowledge of the industrial arts," but I will explore a few of the pertinent elements of technology, starting with the basic problem of wood machining. There is hardly a subject of more importance to the woodworker than machining. At the same time, I cannot imagine an aspect of woodworking that is more complex. Several basic factors affect machining quality.

Machining is a stress-failure process. By hand or machine power, force is transmitted to the wood by means of a cutting tool. The orientation and direction of the force are controlled by the design of the tool and/or by the hand of the woodworker. The tool has pertinent geometry, and the wood has pertinent physical and mechanical properties. The direction of motion and configuration of the tool determine the way stress develops and is resisted by the wood, and therefore the manner of failure or "cutting" that occurs. Two concepts are important in this regard. One is the idea of sharpness, in which the cutting area (A) of the tool edge is small enough so that the force (P) applied to the tool will cause a stress (P/A) greater than the strength of the wood. The second factor is the condition of the wood—its moisture content, temperature, defects, etc.

It is convenient to analyze machining as the action of a **cutting tool** on a piece of wood or **workpiece**, with the cutting action that takes place referred to as **chip formation,** wherein a portion of wood called the chip is separated from the workpiece. Chip formation involves the geometry of the tool, the condition of the wood and the motion of the tool relative to the orientation of the structure of the wood.

The objective of machining always should be the paramount consideration. The approaches used may differ drastically, depending on the objectives sought. These objectives may be classified as:

- Severing: To make two or more pieces from one, for example, splitting firewood or bandsawing rough parts from a plank.
- Shaping: To impart a specific shape to the workpiece, in some cases a flat-planed surface, in others some specific contour. Jointing a flat surface on a cupped board is one example, milling an ogee molding is another.
- Surfacing: To create a surface of prescribed quality, for example, sanding a surface prior to finishing, or jointing edges that are suitable for gluing.

In most cases, two or even all three of the above are involved concurrently. In ripping boards into strips, for example, one might want the resulting surface to be true enough and of appropriate condition to be glued. In most machining the objective is the workpiece, and the nature of the chip removed is irrelevant. An interesting exception is knife-cut veneer, where the chip itself, the veneer, is of primary interest, and the surface left on the workpiece becomes one face of the veneer that will be removed by the next cut. Although the average woodworker is not involved in making veneer, the user of veneer should understand the machining process involved in its manufacture.

Let's survey the interrelationship of the workpiece, the tool and the chip formation in machining.

1—A Type I chip produced by 90°-0° cutting, in this case pared from the edge of a board by a jointer plane with sharp iron and closely set cap.

The workpiece

The aspects of wood that affect machining have already been covered in previous chapters. The structural nature of wood in terms of its three-dimensional properties is particularly important. Density variation among and within species is also of obvious importance, as is unevenness of grain, especially in ring-porous hardwoods and uneven-grained softwoods. Heartwood extractives in some species are particularly abrasive and contribute to tool dulling. Defects such as knots create both irregularities of grain direction and variations in density. Structural irregularities such as wavy or interlocked grain cause special machining problems. Moisture content influences machining as it affects the strength of wood, and so do stresses or checks developed in drying.

Strength of wood is, of course, the bottom line. The relationship between the strength of wood parallel to the grain and perpendicular to the grain is perhaps the most important part, although every other factor affecting strength in turn affects machining.

The tool

At the business end of a cutting tool, where chips are being formed, the tool geometry can be described in terms of a **cutting edge** formed by its intersecting **face** and **back** surfaces, or planes *(1)*. The critical geometry of the cutting edge is usually defined in terms of its direction of motion:

α = the **rake angle** (also called the **cutting angle,** the **hook angle,** the **chip angle** and the **angle of attack**) is the angle between the tool face and a line perpendicular to the direction of travel of the edge.

β = the **sharpness angle,** the angle between the face and back of the knife.

γ = the **clearance** angle, the angle between the back of the knife and the direction of travel of the edge.

As required by circumstance, cutting-tool geometry can be varied considerably. The sharpness angle will always be a positive value. The cutting angle and clearance angle can be negative values. In the case of the clearance angles, negative values indicate interference between tool and workpiece.

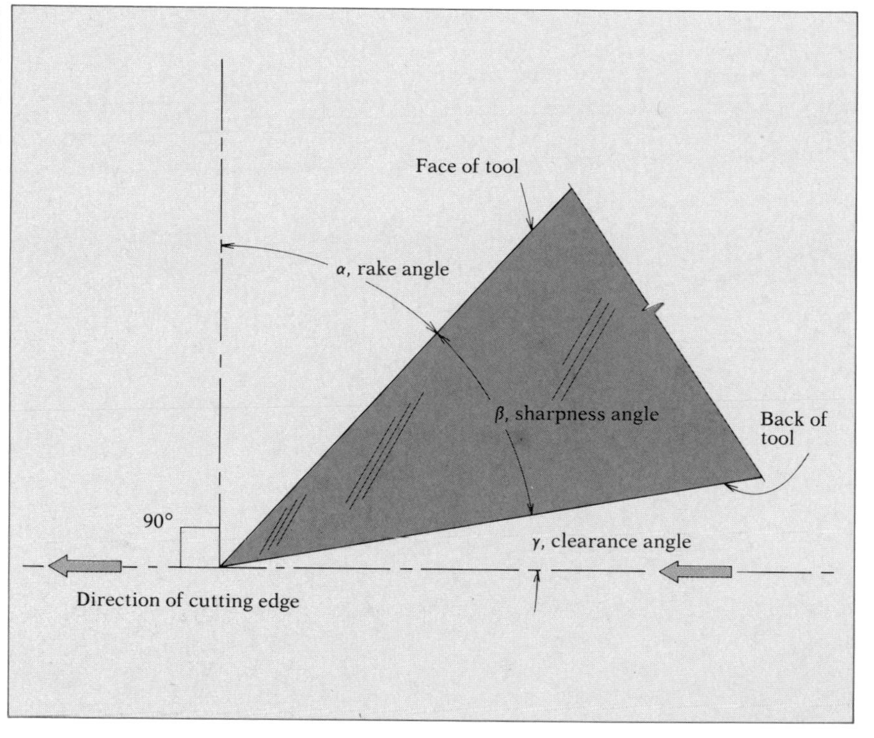

1—The business end of a cutting tool consists of an edge formed by its intersecting face and back surfaces. Its geometry can be described by the rake angle, α (alpha), measured from a line perpendicular to the direction of travel to the tool face; the sharpness angle, β (beta), measured between the face and back of the tool; and the clearance angle, γ (gamma), measured between the back of the knife and its direction of travel.

Chip formation

Figure *2* shows an ideal cutting action, which will vary according to the resisting grain orientation of the wood and the tool geometry. Energy is consumed in severing or separating the wood structure to form the chip, in deforming or rotating the chip, and in the frictional resistance of the tool face against the chip and the tool back against the newly formed surface of the workpiece. Each element varies in importance, depending on the type of cutting. In riving shakes, for example, after initial entry of the cutting edge and initiation of the split, the chip forms by fiber separation well ahead of the knife edge, with energy being expended to overcome friction on both sides of the knife and to deform (usually by bending) the chip being separated. In planing wood across the grain, the edge of the iron severs cell structure with minimal frictional resistance from the weak chip being separated. When the resulting shape and surface quality of the workpiece are important, it is critical to keep chip formation close to the tool edge itself. When the chip is being formed well ahead of the tool, as in splitting wood, neither its shape nor its surface quality can be well controlled.

In cutting-tool design, a compromise must often be made. It would seem advantageous to increase the cutting angle and reduce the sharpness angle, thereby reducing the amount of distortion of the chip and the resulting forces against the tool. In practice, however, one soon runs up against the limitations of steel. Such an idealized cutting edge would have little strength, and would soon break or become dull. Increasing the sharpness angle of the tool makes the edge more durable, but eventually leads to excessive frictional resistance or to uncontrollable chip formation.

When severing wood tissue with an edge, two points must be kept in mind: First, failure occurs only when ultimate stress is reached; second, stress is *always* accompanied by strain. This means that contrary to the idealized cutting action shown in Figure *2*, we must imagine something else, where in order to develop enough stress to produce failure, the wood must first deform *(3)*. Considered another way, since stress is load divided by area, then the smaller the area of application, the higher the stress that will be produced by a given load. We should therefore try to concentrate cutting force on the smallest possible area.

This, of course, is what we commonly call sharpness. Most people think of sharpness as the minuteness of the cutting edge, which influences the relative force required in cutting. But one must also think of sharpness or dullness in terms of the deformation of the wood tissue in both the chip and the workpiece. A helpful model for visualizing this relationship is to try to take a ⅛-in. slice from the edge of a wet cellulose sponge with an ordinary table knife. The sponge simply moves out of the way, and no cutting is done. With a very sharp knife or razor blade the sponge can be cut, but only after it deflects noticeably ahead of the knife. "Chip formation" is irregular. Some sponge may tear away in places other than at the exact blade edge. When the "chip" of sponge is finally severed, the "workpiece" springs back, and an irregular surface is the result, because of the irregular deflection. The springback of material after the cutting edge passes also illustrates the need for an appropriate clearance angle.

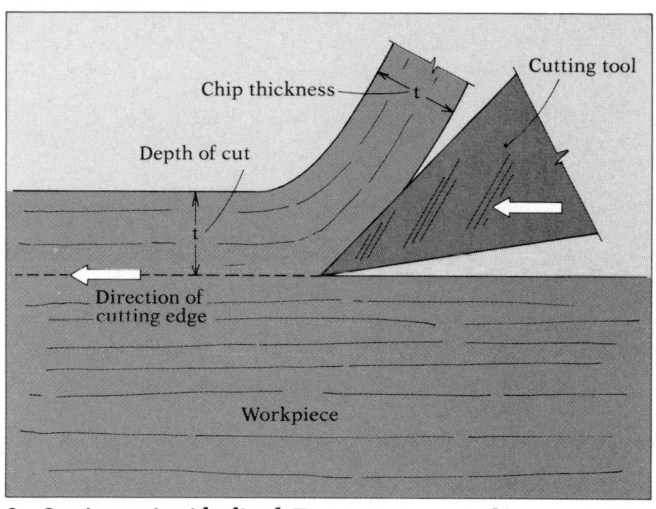

2—Cutting action idealized. Energy is consumed in severing the wood to form the chip, in deforming or rotating the chip, and in friction of the tool face against the chip, plus friction of the tool back against the new surface of the workpiece.

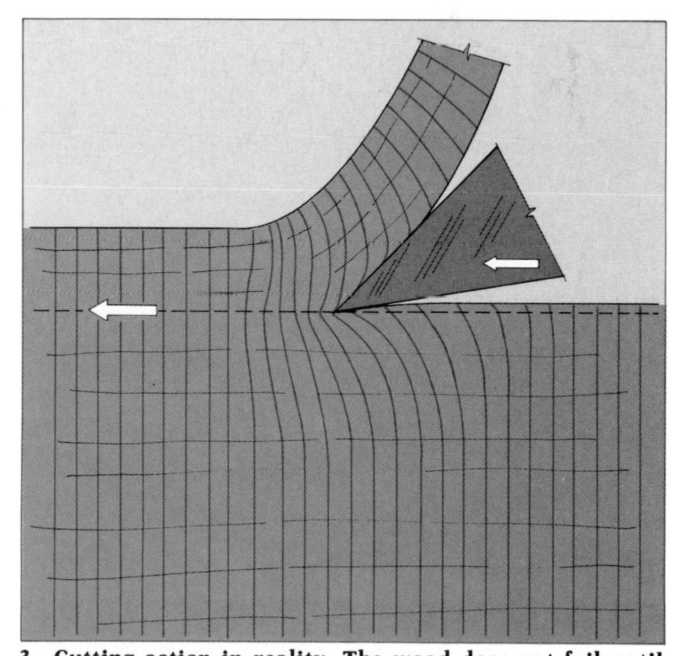

3—Cutting action in reality. The wood does not fail until ultimate stress is reached, and stress is always accompanied by strain. As the cut proceeds, the workpiece deforms ahead of the tool, severs, and then both workpiece and chip spring back to some extent.

Types of cutting action

There are two basic types of cutting action. The first is called **orthogonal cutting,** in which the tool edge is more or less perpendicular to its direction of motion and where the cut is in a plane parallel to the original surface of the workpiece, with removal of a continuous chip. An ordinary plane peeling a shaving from the edge of a board is one example. The second type is referred to as **peripheral milling,** in which a rotary cutterhead carrying one or more cutting edges intermittently contacts the work surface. Each cutter proceeds on a curved path, and removes a single chip. Virtually every cutting situation can be compared to either orthogonal cutting or peripheral milling. Note that as the cutterhead radius increases in peripheral milling, it approaches orthogonal cutting.

Visualize a cube of wood with its sides oriented in the radial, tangential and longitudinal planes. According to notation reported by W.M. McKenzie, orthogonal cutting is described by two numbers. The first is the angle between the cutting edge and the cellular grain direction and the second the angle between direction of cutting and the grain direction. Thus there are three basic cutting directions: 90°–0° cutting, 90°–90° cutting and 0°–90° cutting *(1)*. By considering each type of orthogonal cutting, some common types of machining can be understood more clearly.

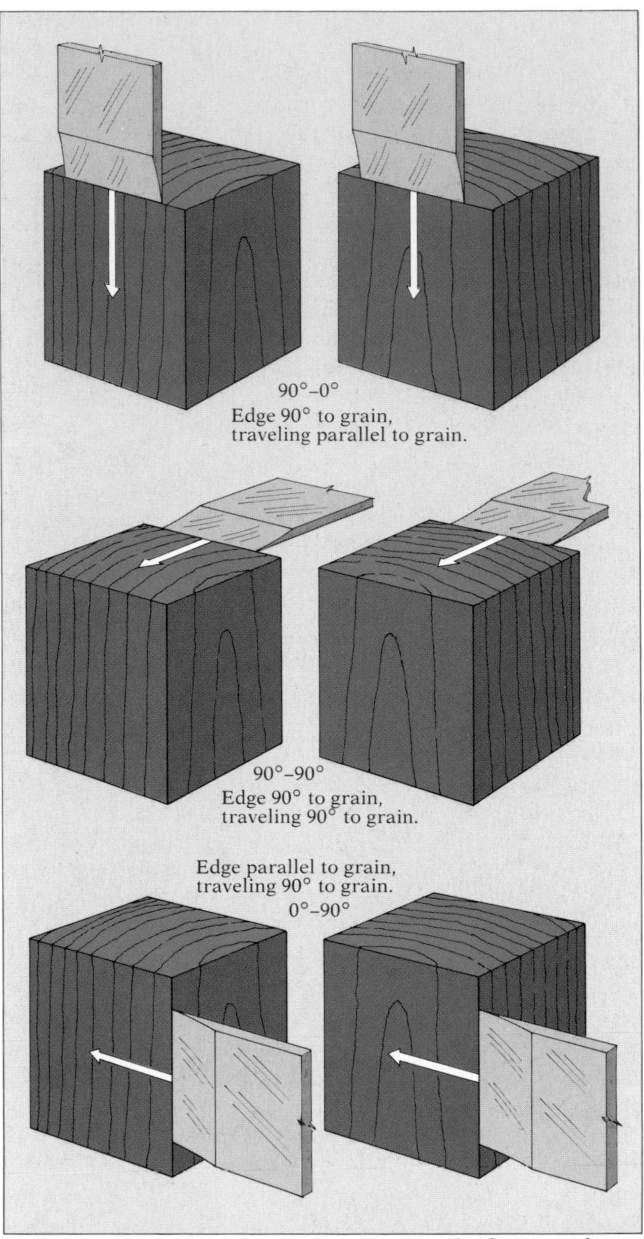

1—The three types of orthogonal cutting. The first number is the angle between cutting edge and grain direction; the second is the angle between direction of cutting and grain direction.

90°–0° cutting (planing along the grain)

Parallel-to-grain cutting is best typified by the standard hand plane. The chip forms as the plane is pushed along the board. The typical cutting action involves a cyclic sequence of events *(2)*. The iron separates fibers lengthwise to begin a chip (A). As the knife advances, the separated chip slides up the iron. The chip is now a cantilever beam that resists bending. It lengthens itself by failure of the wood in tension perpendicular to the grain well ahead of the knife edge (B). Finally, the chip is so long that bending stresses equal the strength of the wood and the chip breaks (C). The iron advances to the fracture point and begins to lift the next segment of chip (D) and so on. The chip, produced in a long jointed curl (see Figure *1*, page 144), is referred to as a Type I chip in 90°–0° cutting.

The typical plane cutter or iron is set at an angle of 45°. Sharpening to an angle of 30° leaves a 15° clearance angle. If the rake angle becomes too great, the friction of the chip upon the iron face would increase and the efficient bending and breaking action would be lost.

At smaller cutting angles, a greater component of forward compression and a smaller component of upward lifting are transmitted to the chip. Failure may occur as a diagonal plane of shear, bending the fiber structure, so chip formation develops as a continuously generated curl of deformed cell structure. This is classified as a Type II chip *(3)*. The cutting edge produces the surface as it dislodges cell structure. Greater force is required because of the compression resistance. Where the tool is well controlled and a reasonably thin chip is taken, chip formation takes place quite uniformly and an excellent surface is produced. Some special hand planes with a cutting angle of only about 30° are designed to take advantage of this type of cutting action.

As small (or even negative) cutting angles are used, force is transmitted mainly as compression parallel to the grain. The knife edge produces the surface as it shears free the cell structure. As the wood fails in compression, the damaged cell structure packs up against the cutting face and may form a wedge that transmits force and causes failure out ahead of the knife edge, often below the projected cutting plane. The failure is erratic and leaves an irregular surface, and is accompanied by an irregular chip of mangled cell structure. This is Type III chip formation *(4)*. With very low cutting angles, a smooth surface and uniform cutting action occur only when a thin enough chip is taken to form Type II chips. This is the cutting action of scrapers.

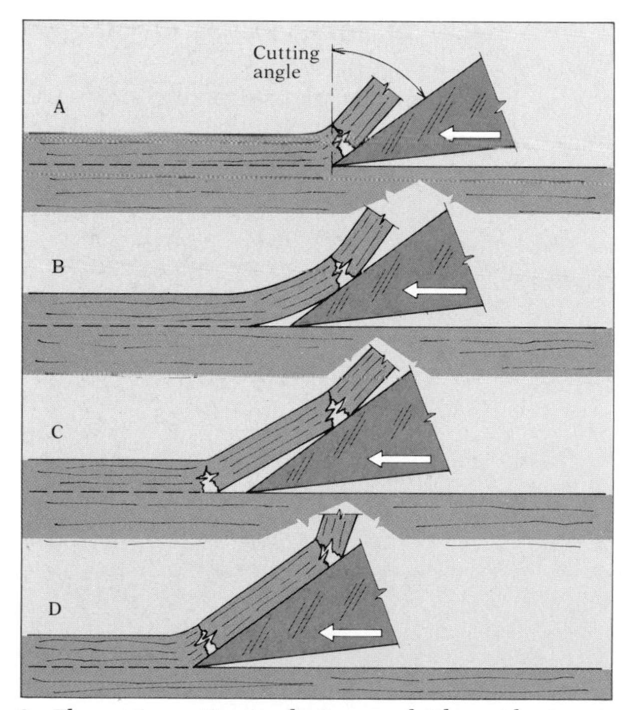

2—The cutting action in planing wood. The cut begins at A. The chip bends as it slides up the knife, and the wood fails ahead of the edge due to tension perpendicular to the grain, B. Finally the chip breaks, C, whereupon the next segment of the cut starts, D. In 90°–0° cutting this is known as a Type I chip, produced by a relatively large cutting angle.

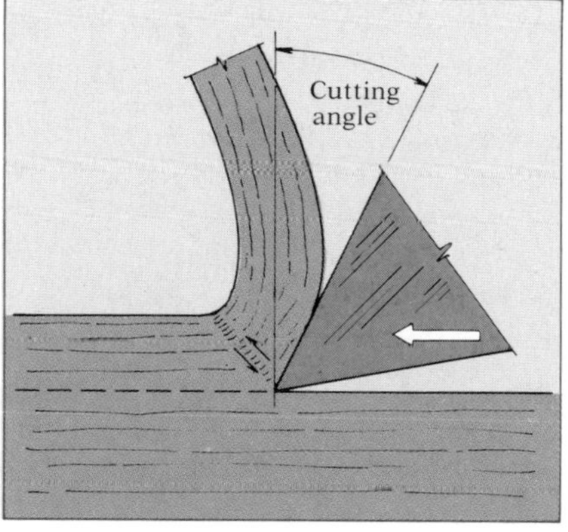

3—At small cutting angles, the face of the knife produces more forward compression than upward lifting. Failure occurs as a diagonal plane of shear right at the cutting edge. With enough force and a thin chip, the workpiece surface can be left in excellent condition. This is a Type II chip in 90°–0° cutting.

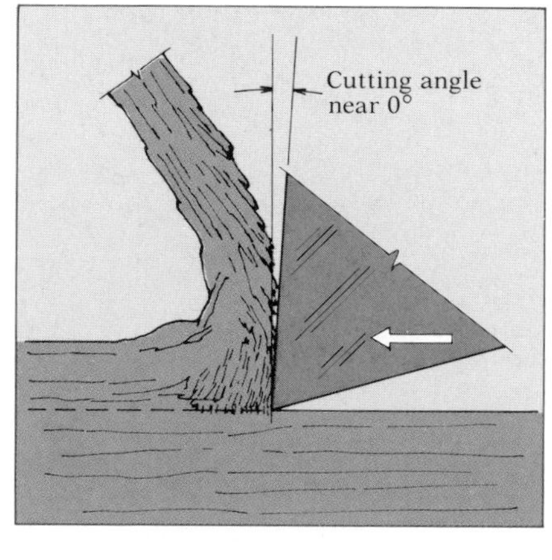

4—At very small (or even negative) cutting angles, force is transmitted mainly as compression parallel to the grain. The damaged cells pack up against the cutting face, often causing erratic failure ahead of and below the edge. This is a Type III chip in 90°–0° cutting. The snowplow effect can only be avoided by taking a very thin chip, whereupon it becomes Type II cutting. Cabinet scrapers work this way.

The quality of cutting depends mainly on two factors: the grain direction and the mechanics of chip breakage. Figure *2* on page 147 assumes perfectly straight grain, which in reality is more the exception than the rule. Usually some degree of cross grain exists wherein the fiber direction either rises ahead of the projected line of cut or leads down below it. The former case is termed cutting **with the grain,** the latter cutting **against the grain** *(1)*. Cutting with the grain is preferable, since the splitting of the wood associated with chip formation projects harmlessly into the next chip segment which subsequently will be removed. The cut produces a new surface generated by the continuous severance at the tool edge. Cutting with the grain is very efficient because most of the chip segments fail readily due to cross grain. The woodcarver will find that cutting with the grain at an acute angle to the grain is a fairly efficient way to remove large amounts of stock. At the same time, 90°–0° Type I chip formation using a hand chisel can be painfully undesirable in carving. The case in point involves carving with the grain, where a splinter-type chip slides all the way up the face of the tool and jabs the carver in the hand *(2)*. I have numerous scars on the outside heel of my left hand thanks to this situation. I collect another scar every time I fail to wear a glove when carving.

By contrast, cutting against the grain can result in chip formation where the splitting projects below the intended plane of cutting. The resulting surface is called **chipped** or **torn grain.** In some cases, as in planing the edge of a flatsawn board with spiral grain or in passing a flatsawn board with diagonal grain through a surface planer, the board can be alternately turned end-for-end so that cutting will occur with the grain. In other cases, however, as with the bulge of grain direction associated with a knot, some cutting must take place against the grain. To minimize the depth of torn grain, the breaking length of the chip segments must be controlled. One approach is to take an extremely thin cut, in which the chip segments break frequently. Otherwise, a "chip-

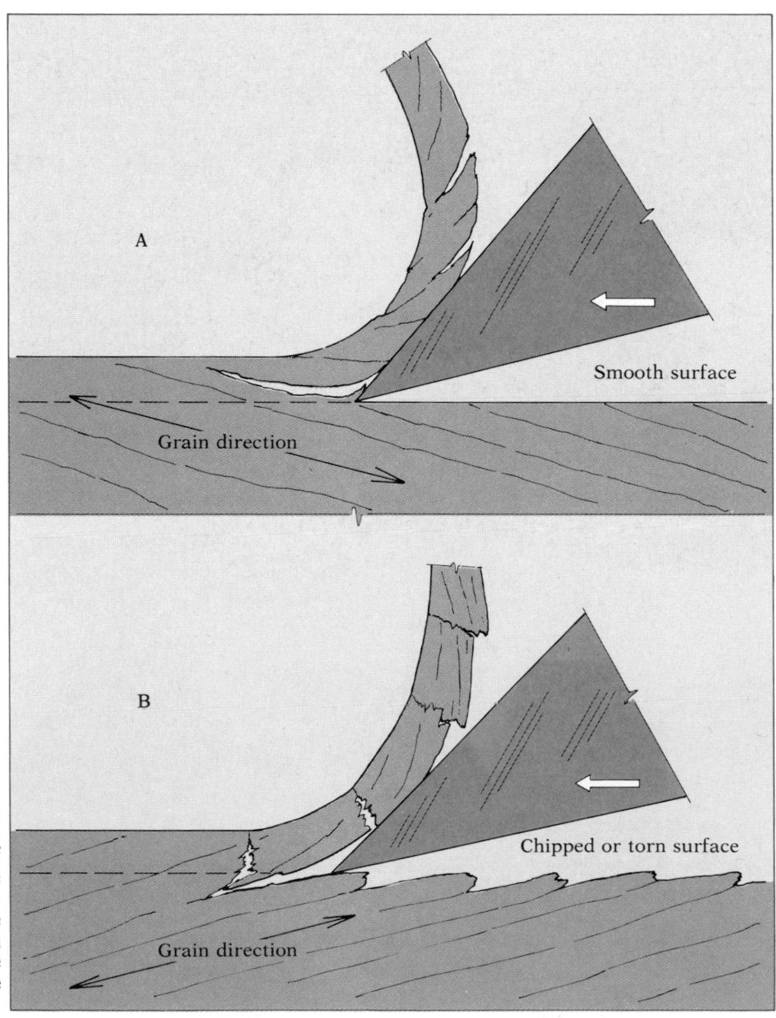

1—In the real world, the wood grain is rarely parallel to the cutting direction. Usually the fibers are rising ahead of the line of cut, and cutting with the grain leaves a very smooth surface (A). When the fibers lead down below the line of cut, cutting against the grain leaves a chipped surface (B). We usually reverse either the work or the tool, to go with the grain.

breaker" must be introduced, such as the cap iron on a hand plane *(3)*. The cap must be located suitably close to the cutting edge and must fit tightly enough to the face of the iron so the chip will not lodge but will slide up easily and bend beyond its breaking point quickly. The face and mating edge of the cap should be shaped as precisely as the cutting edge of the iron itself, for the cap iron is an integral part of the cutting mechanism.

The importance of the clearance angle in the cutting process should be appreciated. As with the "springback" of the wet sponge, it is also true that some deflected cell structure will recover after the chip forms. In order for the back of the cutting edge to clear this material, frictional drag and pressure against the back of the knife must be eliminated. If the cutter were infinitely sharp, of course, little clearance would be needed, and recommended clearance angles of up to 15° may seem excessive. However, such large clearance angles are probably safeguards against less-than-perfect sharpening. If the back (beveled side) of the cutter is not perfectly flat, the clearance angle is reduced. As will be pointed out in discussing sharpening, it is crucial that no portion of the cutter be deeper than the cutting edge itself.

In 90°–0° cutting with a hand chisel, the back of the tool itself guides the cutting direction. The clearance angle is effectively zero, and springback is automatically compensated by the angle at which the tool is held. Frictional resistance is overcome by whatever force is applied to advance the chisel.

Most hand planes used along the grain involve 90°–0° cutting, whether they operate on the whole surface of the workpiece (as in planing a flat surface) or whether they plow a groove or form a rabbet. Spokeshaving down a canoe paddle is another example of 90°–0° cutting where tool geometry and depth of cut are fixed by tool design and adjustment. Rough-shaving an ax handle with a drawknife and taking long shavings along the grain with a pocketknife are also 90°–0° cutting; in these cases the cutting angle, clearance angle and depth of cut are controlled by the way the tool is held.

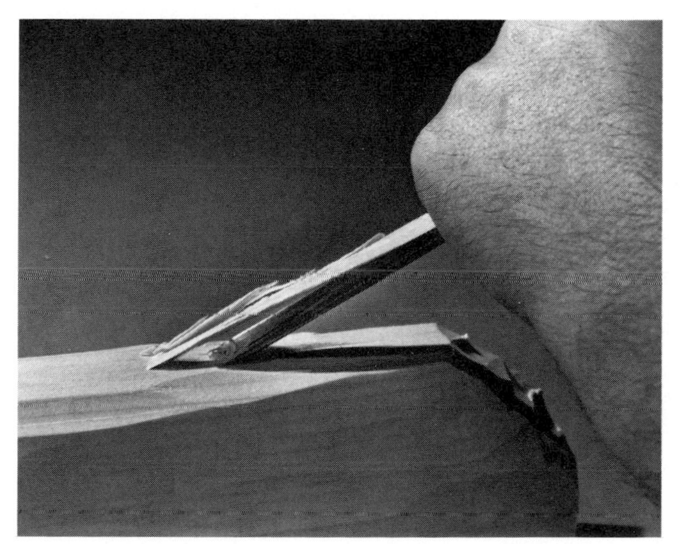

2—In woodcarving, the splinter-type chip resulting from 90°–0° cutting with the grain can slide up the tool face and jab an unwary carver in the hand.

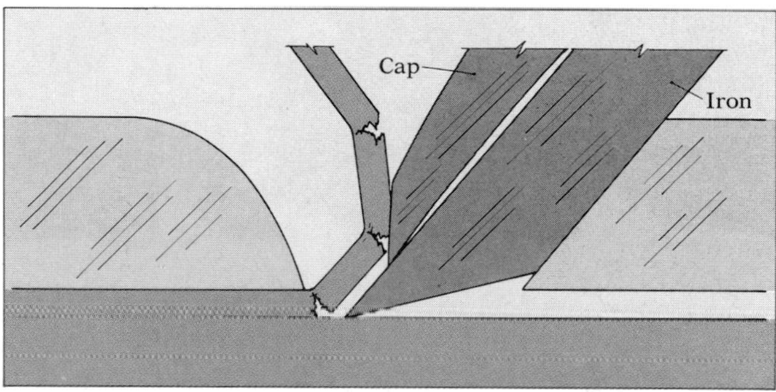

3—In 90°–0° cutting with a hand plane, the cap iron minimizes torn grain by breaking the chip near the cutting edge.

Peripheral milling (machine planing)

Orthogonal cutting in the 90°–0° mode has a counterpart in peripheral milling in the cases of the typical jointer, single surfacer, spindle shaper and router—wherever a revolving cutterhead operates along an edge or face of a board.

The cutting action is modified by the path of each cutting edge, which by combined revolution of the cutterhead along the surface of the workpiece follows a trochoidal path *(1)*. Each cutting edge takes a curved chip from the workpiece. Customarily, rotation of the cutterhead moves each knife in a direction opposite the relative direction of the workpiece, representing the **up-milling** condition *(1, inset)*. In most rotary cutterhead designs the cutting angle is decreased to between 10° and 30°. This requires more power, but the chip type produced approaches a scraping Type II or Type III chip rather than a splitting action, as in Type I chips, and there is less uncontrolled splitting ahead of the knife edge. The surface generated by the overlapping cutting arcs of successive edges is wavelike. These waves are often visible and are known as **knife marks.** Figure *2* shows an extreme case of knife marks in crudely planed structural lumber—only four to the inch. The marring of the surface is plainly visible. However, in finish lumber the best surfaces are produced by 12 to 25 knife marks per inch. In this case, the height of the waves is typically quite small, and may not be seen easily with the naked eye. Their visibility is the result of crushed or buckled cells, rather than the actual surface irregularity of the waves *(3)*. When the number of knife marks per inch exceeds 30, unless the cutting edges are *extremely* sharp, the surface may actually get worse. The chip gets so small that each cutting edge does not bite, but rather rides over the surface, as with the table knife and the wet sponge. Frictional heat also may be produced and the resulting surface, although apparently smooth,

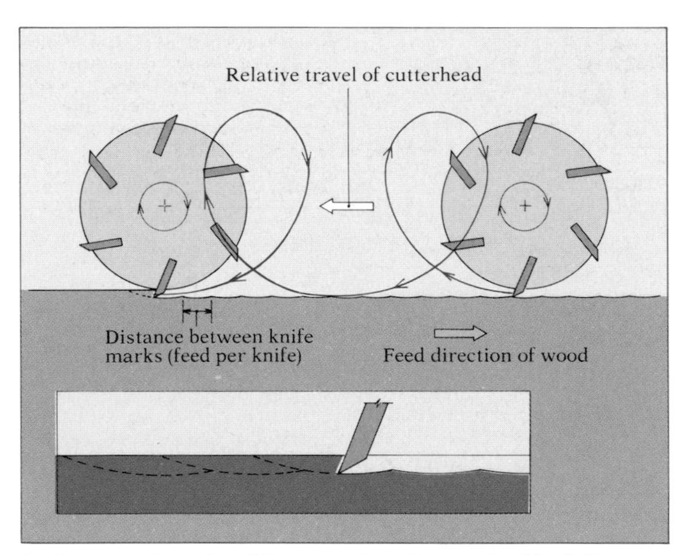

1—In peripheral milling, each cutter actually follows a trochoidal path relative to the workpiece, the result of cutterhead rotation plus feed. Each cutter takes a curved chip from the workpiece, usually by up-milling (inset).

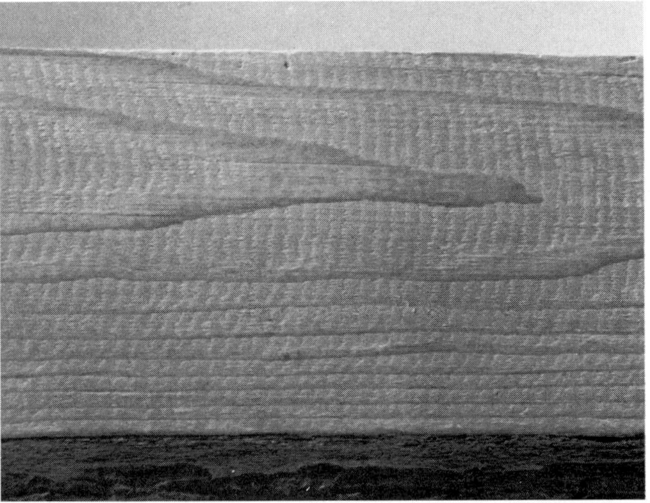

2—Crushed cells delineate knife marks—about four to the inch—on this eastern hemlock board.

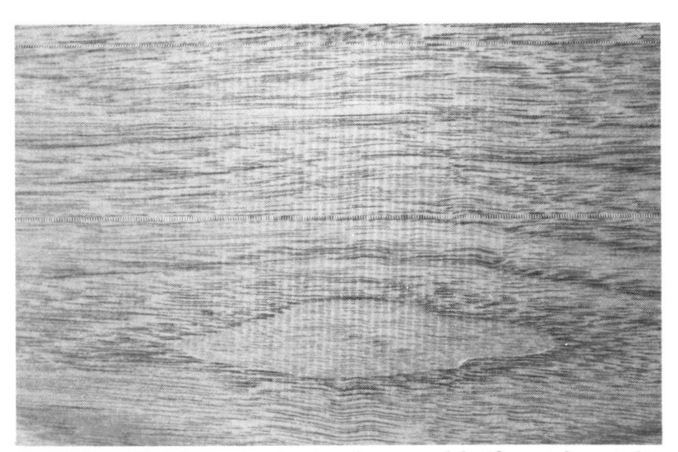

3—Light reflection reveals closely spaced knife marks on this butternut board.

simply may be glazed by the crushed cell structure and chemically altered by heating. Extremely slow feeds actually may scorch the surface. If a board sticks or pauses in a thickness planer, a glazed or scorched band often is produced across the board. Such glazed surfaces on thickness-planed boards are not acceptable for gluing.

Another problem with chip formation in peripheral milling is related to the practice of jointing the knives. Were it possible to set every knife perfectly in the cutterhead, jointing would not be necessary. Jointing is done by passing a stone along the revolving cutterhead so that it at least touches every knife. Any high spots are therefore ground back, ensuring that all the edges lie on the same cutting circle and that each knife will take a chip of equal depth. The jointing produces a flat area or land behind the knife edge, which has a 0° clearance *(4)*. Jointed knives have both advantages and disadvantages. A very narrow land less than 0.01 in. will not create problems because the area of zero clearance is very small; at the same time, the sharpness angle will be increased, thus making the cutting edge more durable. Every woodworker quickly discovers that the cutterhead of a planer can be resharpened by a little extra jointing, which effectively restores to sharpness a slightly rounded or dulled edge. But of course, the width of the land is thereby increased. As the cutting circle moves forward along the workpiece, a jointed back surface actually becomes a negative clearance angle, and as the land becomes excessively wide, say 0.03 in. to 0.04 in., the wood surface is pounded and heated. Dull or rounded cutting edges also produce a negative clearance angle that compresses the wood surface *(5)*.

Pounding caused by dull knives and excessive jointing can produce serious defects in the wood. **Raised grain** results when wood is unevenly compressed during cut-

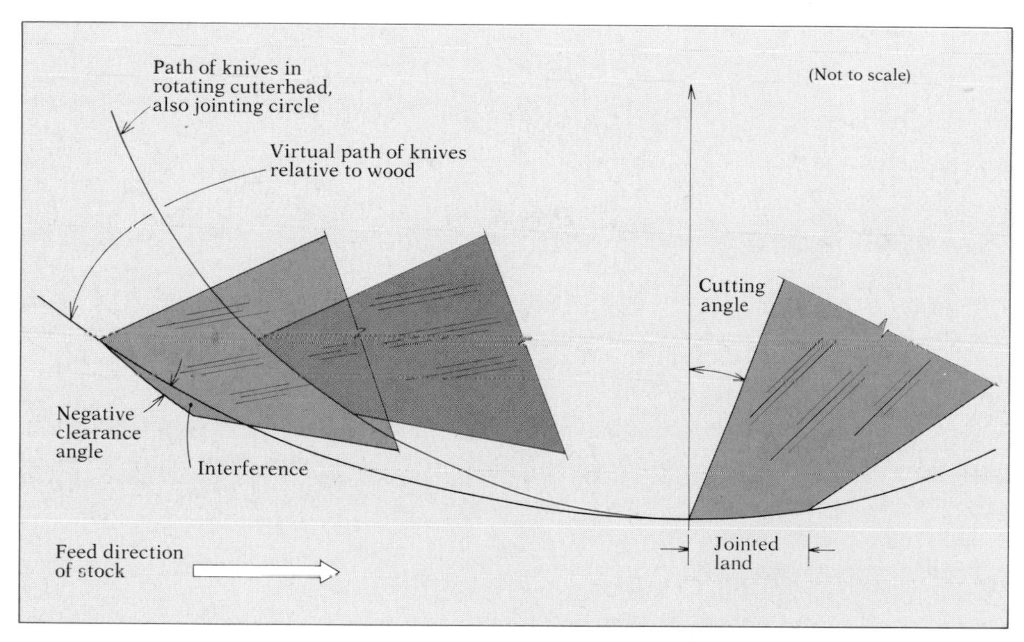

4—The jointed land that forms the back surface of the cutting edge is ground to a 0° clearance angle, but the relative feed of the work produces a negative clearance angle. The land should therefore be as narrow as possible to minimize the interference that causes frictional heating.

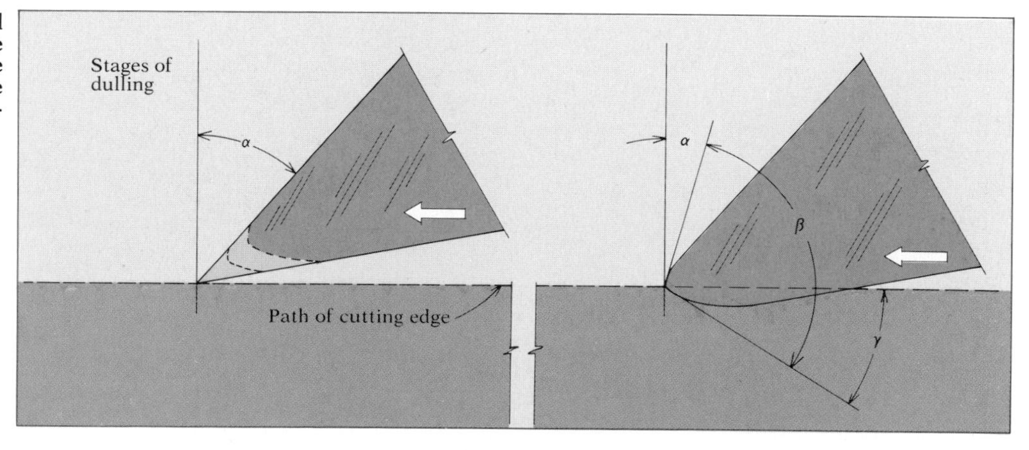

5—A sharp edge becomes dull with breakage and wear. The profile becomes blunt, and the leading edge is no longer the lowest point, creating a negative clearance angle (γ).

ting, with resulting uneven springback. The extreme situation occurs on the pith side of flatsawn boards of uneven-grained conifers *(1)*. The acutely angled layers of latewood are driven down into the weak supporting layers of earlywood as they are cut and pass under the knife. Then they spring back and rise up above the machined surface. In extreme cases, the cell structure of the supporting earlywood is so badly damaged by compression that upon springback, the layers of latewood actually separate. This is referred to as **loosened grain** or **shelled grain**. When wood is at moisture contents above 12% to 15%, raised grain may also occur, more as a result of soft earlywood than of dull knives.

Raised or loosened grain usually is apparent immediately. In some cases, however, additional strain recovery or separation takes place at a slower rate. Even boards that are sanded smooth may continue to recover for years afterward. This effect is doubtless amplified by variations in moisture content. The "grain" will often "show through" a painted board years later. The worst effects of raised grain in softwoods can be avoided by using a board so that the bark side is visible and the pith side is concealed. (This memory crutch may help: B stands for better and bark side, P stands for poor and pith side.)

Another defect that commonly results from faulty surface planing is **chip marks** *(2)*. This problem is machine-related, not due to inherent flaws in the wood. It occurs when chips are not being cleared from the cutterhead because of insufficient air flow or static electricity. Instead, the chips are caught by the knives and then dragged through the region of chip formation, where they cause compression on the surface that has already been produced.

Woolly or **fuzzy grain** *(3)* may result from planing material with a high moisture content, especially at low rake angles. Tension wood in hardwoods also tends to produce woolly surfaces.

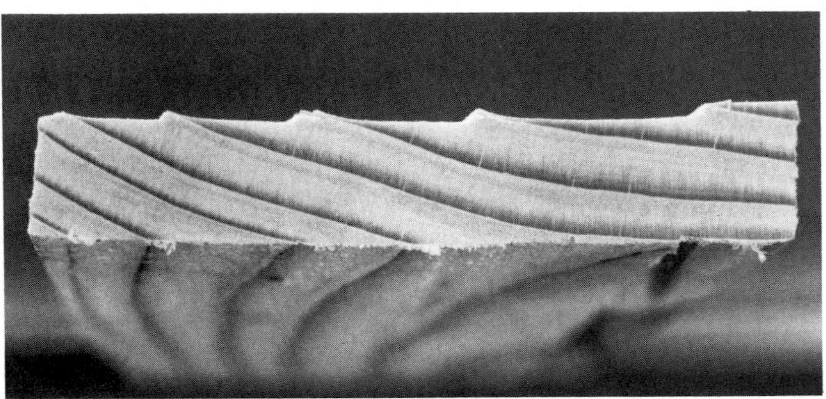

1—Raised grain (left) occurs during peripheral milling when layers of harder latewood are driven down into soft earlywood, then spring back unevenly. It's usually worst on the pith side of a flatsawn board (above), and in extreme cases the earlywood layers may actually fracture, resulting in loosened grain.

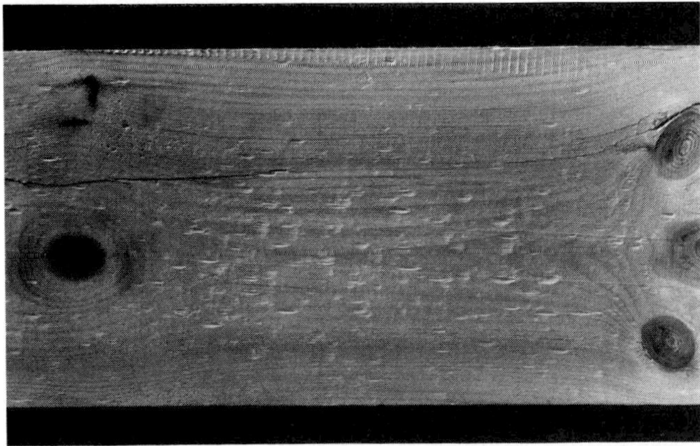

2—Chip marks occur when a planer's exhaust system isn't able to clear debris from the cutterhead. The knives catch the chips and drag them across the wood.

3—Woolly grain is liable to result when planing green material, or in machining tension wood in hardwoods.

90°–90° cutting (planing end grain)

Planing across the end-grain surface of a board is a common example of 90°–90° cutting. The cutting edge must sever the chip by cutting longitudinal cell structure across the grain *(4)*. The chip is displaced as much by shear deformation and failure as it is by bending, and it often moves up the face of the cutting edge as a partially connected string of rectangular chip segments.

Since the cutting tool must sever fibers across the grain, a dull edge (or a low rake angle) will drastically deform the wood in compression perpendicular to the grain during cutting. This may result in severely bent-over fiber ends and even splits down into the surface of the resulting cut. For this reason, high rake angles, low sharpness angles and sharp cutting edges are essential to minimize damage on end-grain surfaces.

Ripsawing with a band saw or frame saw is a special case of 90°–90° cutting. Unlike planing across the end of the board, when the blade is usually wider than the cut, ripsawing incorporates a cutting-edge width, or **saw kerf**, that is narrower than the workpiece. Since the objective of sawing is often simply to cut one piece into two separate parts, we may tend to think of it as similar to slicing a loaf of bread. However, we should keep in mind that in sawing we are working primarily at the surface of the bottom of the kerf. Therefore, in addition to forming the chip, we must sever it along its lateral faces in order to free it from the bottom of the kerf. But since the shear strength parallel to the grain is so low compared to the stress developed by the rake angle of the cutter face, the chip is easily sheared free along its sides. To avoid frictional contact of the saw against the sides of the cut, its teeth must have **set**—that is, the saw must be wider across its cutting edges than the sawblade thickness *(5)*. This is usually accomplished either by **swage-setting** or **spring-setting**. If ripsaw teeth have the set developed by swaging or spring-setting at the very tip, the sawn surface they produce is characterized by distinct tooth lines that stand out against the somewhat roughened plane produced when the chips were sheared free from the side walls of the kerf. Slightly jointing the sawteeth (back as far along the sides of the teeth as the depth of the chip) will clean up the sheared wood and will leave a smoother surface. However, excessive side jointing is counter-productive to set and will result in frictional heating. Chopping out the ends of a rectangular mortise is another example of 90°–90° cutting accompanied by lateral shearing out of the severed chip. Here also the sharpness angle should be as low as it can be and still survive.

In circular ripsaws on a standard saw bench, the cutting action approaches 90°–90° when the blade is raised to its fullest height *(6)*. But when the blade is adjusted for making a shallow grooving cut, the rip teeth develop 90°–0° cutting and shear the chips laterally from the cut. The chips removed from the bottom of the kerf tend to be more stringy.

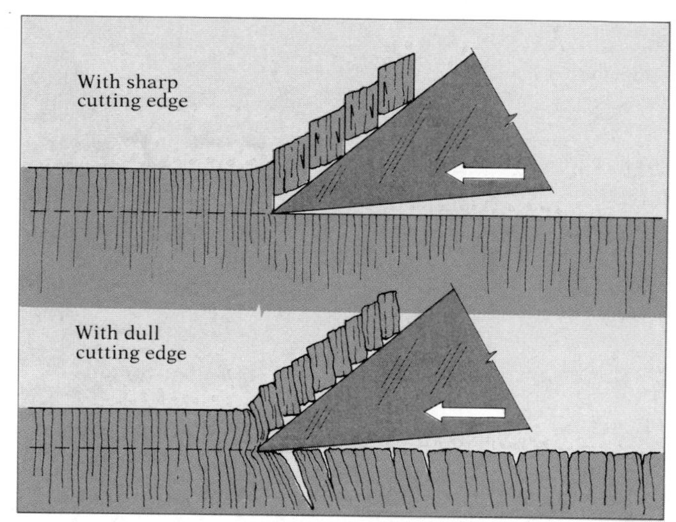

4—90°–90° cutting action: If the tool is dull or a thick chip is taken, damage may extend deep into the end-grain surface.

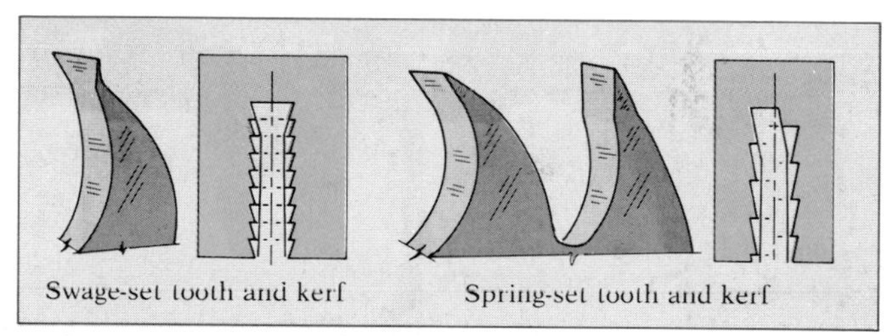

5—Setting a saw, or making the tips of the teeth slightly wider than the gauge of the saw, prevents the saw from binding in the kerf. Swage-set teeth are spread at the tip; spring-set teeth (the more common for ripsaws) are bent alternately to the sides.

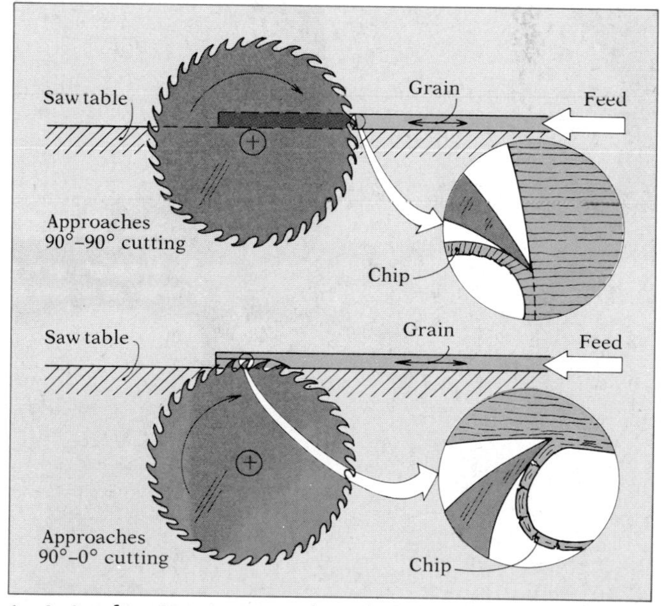

6—A circular ripsaw approaches ideal ripping action (90°–90° cutting) when the blade is raised to its maximum height above the table. It is also a more dangerous working situation. When the saw is lowered for safety, or to make a grooving cut, 90°–0° cutting is involved.

0°–90° cutting (planing across the grain)

While 0°–90° cutting is exemplified by using a hand plane across the side-grain surface of a board, the classic case is veneer cutting. In veneering, the chip itself is as important as the surface generated in the workpiece. In fact, one can produce a small strip of veneer by planing wet wood across the grain. With low-density species such as basswood, in green condition and especially if heated, a fairly respectable ribbon of veneer can be produced by taking a thin cut. However, if a thicker cut is attempted on unheated wood, one experiences a clicking sensation as the chips are cyclicly broken as they bend away from the workpiece. Examining the cutting action *(1)* we see that failures called **knife checks** are produced at regular intervals in the veneer. The face of the veneer having the knife checks is called the **open** or **loose side.** The opposite side is called the **closed** or **tight side** of the veneer. Knife checks are more pronounced in veneer greater than ⅛ in. thick. As veneer is cut thinner and thinner, knife checks tend to become insignificant. In gluing veneer, it is important to determine which side has the knife checks and always to place this side into the glueline. This ensures a check-free surface that will accept finish more uniformly.

In cutting veneer, the maximum possible rake angle is desired to minimize knife checking. It is limited by the minimum sharpness angle that will hold up (about 21° in commercial veneer cutting) plus a clearance angle (0°–2°). A rake angle of about 68° usually results. Knife checking is still a serious problem in most veneer cutting, and so an auxiliary tool component, the **nosebar,** is introduced *(2)* to restrain the veneer as it is cut. By setting the nosebar to give proper pressure, knife checks can be minimized or virtually eliminated without unduly damaging the veneer by over-compression.

In woodcarving, 0°–90° shaving is often an easy way to remove material with very little force. However, the wood tissue yields and easily tears ahead of the cutting edge, both above and below the plane of cut, so some tearing out of the surface may result. And dull spots on the tool cutting edge will be signaled by scuffs along the severed surface.

Planing across the grain is often a desirable compromise on irregular grain, which will be torn by 90°–0° planing no matter which way it is oriented. Other common applications of 0°–90° machining include cleaning up the bottom of a dado with a router plane, shaving the sides of a rectangular mortise, or slotting a lathe turning with a parting tool. Bowl and spindle turning, in fact, involve extremely smooth transitions from one cutting mode to another, as the turner changes the orientation of his tool with respect to the rapidly whirling work. For this reason turners are quite likely to have a highly developed intuitive understanding of the various types of cutting, and often they are able to apply what they have learned at the lathe to other forms of machining as well.

Knife checks
Tight side
Chip (veneer)
Loose side
Nosebar

1—Veneer-slicing is an example of 0°–90° orthogonal cutting. Unbalanced pressure from knife causes checks in loose side of veneer. Counter-balancing nosebar and greatest possible rake angle minimize knife checks.

2—With nosebar retracted, left, veneer checks seriously as it curls over the knife. A little nosebar pressure, center, reduces the amount of checking. When the nosebar pressure approaches 15% to 20% of the veneer thickness, right, checks are nearly eliminated.

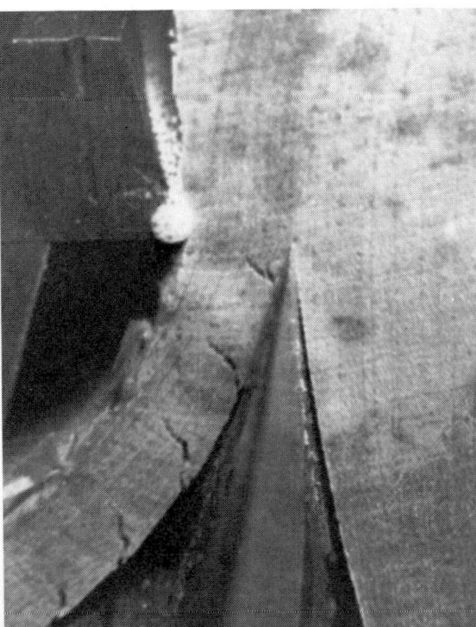

Crosscut sawing

Sawing wood across the grain using a ripsaw is seldom satisfactory. The rip tooth could easily initiate chip formation from the main piece (by 0°–90° cutting) as the result of parallel-to-grain fiber separation. However, the chip remains anchored by longitudinal cell structure, across which shear resistance is extremely high. The forming chip would therefore have to be displaced by shearing across the grain or by simply ripping the cell structure loose. It is possible to crosscut with a saw designed for ripping, but the end-grain surfaces produced will show badly mangled cell structure *(3)*. Where cleanly cut end grain is desirable, an additional cutting action is necessary to sever the chip loose at its end-grain faces. The old-fashioned two-man (or one-man) crosscut log saw is a classic solution to this problem. The crosscut design embodies two kinds of teeth *(4)*. **Scratcher teeth** in even-numbered sets (usually two or four) are filed to a spear-point profile in side view and to a slim, almost knife edge as viewed from the end of the saw. In each set, alternate scratchers are spring-set and filed to opposite sides. Their job is to sever fibers at each edge of the kerf *before* the chip is formed. This is a sort of 90°–90° cutting. Once the ends of the potential chip layer have been separated in this manner, the **raker teeth** roll the wood out of the kerf by 0°–90° cutting. The scratcher teeth must therefore be filed to project $1/64$ in. to $1/32$ in. beyond the cutting plane of the rakers, to ensure that the chips are severed before they are formed.

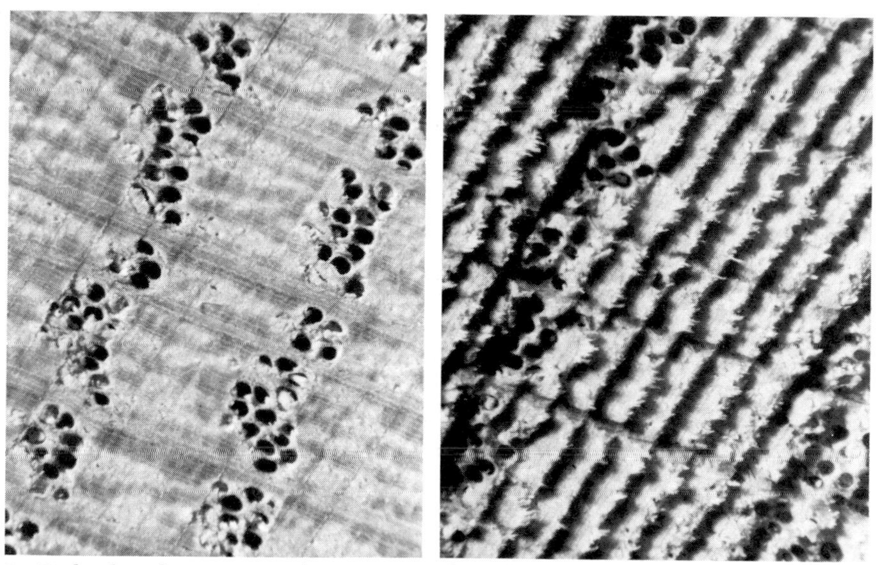

3—Red oak end grain cut with a ripsaw (right), which mangles the cell structure, and with a crosscut saw (left), which severs the fibers cleanly.

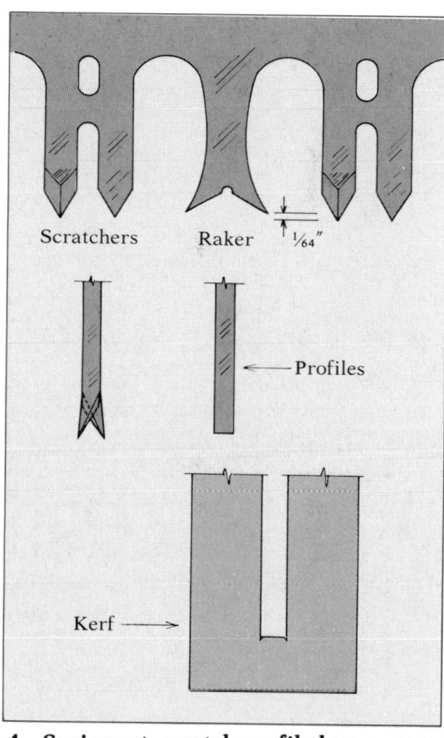

4—Spring-set scratchers filed to a spear point on a two-man crosscut saw cut the fibers free at the edges of the kerf. Then slightly shorter raker teeth come along to roll out the chip. Such a saw cuts when pulled in either direction.

The raker teeth are somewhat forked, with cutting edges facing in both directions, so the saw cuts in both directions. The chip can be compared to a very narrow strip of veneer, although it is quite shattered by severe knife checking because of the rather low rake angle of the rakers. The deep gullets to each side of the tall raker teeth are necessary to roll up and hold the long, stringy chips being "raked" from the kerf. The piles of "caterpillars" that emerge from the cut are the hallmark of a well-fitted crosscut saw. To me, there is magic in the way a well-filed crosscut saw almost effortlessly whisks wood out of a green hardwood log.

Modern circular crosscut saws, which cut in only one direction, are a modification of this design. The periphery of the blade has sets of teeth interrupted by large gullets. Behind each large gullet, the first tooth of a section is a raker tooth filed straight across, followed by an even number (usually four) of "cutting" teeth, with alternately filed points set to one side or the other. Rakers are filed 1/64 in. below the outer cutting circle.

In crosscut handsaws, the rakers are commonly eliminated *(1)*. The spring-set pointed teeth shear the sides of the kerf, and fragments break loose into particles of sawdust. Crosscutting circular saws also are made in similar styles.

Combination saws are designed to cut as both rip and crosscut saws. One style resembles the alternately pointed crosscut except that the teeth are wider and their cutting edges are closer to straight across, but they are not perfectly straight across as in rip teeth. Another type of circular combination saw is similar to the crosscut, with sets of raker and cutting teeth *(2)*. The cutting

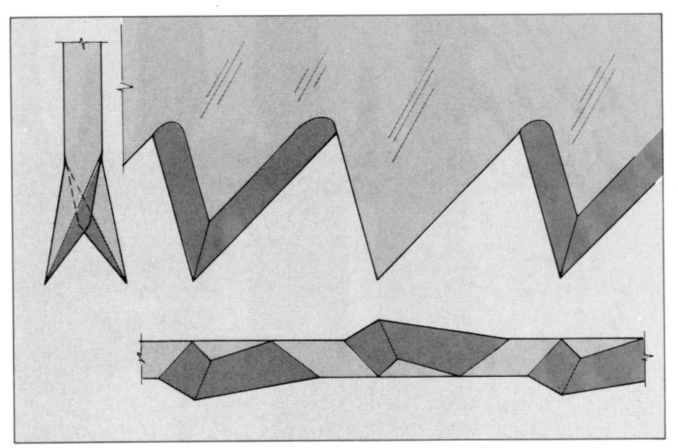

1—The teeth of a crosscut handsaw are usually fine enough to eliminate the need for rakers. The spear-point teeth are spring-set alternately left and right.

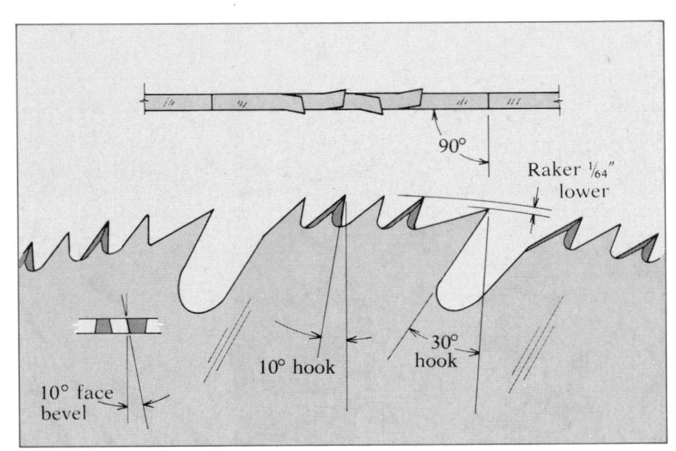

2—Circular combination saws can be used either as rip or crosscut saws. The style shown, similar to a crosscut saw, has four cutting teeth per raker.

teeth are not as pointed and there are often only two cutter-type teeth per raker.

Boring bits are another example of rather complicated combinations of cutting action. The typical wood-boring bit *(3)*, is of the double-spur, double-twist type. For hand-augering it usually has a screw-threaded point to control the feed into the wood. Bits for machine-boring usually have only a brad point or no point at all. The spur does the initial cutting, forming the side surfaces of the hole. In the typical hole bored across the grain, during one revolution the spur cycles through two complete transitions of 90°–0° to 90°–90° and back to 90°–0° *(4)*. Because of the variable resistance to cutting—across, with, along and against the grain—the hole may be diagonally oval, not round. The spur has a fairly large rake angle and slim sharpness angle with little if any clearance angle. Once the hole is generated by the spurs, the lips remove stock from the bottom of the hole; in so doing they vary between 90°–0° cutting and 0°–90° cutting. The lips are typically filed to a rake angle of 40° with a clearance angle of 10° to 15°.

In boring parallel to the grain, the spurs are cutting in the 0°–90° mode, the lips 90°–90°.

The standard twist drill *(5)*, commonly used for boring metals, is sharpened by grinding the clearance angle on the back face of the cutting edge on the lips. The bit has no spurs or point; the surface of the hole is generated by the cutting action of the outer corner of the lip. Although the quality of hole produced is inferior to that which can be drilled with a standard auger bit, the simplicity of this bit makes it well suited to small-diameter holes, as for screws and dowels.

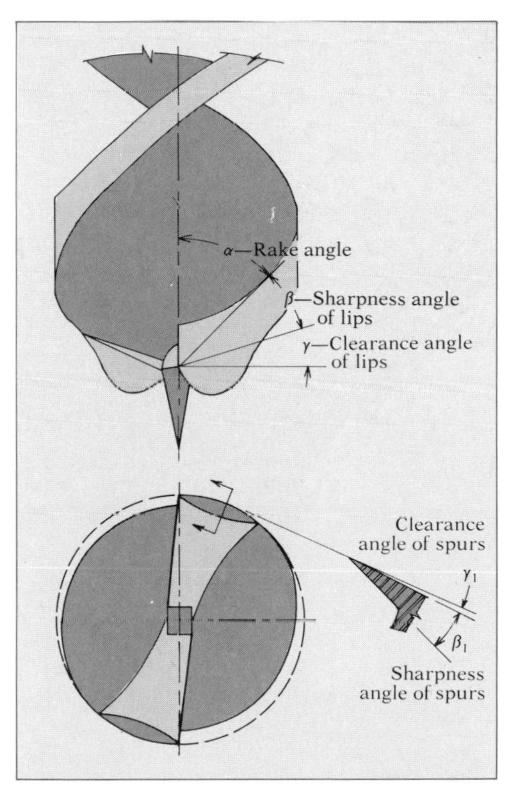

3—The standard wood-boring bit has double spurs and double twist. The spurs, which have a fairly large rake angle and a slim sharpness angle, form the sides of the hole. The lips, with a rake angle around 40° and a clearance angle of 10° to 15°, follow behind the spurs to clear chips from the bottom of the hole.

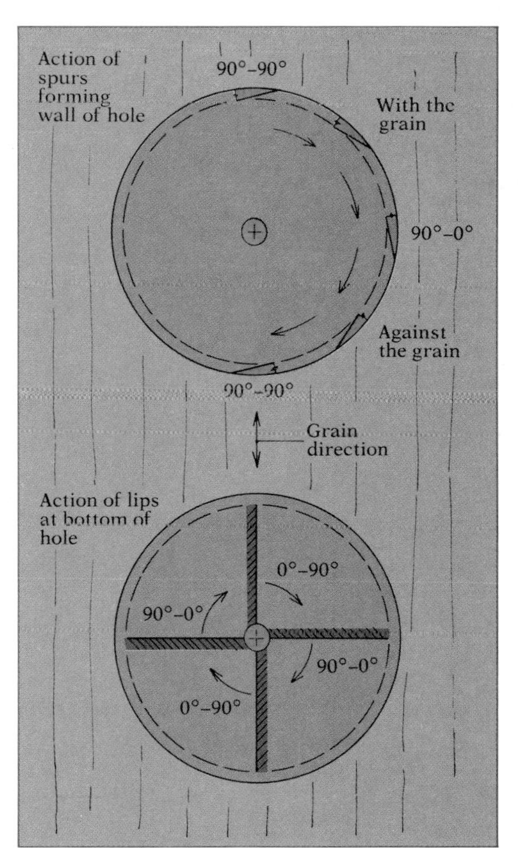

4—In boring a hole perpendicular to the grain, the spurs alternate through a sequence of 90°–90° and 90°–0° cutting. The quality of these cuts determines the quality of the hole. At the bottom of the hole, the lips cycle between 0°–90° and 90°–0° cutting action as they clear the chips.

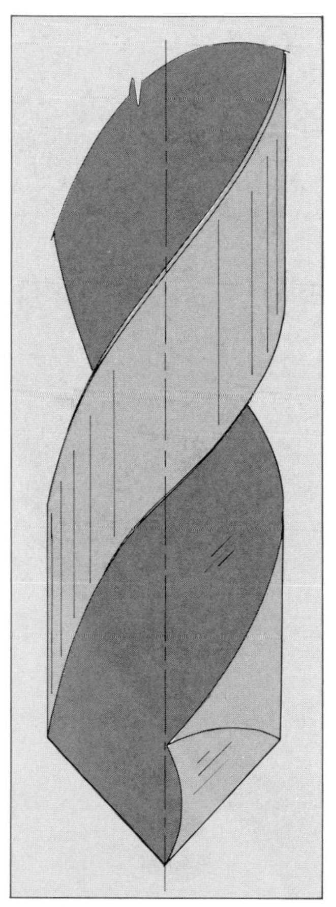

5—The metalworker's twist drill is commonly used for small holes. It has neither spurs nor point; the surface of the hole is made by the outer corners of the lips.

Combination and compromise

In summary *(1)*, some forms of machining such as veneer cutting, hand-planing and ripsawing are virtually pure cases of orthogonal cutting. Some tools, such as the two-man crosscut log saw, involve a combination of cutting modes. But many involve cutting action that is a compromise between two types (for example, planing the bevel across the end of a raised panel) or which varies between two types (as in ripsawing with a circular blade raised to different heights, or the spurs of an auger through each half-revolution).

Some aspects of woodworking involve all conceivable interactions and combinations. Woodturning and woodcarving are good examples where every imaginable combination of wood structure and cutting direction seems eventually to present itself. I realize this every time I scoop out a wooden bowl or hollow out a decoy body. By experience the woodcarver develops an intuition for the best cutting action or compromise to use in a given situation. By intuition, the workpiece is turned this way and that from cut to cut, to favor 90°–0° cutting with the grain and to avoid the costly splits of carving against the grain. There would be no value in attempting to analyze every cut in terms of orthogonal cutting or tool geometry. But the better the various types of cutting action are understood, the more successfully intuition will be developed. And oftentimes, when a chronic machining problem such as torn side grain or scuffed end grain seems insoluble, careful analysis of the cutting action will suggest whether the problem is one of depth of cut, cutting angle or tool sharpness.

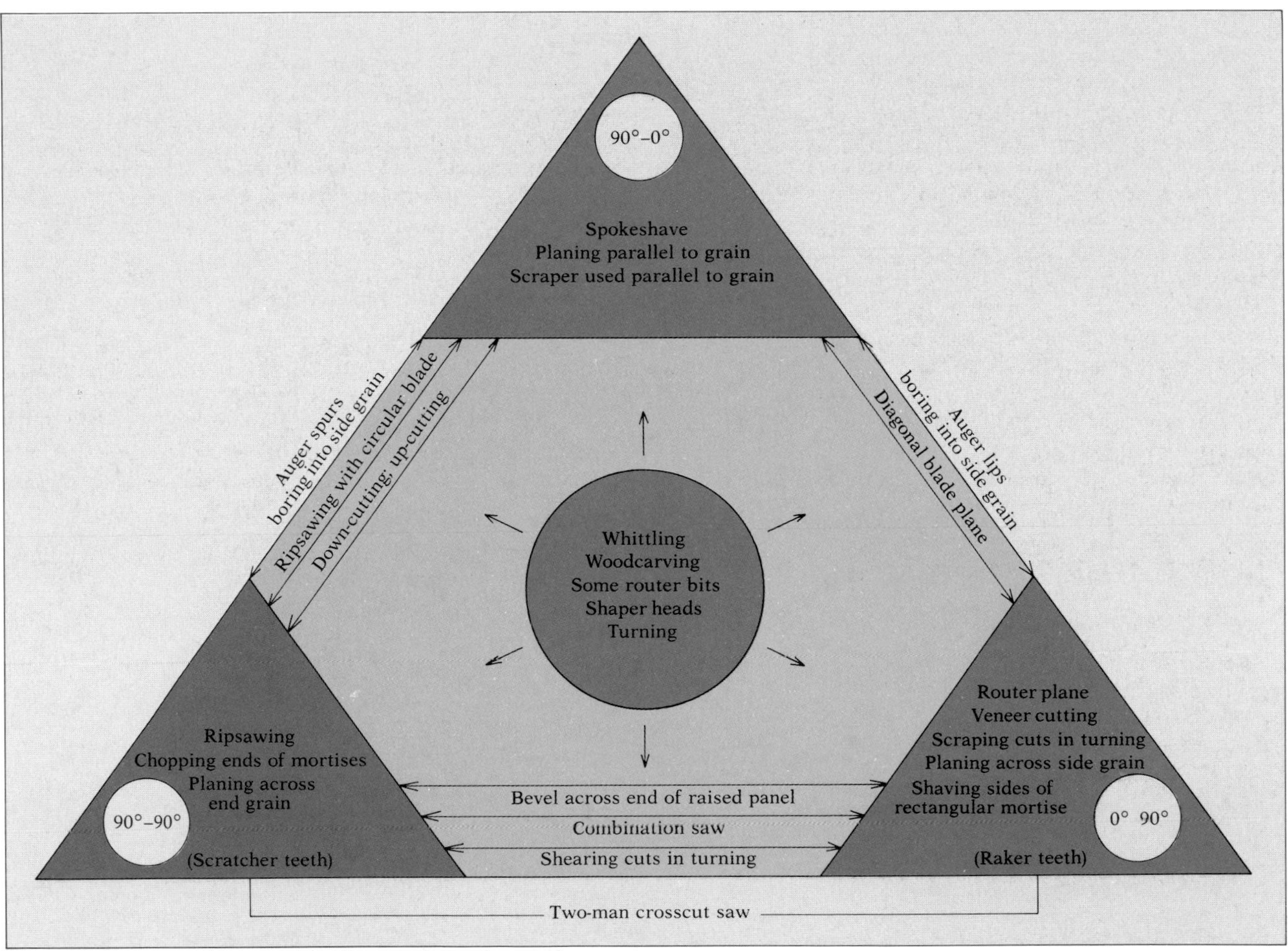

1—Summary of machining classification by orthogonal cutting type.

Sharpening

To me, the term **sharpness** includes two things. The first is the condition of the cutting edge, that is, how well the face and back surfaces combine to form a line of intersection. The closer the face and back approach true planes, the more closely the cutting edge approaches a straight, not irregular, line. Of course, no matter how well the edge is formed, if magnified enough, it will look like a mountain range *(2)*. The second aspect of sharpness is correct tool geometry. A sharp edge is of little use if it is not associated with suitable cutting, sharpness and clearance angles.

A common example is the ordinary pocketknife *(3)*. It is usually sold with polished blades, with narrow factory-ground faces forming a cutting edge that includes a large sharpness angle. The cutting edge may be well formed, and it may readily cut your finger. But the tool geometry is all wrong for cutting wood. Try a typical paring cut and the knife won't work at all. It just slides along the wood surface. Held in the usual position, the negative clearance angle prevents cutting. If the blade is rotated enough to eliminate the negative clearance, the rake angle becomes so small that the knife cuts poorly, with too much force required. The sharpness angle is simply too great. Its geometry must be corrected so that the cutting edge contacts the wood (that is, the clearance angle is zero or greater) when the knife is held in a comfortable position, and the sharpness angle must be reduced. For utility cutting toward and away from you, lengthening both faces is one compromise. For whittling with a jackknife in the traditional draw-grip position, I like to keep the side of the knife that contacts the wood flat, and do all my sharpening on a bevel on the other side.

One of the greatest difficulties in sharpening is keeping flat faces. Since two problems are worse than one, I always try to keep one face of an edge flat, and sharpen the other to the correct geometry. This way I have only one side to work on and one side to correct if I get it

2—Under sufficient magnification, even a razor-sharp plane iron has a jagged edge.

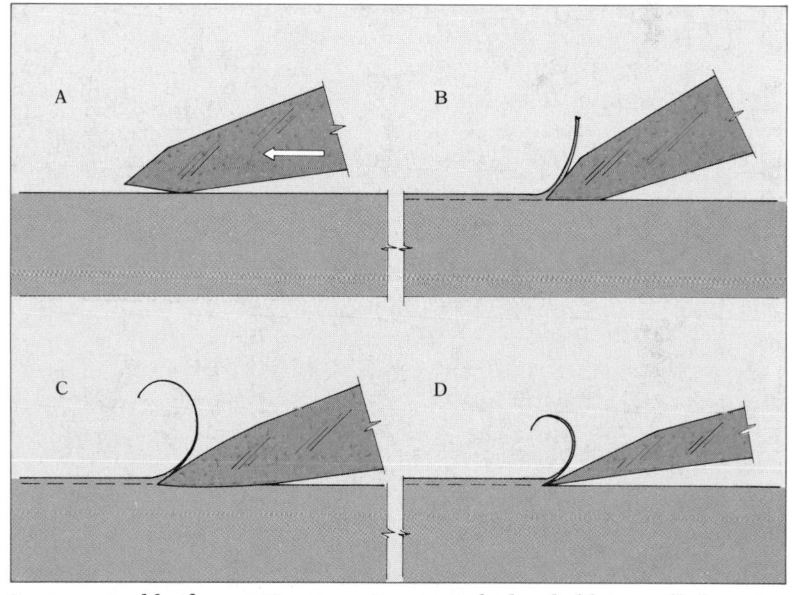

3—A new jackknife sometimes won't cut wood when held normally because its large sharpness angle results in a negative clearance angle (A and in the photo below). When rotated to eliminate the negative clearance angle (B), the cutting angle is so small that chip formation is difficult. Both faces of the blade must be lengthened to reduce the sharpness angle (C), or all the sharpening must be done on one face, keeping the other flat (D).

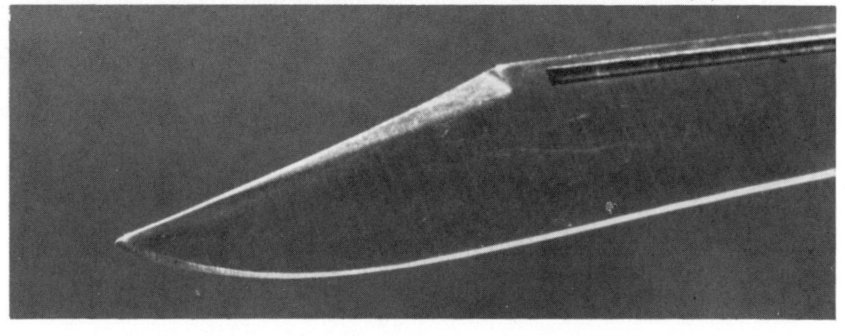

fouled up. (Some tools, such as a standard ax, for example, must be sharpened on both faces.)

Woodcarving chisels usually form the cutting edge with a single beveled surface, as is the case on the outside bevel of a typical gouge. The bevel thus becomes the back surface as used in carving. The chisel is actually guided by its back surface, so an effective 0° clearance angle results. It is not critical that the back bevel be absolutely flat, and in hand-sharpening a slight belly invariably develops. Common sense will suggest the degree of flatness the bevel must have as it merges with the edge to form the actual sharpness angle.

Another concept the woodworker can use to advantage is **microsharpening**, that is, increasing the sharpness angle near the very edge of the cutting knife. Most cutting geometry (a carving chisel and a jackknife are good examples) is a compromise between maximizing the rake angle for ease of cutting and maximizing the sharpness angle for a durable edge. Further, degradation of the cutting edge (dulling) begins at the very apex of the edge and works back from there. Microsharpening offers an interesting tradeoff by increasing the sharpness angle at the very edge, where it is needed for durability, without appreciably changing the rake angle in the area associated with rotation of the chip. On a woodcarving gouge, for example, a sharpness angle of 15° can be strengthened to 30° by a microbevel of 0.005 in. with almost imperceptible change in the cutting resistance *(1)*. With a standard bench plane whose iron is set at a rake angle of 45° and sharpened to about 30°, the sharpness angle can be microbeveled to about 40°, still leaving a clearance angle of 5° or so for good measure. Plane-sharpening instructions often refer to this as the "honing angle." The moderate jointing of

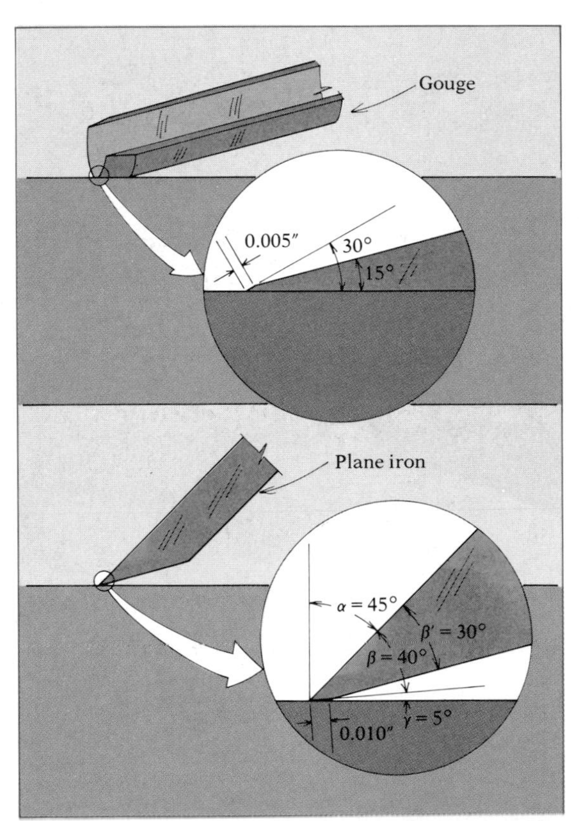

1—Microsharpening creates a larger sharpness angle near the edge (for greater durability) without significantly affecting chip formation. Shown are a microsharpened gouge and a microsharpened plane iron.

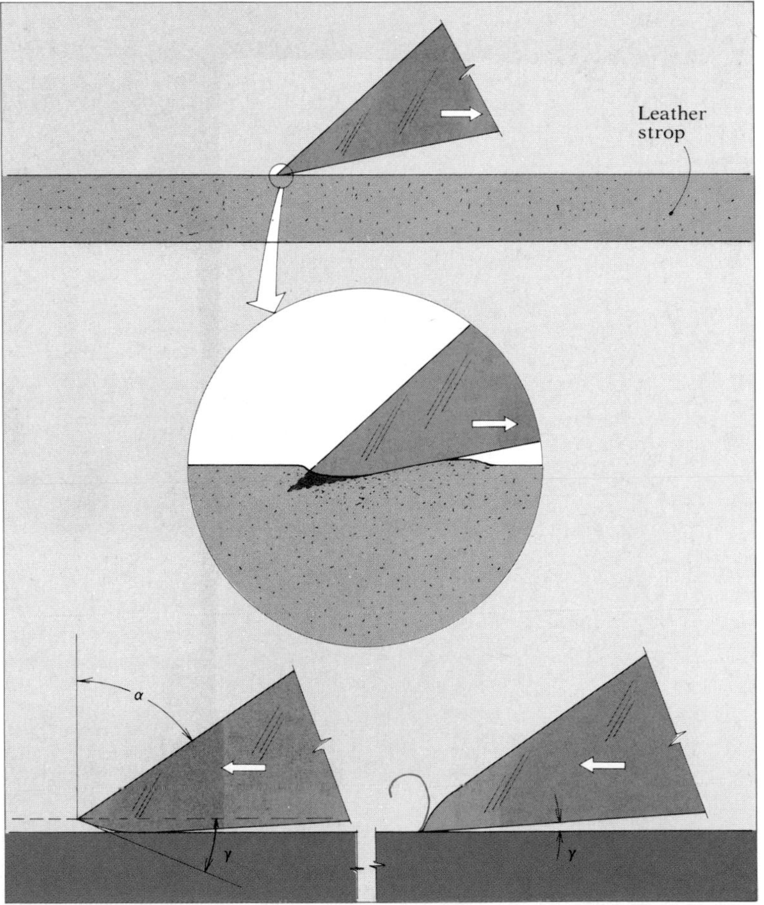

2—Stropping the cutting edge against a non-rigid surface like leather probably causes a slight rounding near the edge. Stropping on the face bevel instead of on the back ensures that the cutting edge remains the lowest point on the tool, and that the clearance angle is never less than 0°.

planer knives is also microsharpening. Once again, microsharpening must not produce a negative clearance angle; the cutting edge must always be the deepest part of the tool.

When a blade is stropped on a non-rigid surface like leather, I can't believe the faces are not altered at the very edge. I visualize something such as shown in Figure *2*. This slightly rounded spear-pointing may then in effect be equivalent to microsharpening. With this in mind, I try to bias my stropping to favor heavier removal on the face than on the back of the edge.

While it is tempting to let this discussion wander over into techniques of sharpening, that subject is beyond the scope of this book. But I can't resist adding a few comments. First, the quality of the tool face and back, and therefore of the edge formed at their intersection, is closely related to the scratch pattern left by whatever abrasive was used to do the sharpening. Needless to say, deep scratches leave larger, fragile projections at the edge, and breakage of larger projections causes faster dulling. The finest abrasives leave the shallowest scratch patterns and a more durable as well as a "sharper" edge. Most woodworkers agree that Arkansas stones produce the finest and truest possible edge surfaces.

A certain mystique has been created around the notion of **hollow grinding**. I suggest it be looked at in terms of tool geometry. Hollow grinding gives a greater clearance where the back bevel is restricted, as with a planer knife. But it also reduces the sharpness angle, sometimes too much. In the case of a planer knife, a slight jointing takes care of this. I see hollow grinding as a convenient partner to microsharpening. In sharpening carving tools, for example, I like to hollow-grind the bevel *(3)*. This establishes a basis for the sharpening and honing angles which I like to do by hand on flat stones, adding a final microbevel with a hard Arkansas pencil and stropping lightly with a piece of notebook paper. The hollow grinding narrows the area to be stone-sharpened to the edge and heel of the back. It also reduces friction in cutting *(4)*.

Experienced woodworkers agree that sharpening is important. The beginner too often is slow to acknowledge this reality. The beginning woodcarver always asks "Which tools should I buy first?" My stock answer, "A set of good stones," is seldom received with joy. Experience with dull and really sharp tools, however, quickly confirms that tools are almost worthless if they cannot produce the desired cutting action.

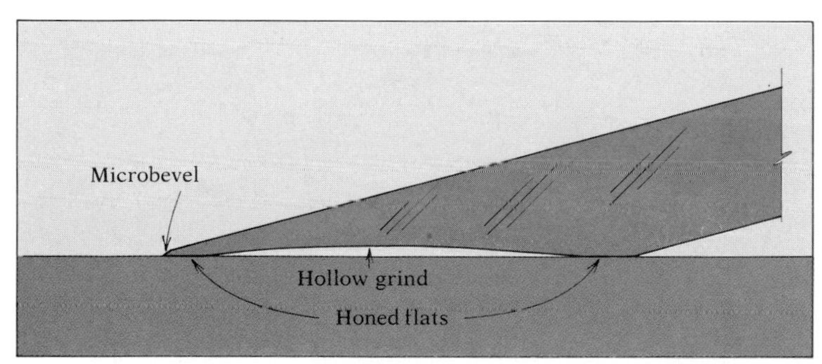

3—To sharpen a carving chisel, Hoadley first hollow-grinds the bevel against an abrasive wheel, then hones the bevel, first with a fine India stone followed by a hard Arkansas stone, to produce flats at the tip and heel of the bevel. Next the face is microbeveled, followed by a light stropping, mainly of the face microbevel.

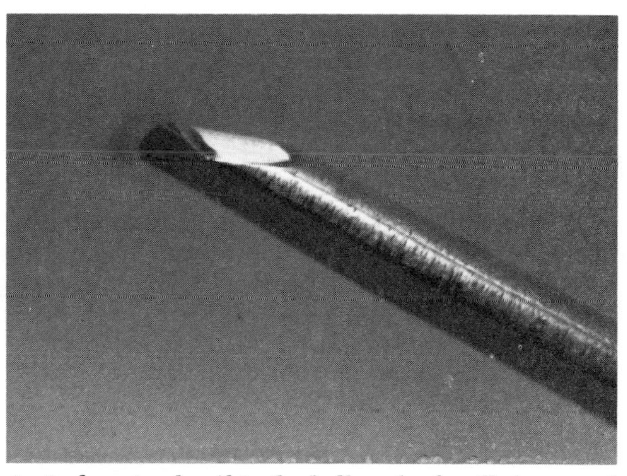

4—A gleaming bevel is the hallmark of well-sharpened carving tools.

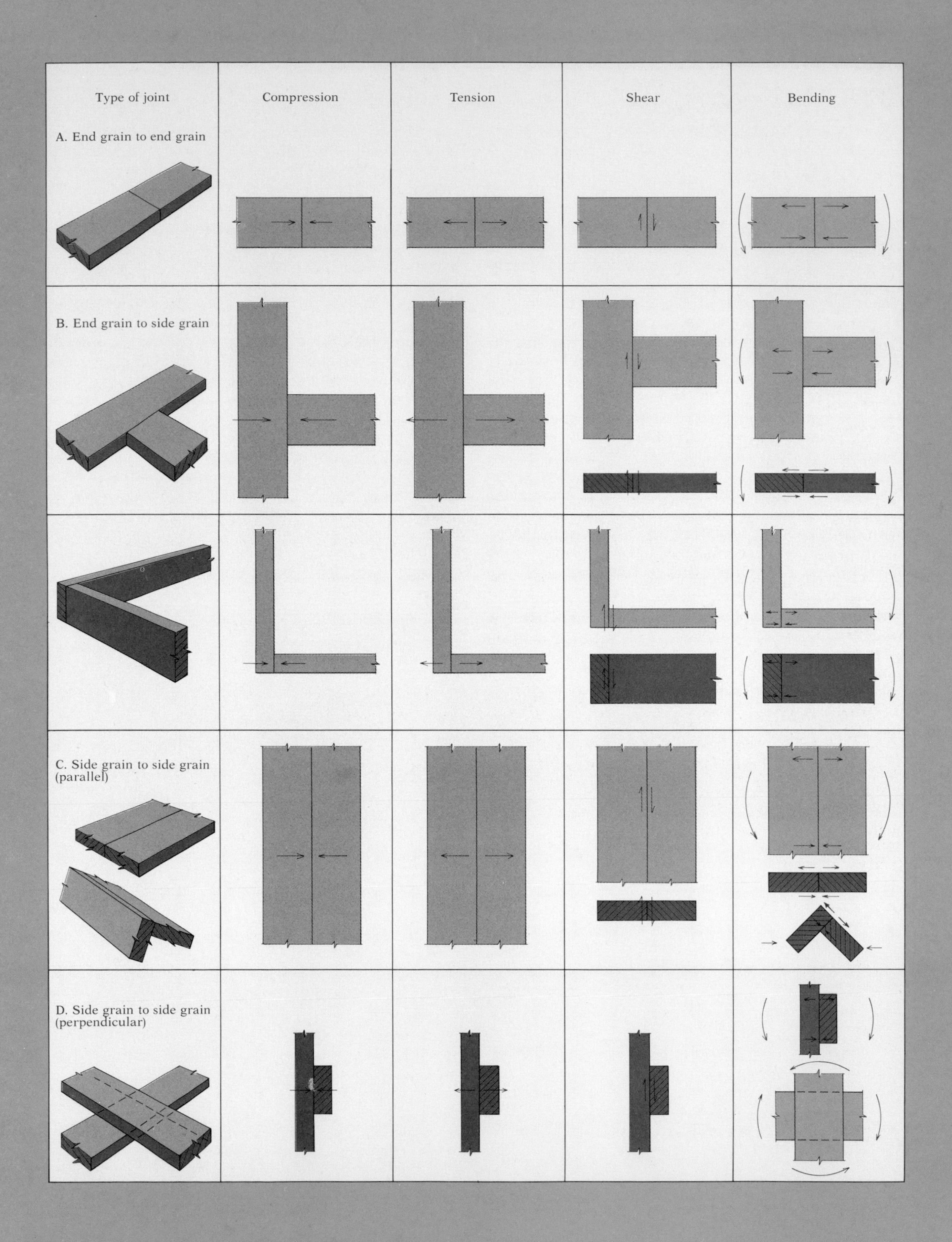
Type of joint
Compression
Tension
Shear
Bending
A. End grain to end grain
B. End grain to side grain
C. Side grain to side grain (parallel)
D. Side grain to side grain (perpendicular)

Joining Wood

CHAPTER 9

How many things are single pieces of wood? I can quickly visualize salad bowls and spoons, baseball bats, rulers, some sculpture and carvings, ice-cream sticks. . . it's not a very long list. As you compose such a list, you soon realize how many products are made of more than one piece of wood, somehow joined, held or stuck together, or of pieces of wood attached by materials other than wood. The list of products made of joined pieces is endless, and the more functionally elaborate and important the things—furniture, boats, houses, pianos—the more complex and varied are the means used to join their components.

More often than not, performance depends not so much on the physical or mechanical properties of the separate parts, but upon how successfully they are fastened together. Wooden items are rather more prone to fall apart than to break. It is the rule rather than the exception that the weakest point in any wooden construction is at a joint. The successful woodworker focuses not just on each piece of wood, but on the interrelationship of pieces of wood and especially on how they are joined together.

In our modern world, it is important to sense the historical evolution of wood use. Some functional forms, like the walking stick, need no improvement, nor are they liable to become obsolete. But new and more complex forms of wood are constantly being invented. These new materials and products owe their success mostly to some kind of marriage between wood and other materials. The evolution of products is paralleled by the evolution of systems for fastening the parts involved. New methods of fastening evolve in response to our changing resources and as a reflection of changing technology, and each advance in turn paves the way for a new array of products.

1—Some typical joints under loading stress.

The elements of joints

I shall not attempt to catalog or discuss all the joints in woodworking. Rather, I shall discuss the four critical considerations that determine the success of any given joint. One of the four may be of overriding importance in some particular situation, or any one may be interrelated with one or more of the others. These four basic considerations are the stress system involved, the grain direction of the joined parts, wood movement in response to moisture, and the surface quality of the mating parts.

The **stress system** is what the joint is being asked to do mechanically, as a consequence of its being part of a structure for use. Joints may be in compression, tension, shear or racking (bending), and usually the great difference in the strength of various joints depends on the stress situation. Most compression joints give little trouble, and shear stress is not too difficult to overcome. Joints subject to tension and racking are usually the most troublesome. Figure *1* shows joints under representative stress systems. Although it is not usually necessary to figure the loads precisely, one must have a general grasp of the direction and relative magnitude of stress in order to design and construct good joints. This analysis can come only from realistic examination of the structure in which the joint will be used, and the loads it is likely to encounter in service.

The second element is the **grain direction** in each mating surface of the joint, as related to the stresses involved. For example, the most difficult surface combination to fasten is end grain to end grain *(1A)*. It would not matter if the load were exclusively compression, but this is most uncommon without some racking stress also. As a result, timber framers who must often lengthen stock to span a space have evolved an elaborate system of scarf joints, many with mechanical interlocks, whose principal purpose is to convert mating end-grain surfaces to long-grain surfaces. End grain to side grain *(1B)* is a very common situation, which can be accomplished quite satisfactorily if all factors are considered. When stressed in compression, such a joint is usually limited by the perpendicular-to-grain compression strength of the side-grain piece. When stressed in tension, the fastening to the end grain may be difficult, and when under racking stresses, either part may be the limiting factor. The solution usually involves a mechanical interlock formed on the end-grain piece or made by adding a third piece of wood to cross the joint. Side-grain to side-grain joints *(1C,D)* can be as strong as the wood itself when they are adhesive-bonded, even when the grain directions of adjacent members are not parallel. But here the third element comes strongly into play, the dimensional properties of the wood in response to changing moisture conditions.

Dimensional change in response to moisture is usually no problem in the case of end-grain to end-grain *(1A)* or parallel side-grain to side-grain joints *(1C)* because the orientation of the growth rings can be the same in both pieces. In these same joints, if the growth rings are not similarly oriented, the difference between radial and tangential movement might cause visual difficulties, if not structural problems.

In perpendicular side-grain to side-grain joints *(1D)* and in end-grain to side-grain joints *(1B)*, the conflict between dimensional change along the grain and across the grain (especially where tangential direction opposes longitudinal direction) may become more important than the stress/strength of the original joint. *The potential self-destructiveness of such joints should always be anticipated.* A lap joint *(1D)*, for example, might be very strong when glued, but it could self-destruct as a result of dimensional change.

The last element is the **surface condition** of the mating parts, including the precision of fit and evenness of bearing, the trueness of the surfaces, and the severity and extent of damage to cell structure resulting from the surfacing process. Uneven surfaces may concentrate enough stress to overcome the strength of wood or glue, while the same joint would survive very well if it had fit properly. Poorly fitted parts may also allow unintended motion, ending in destruction. It is quite common for joints to fail not along a glueline, but in adjacent wood tissue that had been mangled by poorly sharpened tools or bad woodworking technique while the joint was being cut. In joinery, the combinations of stress, strength, dimensional change and surface quality are endless. But careful analysis of the factors involved in each joint will develop your judgment and minimize your mistakes. I'll discuss a few of the more common joints, to suggest how you might approach your particular joinery problems.

Basic types of joints

The term joint has various and sometimes overlapping meanings. In its broadest sense, it refers to any junction between two components or materials. Without reference to any accompanying means of fastening, joints can be characterized on the basis of the grain orientation of the mating surfaces as end to end, and end to side or side to side. Flat mating surfaces are loosely termed butt joints, although this term usually suggests either end-to-end butt joints *(1A)* or end-to-side butt joints *(1B)*. Side-grain to side-grain joints are more clearly designated as edge joints when the mating grain directions are parallel *(1C)* or as lap joints when the grain directions are perpendicular *(1D)*. (The special case for miters and scarf joints would be termed cross grain to cross grain.)

Obviously, such joints have no structural integrity without some means of holding or fastening the pieces together. I prefer to consider three basic types of joints or a combination thereof:

Worked joints, where the wood is physically interlocked or fitted together;

Fastened joints, where a "third party," the fastener, is attached mechanically to both components;

Glued joints, where an adhesive forms a continuous bond between two pieces by surface attachment.

Each type has its ancestor far back in history. The first worked joints might have been tree stems with forks or splits, interlocked with others, or perhaps circular branches inserted into knotholes. The first fasteners were probably thongs of hide or vines, used to lash wooden parts together. Who can say when the first crude metallic nail replaced a wooden pin to hold two pieces of wood together? Adhesives were probably discovered when residues from cooking meat accidentally stuck two pieces of wood together. Modern fastening systems are refinements of each of these, in complex and ingenious combinations.

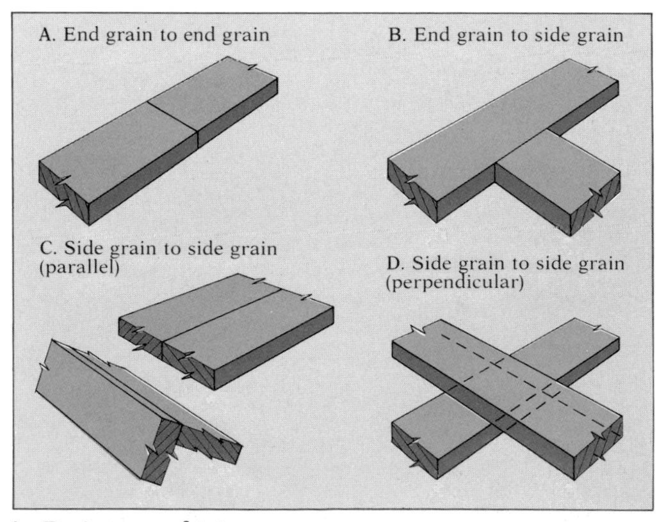

1—Basic types of joints.

Worked joints

Creating interfitting or interlocking shapes to provide strength and integrity in a joint has been a hallmark of the woodworking tradition. Although modern machines can simplify the making of fitting parts, the pride of accomplishment in hand-execution of difficult and beautiful joints will always be among the challenges, pleasures and rewards of woodworking.

The mortise and tenon—Fastening of end-grain to side-grain joints can be accomplished with a high level of success using the mortise and tenon. The basic joint is fashioned by forming the end-grain component, the tenon, into a round or rectangular cross section and inserting it into a hole, or mortise, of the same size and shape in the side-grain component. By closeness of fit alone, this joint can have positive resistance in compression, shear and racking, in which cases the strength of the wood in compression perpendicular to the grain limits movement in the joint *(2)*. The mortise and tenon is commonly associated with frame construction. In chairs, round shapes are usually used. In window frames, paneled doors and other squared frames, the rectangular form is common.

The mortise-and-tenon joint has mechanical restraint in every direction except direct withdrawal of the tenon from the hole. Although this is the way a joint usually comes apart, it most often does so only after damage due to racking. A racking load on a rectangular frame acts to deform it diagonally. Under racking loads, the tenon pivots in the mortise.

The basic "dry" joint can be improved in several ways. The most obvious is to glue it, thus adding side-grain to side-grain shear resistance along the mating mortise-and-tenon cheek surfaces to oppose the rotational effect. A second approach *(3)* adds a shoulder to the tenon, giving additional bearing surface to share the compressive resistance on the outside of the mortise.

In our earlier discussion of beam strength (Chapter 6), it is pointed out that the greater the depth of a beam, the lower the axial stress developed when the beam is loaded. The mortise and tenon (end-grain to side-grain joint) can be thought of as a cantilever-beam attachment, so increasing the height of the "beam" will reduce stress *(4)*. This also increases the surface area of the cheeks, and thus the gluing area. Lengthening the insertion depth will further increase the resisting glue-shear areas.

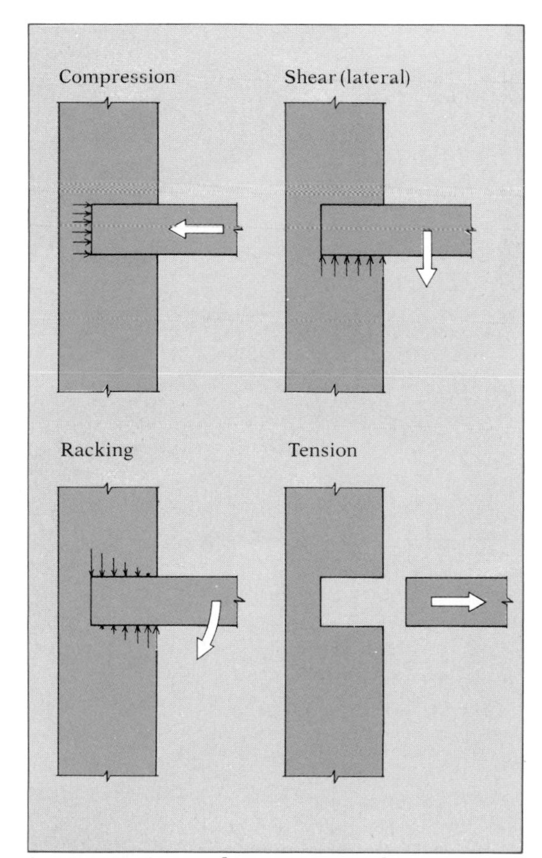

2—A mortise-and-tenon joint has positive resistance to compression, shear and racking, even without glue, but no resistance to tension.

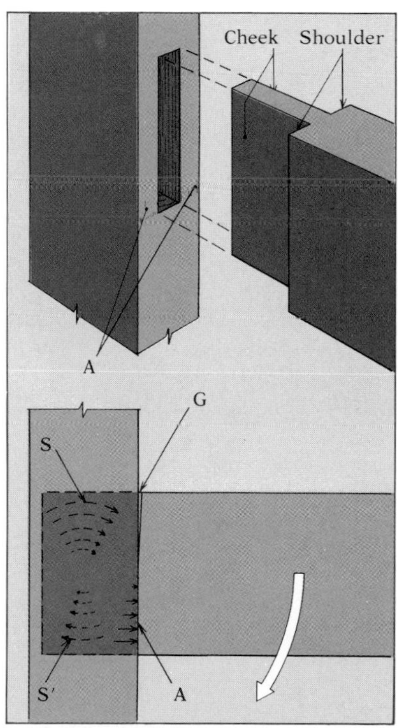

3—In a shouldered tenon subjected to racking, rotation is restricted by shear strength in the glueline bonding the cheeks of the tenon. The bearing surface of the shoulder carries compressive stress at A while a gap (G) may open so that shear in the upper area of the tenon (S) will be greater than below (S′).

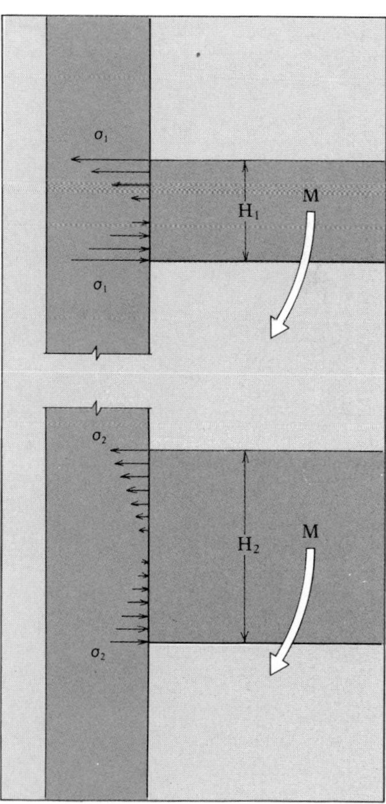

4—A mortise-and-tenon joint functions like a cantilever beam, so increasing the height of the tenon will reduce the stress on the joint, if the force (M) remains the same.

At the same time, the improvement in mechanical advantage obtained by increasing height is offset by increased dimensional conflict between longitudinal and transverse grain orientation. Some careful compromises must therefore be worked out. For example, since tangential shrinkage (and swelling) is about twice the radial movement, it is best to have the radial (rather than tangential) grain direction of the tenon matched to the long dimension of the mortise. It is also better to have the radial direction of the mortise matched to the longitudinal direction of the tenon. In Figure *1*, joint *A* would be best, since radial/longitudinal grain direction is matched along the mortise cheeks both vertically and horizontally, and tangential grain is matched perpendicularly to the plane of the mortise. Joint *D* has the worst dimensional conflict. As the height of the mortise (along the grain) is increased, joint survival increasingly depends upon moisture control.

The usual solution to dimensional conflict is to divide the joint into multiple sections *(2)*. By keeping the dimensions of each tenon within limits, dimensional conflict can be reconciled by mechanical restraint. Thus there is considerable advantage, especially in wide or thick stock, in multiple tenons or multiple splines. These considerations also emphasize the importance of well-made, well-glued joints designed for mechanical restraint of the dimensional conflict as the wood moves, as well as for initial strength.

In summary, the rectangular mortise and tenon seeks to join side-grain to side-grain gluing surfaces and to offer optimum mechanical resistance by maximizing the depth of the joint while still surviving dimensional conflict. For example, in Figure *3*, joint *B* offers three distinct advantages over joint *A*. First, the depth of the individual tenons more than offsets the loss of width. Second, the height of the glued side-grain surfaces is increased. Third, the number of side-grain to side-grain interfaces is multiplied.

The round mortise-and-tenon joint has advantages and disadvantages. By turning the tenon on a lathe and by drilling the mortise, it can be produced with a high level of precision. However, poor tool geometry or poor sharpening commonly leaves drilled hole surfaces and tenon surfaces in poor condition. The joint may therefore have weak mating surfaces. Also, the proportion of side-grain to side-grain gluing surface is somewhat

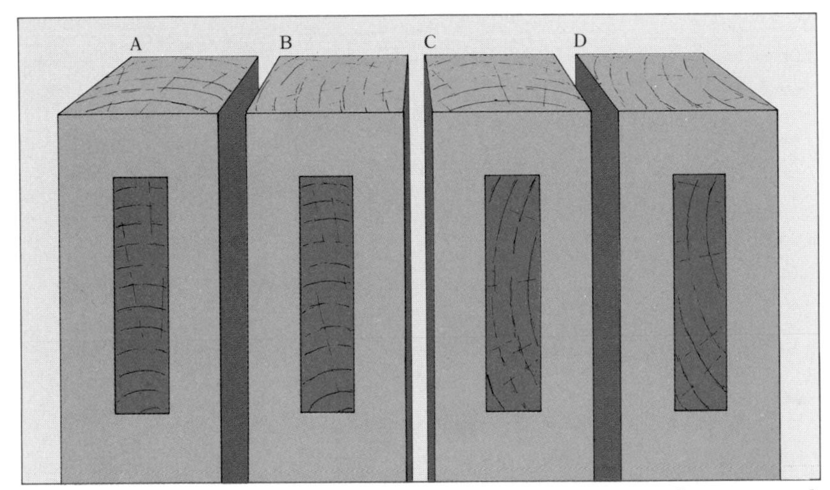

1—The best possible orientation of growth rings in a mortise and tenon is with radial/longitudinal grain direction matched along the mortise cheeks both vertically and horizontally, as in A. Joint D, the worst orientation, is apt to split.

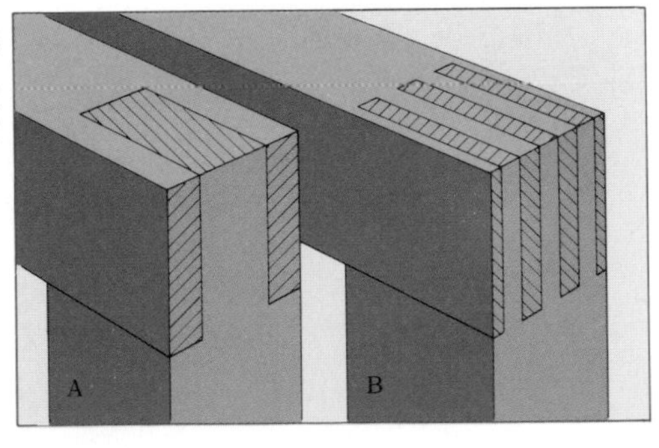

2—Although joints A and B have the same amount of wood in each component, joint B has triple the bonding surface and more balanced dimensional restraint.

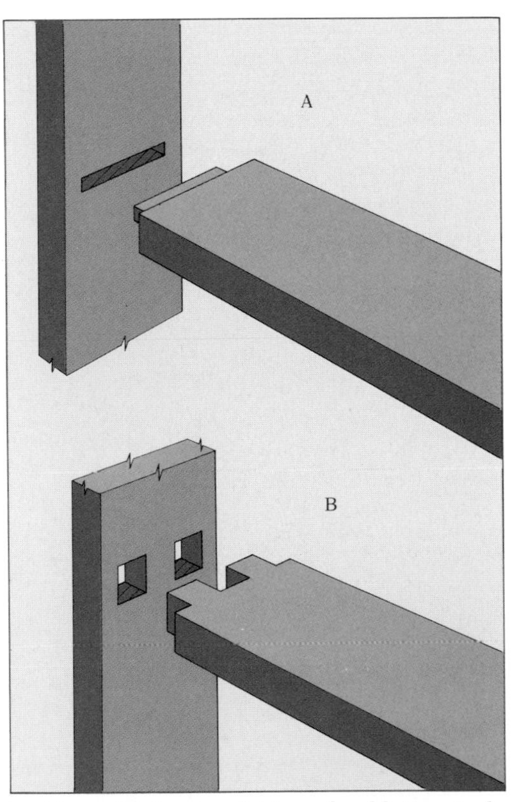

3—The multiple tenon (B) is preferable to a single wide tenon (A). Tenon depth more than offsets the loss of width, and the increase in tenon height and the number of side-grain to side-grain interfaces greatly improves the strength of the joint.

limited and cannot be improved by increasing the dowel diameter. The best side-grain to side-grain gluing area is located at the mid-depth of the dowel, where it can do the least for racking resistance. Moreover, it seems apparent that gluing can do little to improve lateral shear strength. Much of the integrity of a round mortise-and-tenon joint depends on racking strength (that is, resistance to pivoting) from surface bearing. It is therefore critical to maintain the depth-to-diameter ratio above a certain level (about 3:2) to distribute the stress as much as possible. Most joint failures seem to be associated with shallow insertions.

A further complication of the round joint, the result of both poor gluing characteristics and dimensional conflict, is the development of compression-set loosening at the top and bottom edges. This emphasizes the importance of well-machined surfaces and of matching tangential-grain to tangential-grain directions. Experiments with the moisture-cycling performance of joints indicate the high degree of success achieved by using a good glue to restrain mechanically at least some of the dimensional conflict. Some additional reduction of compression-shrinkage loosening of tenons may be afforded by pre-splitting the tenon *(4)*. As compression shrinkage develops in moisture cycling, the split can open to further relieve stress on the glueline. Perhaps this mechanism contributes to the success of wedged tenons. Although the wedge is primarily intended to produce lateral pressure to the glue surfaces and perhaps also to splay the tenon for a dovetail-style mechanical lock, it may play the more important role of providing a stress-relief slot that helps the glueline survive.

4—In this yellow birch joint, the tenon was split radially and tangentially before assembly. After moisture cycling, compression shrinkage has developed entirely in one direction, opening the radial split, while the tangential split remains tight, and the glueline has remained intact.

Dovetail joints

Nothing is more symbolic of the woodworking tradition than the dovetail joint *(5)*. It is a strong and beautiful way to execute the corner side-grain to end-grain joint and is commonly used in carcase construction. The joint consists of interlocking tails and pins, giving it strength in tension along the tail member but not along the pin. It should therefore be oriented to resist tension against the tails. In a drawer, for example, the tails should be in the drawer sides, the pins in the drawer front. In case construction, the pins should be in the sides and the tails in the top to prevent the sides from spreading. Although the joint strength results from the wedging action of the tails against the pin faces, the joint is held in place principally by gluing the side-grain to side-grain mating faces of the tails and pins. In designing the joint, the slope of the tails must be a compromise. If the angle is not great enough, the wedging-locking action will be lost. If the angle is too great, the splayed tips of the tail will be too fragile, and a component of end grain will be introduced. This impairs the side-grain integrity of the gluing surfaces. An angle of 11° to 12° has proven satisfactory. The joint strength depends largely on the success of the glue bond between the side-grain faces (the end-grain areas behind the pins and tails cannot be depended upon for any substantial contribution to the strength) and the shear strength (parallel to the grain) of the wood of the tails. Joint strength therefore increases as the number of tails increases, as long as enough wood remains across the narrow part of the tails. However, if the joints are cut by hand, the added labor should also be considered in determining the number of pins and tails per joint.

When a large number of pins and tails can be cut with precision, as is possible with machines, the glued surfaces alone can develop adequate strength. In fact, the joint can relinquish the wedging taper and have straight tails. Thus evolved the finger joint or box joint, which gains its strength from the many closely fitting side-grain to side-grain gluelines.

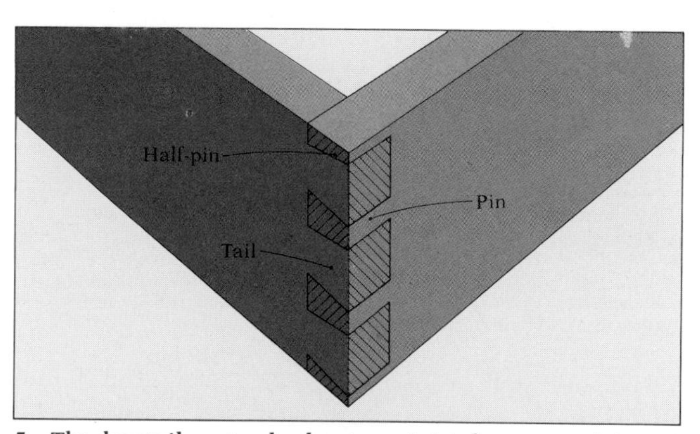

5—The dovetail, a standard carcase joint, has strength in tension along the piece with the tails.

Miter joints

Parallel side-grain to side-grain miters *(1C)* make a very efficient corner joint. Miter joints where the grain direction meets at a 90° angle are attractive but present serious technical problems. Because of the difference between dimensional change along and across the grain, the joint may open if the moisture-content change is great or if the members are wide. If nailed perpendicular to the outside face of either member, nailing into the end grain must penetrate deeply to ensure adequate holding. Gluing miter joints is of marginal effectiveness because of the large component of exposed end grain. Doweled miters, splined miters or combination lap-and-miter joints can improve strength by providing side-grain to side-grain gluing surface.

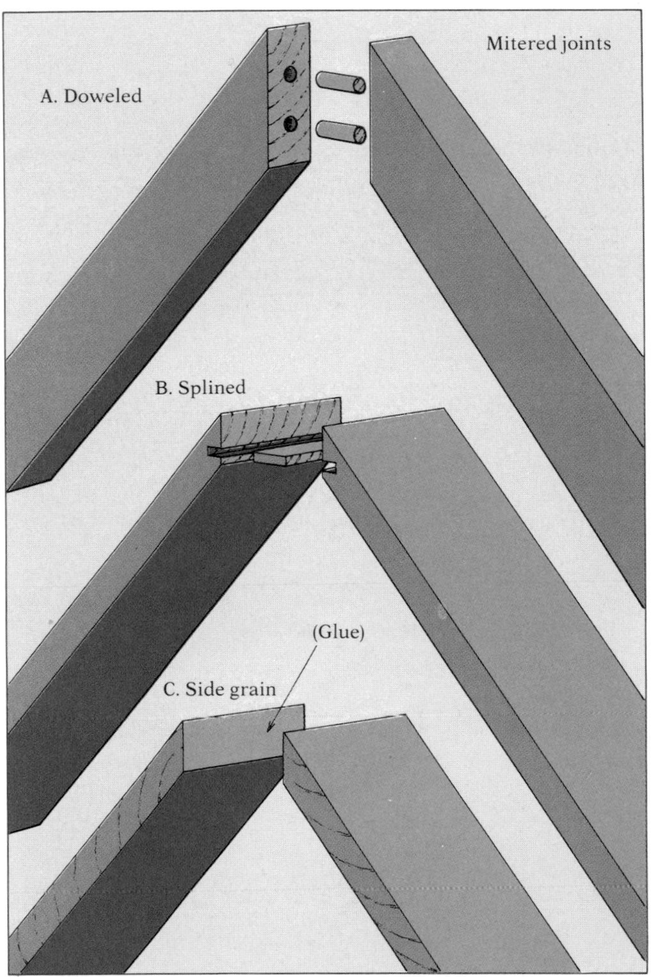

1—End-grain mitered joints can be strengthened by dowels (A) or by splines (B). A properly glued side-grain miter (C) will have adequate strength by itself.

Doweled joints

Dowels are cylindrical wooden rods used in a number of ways to fasten and strengthen joints. I think of dowels as falling into three categories: as tenons, as pins and as gluing accessories.

In cases where dowels are used to modify end-grain to side-grain joints *(2)* a dowel is inserted into a hole in the end of the perpendicular member. Since the piece has side-grain to side-grain contact, a high degree of integrity can be expected. The dowel becomes a tenon extension of the piece (multiple dowels, of course, can also be used). The mating hole in the side-grain surface of the joint becomes a mortise into which the tenon is fitted. In double-dowel joints that are subjected to racking, one of the dowels carries a critical share of the load in tension. The remaining load is transferred as surface compression. In designing such a joint, increasing the height of the member is advantageous, because it enables the dowels to be spaced as far apart as possible *(3)*. The design should incorporate dowels that are large enough in diameter to carry tensile load and deep enough to resist pullout. The dowels should also be able to carry and transfer the shear load parallel to the side-grain member of the joint.

This design, when slightly modified, becomes a doweled miter joint, as in Figure *1A*.

Dowels as pins provide physical constraint to a joint without glue (although glue may still be applied). Examples might include a pinned slip joint, a wooden hinge pin, the guide pins in a table leaf, flooring pins, etc. Historically, large wooden dowels called trunnels (tree nails) pinned framing members together in build-

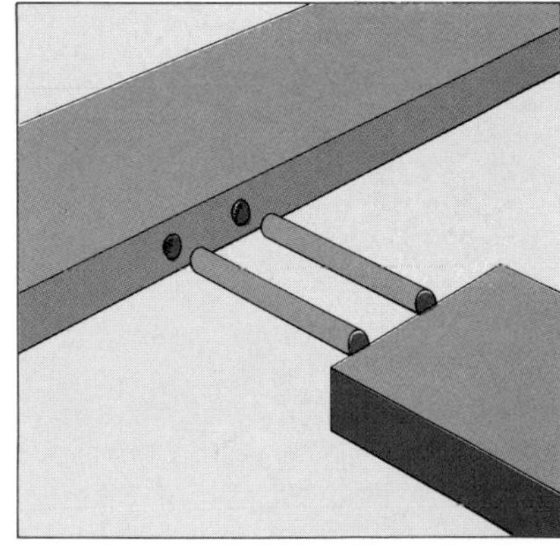

2—Dowels used in end-grain to side-grain joints in effect become tenons; the mating holes become mortises.

ings. The pin usually has its grain direction perpendicular to the grain direction of both parts being joined. These parts may have parallel grain directions but they are more often perpendicular to one another. Pins are sometimes added as a fail-safe measure or as a way of providing clamping pressure to draw a joint home. An example of the former is the pin usually concealed in the neck of a duck decoy, which will hold the head on in case the neck breaks because of weak grain direction. An example of the latter is the draw-bored mortise and tenon, where the hole drilled through the tenon is slightly offset from the one through the mortise, so pounding a pin home pulls the joint tightly together.

Despite the obvious success of the draw-bored mortise and tenon, dowels are most often misused as gluing accessories to hold parts in alignment. For example, in making a tabletop, boards might be edge-glued and held with a series of bar clamps. To ensure alignment of board surfaces at the joints, dowels are sometimes used. However, if gluing is correctly done, full wood strength can be developed by a plain side-grain to side-grain joint—no reinforcement is necessary. Because they do not provide strength, the pins therefore need only be long enough and numerous enough to ensure alignment. For edge-gluing 1-in. lumber, ⅜-in. or ¼-in. dowels that are 1 in. long are plenty. Needless to say, the holes should be bored a little deeper than the length of the dowels. Dowels should fit snugly into accurately positioned holes. When the joints are clamped, no attempt need be made to glue the dowels into the holes in the mating edges. The loss of glueline due to the dowels is negligible. For example, in edge-gluing ¾-in. lumber, a ⅜-in. dowel placed every 8 in. along the joint reduces the glueline area less than 2%. Although it might seem advantageous to make the dowels "good and long" and glue them in "good and tight," a negative effect can actually result. The restraint to normal shrinkage and swelling may cause the wood to fail at or near the glue joint *(4)*. If gluelines fail at edge joints, the problem should be rectified by trouble-shooting the gluing procedure rather than by pinning a bad joint with dowels in an attempt to bring it up to standard. If the gluelines are properly made, there is little to gain in trying to reinforce the joints, since the strength of the wood on either side of the joint is still the limiting factor.

Edge joints also are modified by various tongue-and-groove configurations to assist in alignment. The logic that such joints are stronger because of greater surface area is questionable. If the quality of gluing is up to standard, the glueline is as strong as the adjoining wood. In joints of end-grain to side-grain combination, the spline may become a tenon or a simple locking device. In edge-gluing, the idea of "strengthening" the joint with a longitudinal spline may be tempting, but is a serious misconception. Since the spline is continuous, the reduction in surface area of the board would be substantial *(5)*. For example, if a ¼-in. spline ran the length of a ¾-in. thick joint, the strength of the joint could be reduced by one-third. If the spline is very thin and the joint will be subject to bending, only slight weakening will occur, providing the spline is centrally located along the neutral axis.

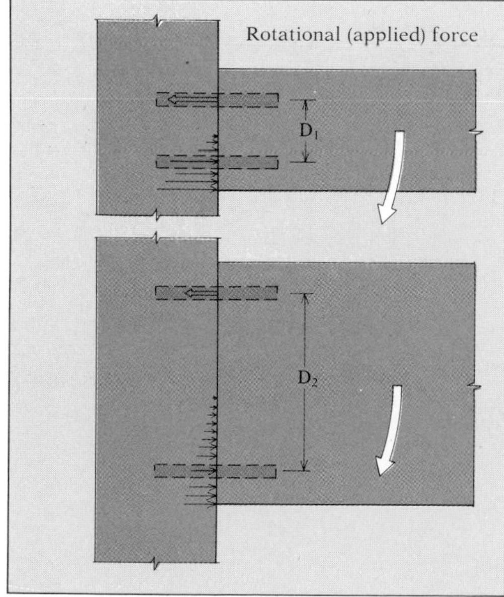

3—In doweled joints subject to racking, increasing the spacing between the dowels reduces the tensile load that each must carry.

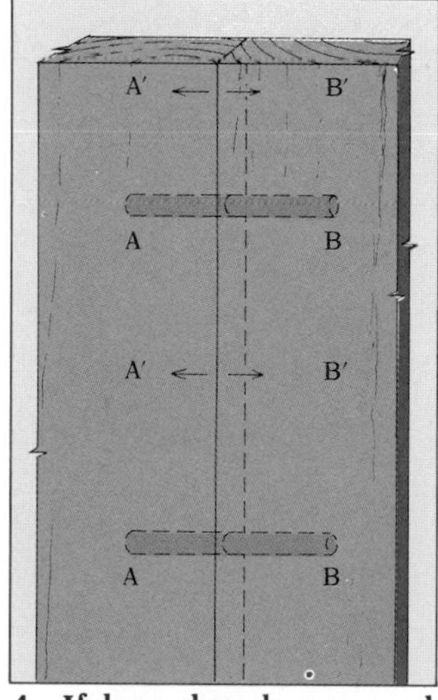

4—If long dowels are used across an edge joint and glued into the holes, shrinkage will be restrained across A-B. Tensile stress will develop across A′-B′ as the boards attempt to shrink.

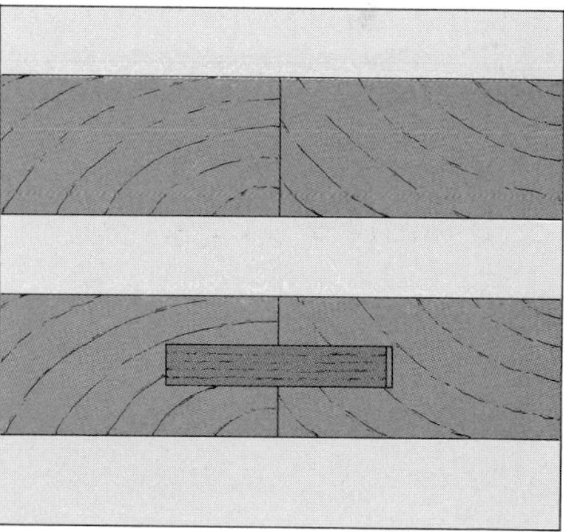

5—Reinforcing an edge-glued joint with a cross-grain spline actually weakens the joint at the margins of the spline.

Fastened joints

The term **fastener** refers to an item that holds together two members being joined. When used in the sense of a pin, wooden dowels are examples. Other wooden components, such as cross-battens, corner blocks and plywood gusset plates, might also be thought of as fasteners. Usually, however, the term fastener suggests nails and screws. It is likely that when civilization learned to extract and shape metals, nails and spikes for wood were among the first items produced. Until the last century, handmade nails had hardly changed and screws were relatively expensive. Today, however, with automated production, improved fastener design, and power installation equipment, mechanical fasteners have become inexpensive and efficient alternatives for assembling wood components. They are likely to remain indispensable to many forms of woodworking.

It has been estimated that some 75,000 fasteners—mostly nails—are used in the average house. Most woodworkers readily appreciate the importance of nails in general carpentry and softwood construction, but also assume a traditional notion which holds that fasteners should be avoided in cabinetmaking. But it has been perceptively observed that wood joints "can be poorly made with considerable ease," and this would certainly apply to many fastened joints. On the other hand, fastened joints can also be well made. Woodworkers should carefully study mechanical fasteners as an alternative means of joining wood.

Since most fasteners are metal and thus have superior strength, failure of the fastener itself need not be a concern. The primary requirement is holding power, which is the ability of fasteners to transfer stress from one member to another without detaching, dislodging or causing failure in either member. Holding power is closely related to the structural strength properties and condition of the wood.

Because of the endless array of styles of modern nails and screws, I will review general considerations making no attempt to summarize the technical data on individual fasteners.

Nails

The common wire nail, with its bright finish, diamond-cut point and flat head, is the most familiar of modern nails. In typical use, it is driven forcefully and rapidly through one or more materials, embedding its point into the side grain of seasoned wood. A general empirical formula for direct withdrawal *(1)* of a bright-finish, common wire nail immediately after driving, is:

$$p = 7{,}850G^{5/2}DL,$$

where

p = maximum withdrawal load, in pounds
G = the specific gravity of the wood (*Table 3*, p. 8)
D = the nail diameter in inches
L = the depth of penetration, in inches, of the nail into the member holding the point.

Under these standard conditions, direct withdrawal varies directly with the diameter and length of the nail; the greater the length or the diameter, the greater the holding power. The formula also reveals that denser woods develop greater holding power.

When a nail is driven into a side-grain surface, the longitudinal cell structure is separated or split apart and also compressed ahead of the point, depending on its taper or bluntness. As the nail progresses, many cells in the path of the nail are broken, and their severed ends are bent and compressed in the direction of driving. The tendency of the cell ends to recover causes them to press against the nail surface, resulting in resistance to withdrawal. Withdrawing the nail restraightens the cells,

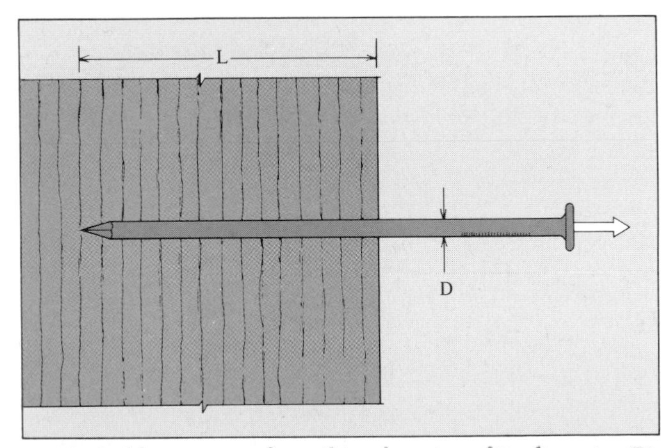

1—The holding power of a nail is a function of its diameter (D), its driven length (L) and the grain direction and density of the wood into which it is driven.

increasing the bearing against the fastener. Only when slippage occurs does the nail finally pull out.

Experiments have shown that in many woods, a spear point with a slim taper results in the greatest holding power. However, the wood fibers also separate, and in some species this type of point causes splitting. A blunt point has less tendency to cause splitting, because a plug of compressed wood structure is torn loose and pushed ahead of the point, rather than being pushed aside to start a split. However, the greater cell damage reduces holding power. The common diamond point, then, is a compromise that seems to afford the greatest holding power with the least splitting in common softwood structural lumber.

The fact that nails can be driven without preboring pilot holes has apparently led to the assumption that they *should* be driven without preboring. An unfortunate corollary seems to be that nails are therefore limited to use where they can be driven without splitting the wood or bending over. In reality, the best holding power develops when nail holes are prebored. While most woodworkers accept the idea of installing woodscrews in prebored holes to prevent splitting and to maximize holding power, they seldom consider preboring nail holes. Pilot holes ranging from 60% of nail shank diameter for low-density woods to 85% of shank diameter for high-density woods give maximum withdrawal resistance. In routine construction and carpentry it is obvious that preboring is not feasible. For cabinetmaking and other woodworking, however, nails installed in prebored holes are extremely effective fasteners and deserve greater consideration.

In nailing two pieces together, driving the nail through the first piece builds up a compression zone that may cause splitting or other disruptions as the nail emerges and enters the second piece. The resulting rupture can keep the pieces from maintaining close contact. Appropriate preboring eliminates this problem.

The holding power of nails diminishes over time following installation. The long-term holding power, especially where extreme moisture variation causes dimensional change in the wood, can be reduced to as little as one-sixth the loads indicated by the formula given.

Holding power can be improved considerably by surface modification of the nail shanks *(2)*. Resin-coated nails (called cement-coated nails) have about double normal holding power but this advantage disappears over time. Withdrawal resistance can be substantially improved by placing annular grooves on the nail shank *(3)*. Annularly threaded nails are understandably harder to drive, but when installed into side-grain prebored holes they provide positive resisting surfaces for bent-over fibers. Spiral-threaded nails have improved holding power that appears least reduced over time, perhaps because of the minimum damage to the holes as the nails "screw" into place.

All nails have less holding power when driven into end grain than into side grain. Test results indicate that the

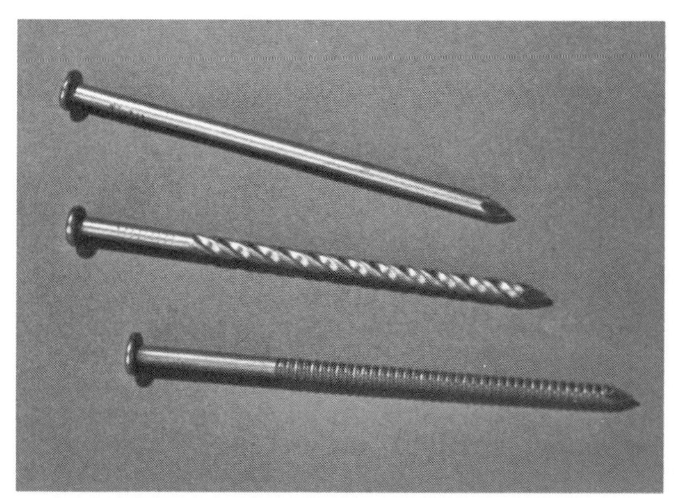

2—From top to bottom, a common nail with smooth, bright finish, a spiral-grooved nail and an annularly grooved nail.

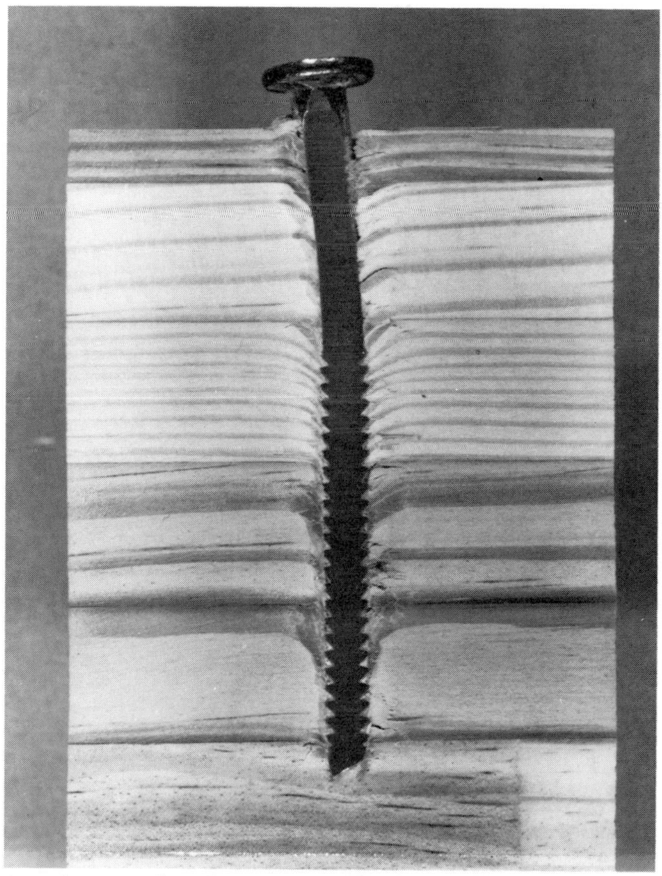

3—When a nail is driven perpendicular to the grain, annular grooves provide added bearing surface for the bent wood fibers, which resist withdrawal of the nail.

difference is smallest with high-density woods, but with low-density woods end-grain holding power may be as little as half of side-grain holding power. Unfortunately, one hears the flat statement that nailing into end grain should be avoided on the grounds that holding power is reduced. Actually, the lower holding power can be compensated for by preboring and increasing the diameter, length or number of nails. Nailing into end grain, in fact, may be beneficial in certain situations. For example, when a nail is driven into side grain, later shrinkage shifts the wood along the nail shank, causing the nail head to protrude rather than pushing the point deeper into the wood. The longer the nail, the greater the protrusion. This is why, in dry-wall work, the shortest nail that will hold the gypsum board in place is recommended. When the wood swells, it moves equally away from each side of the shank center, thereby backing the point still farther from the bottom of the hole. Repeated cycles cause further emergence of the nail. In end grain, however, since wood does not change dimension along its grain direction, using longer nails for greater holding power does not increase nail popping.

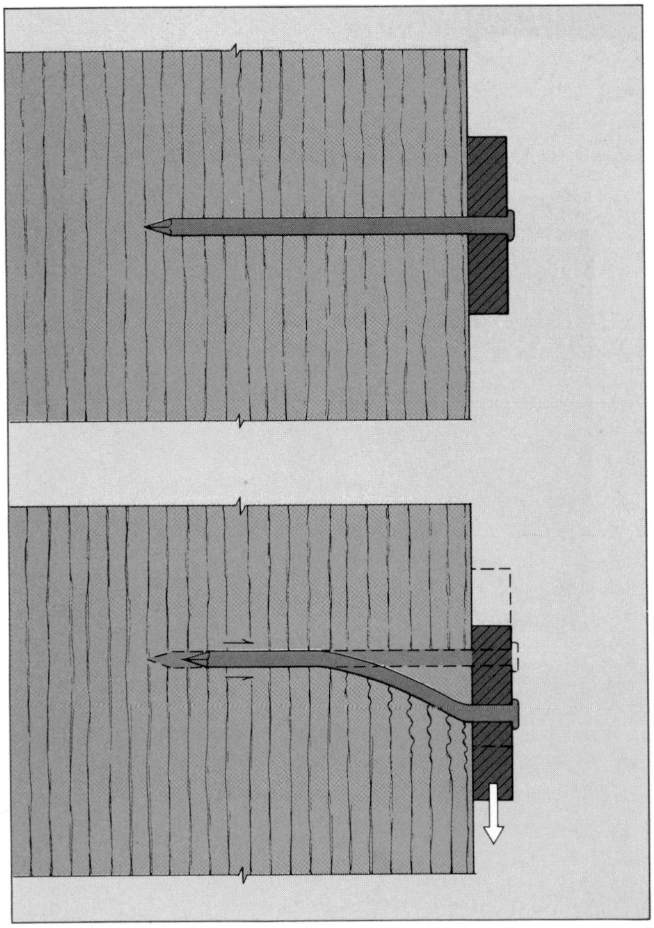

1—When a nail subject to lateral loading fails, it typically crushes the wood near the surface, then bends and pulls out of the hole.

In lateral loading *(1)*, joint slip rather than maximum load is critical. Stouter nails offer increased bearing against the wood so lateral load resistance increases exponentially with nail diameter. (If the diameter is doubled, the lateral resistance is increased by a factor of $2^{3/2} = \sqrt{2^3} = 2.8$.) It is also important that the fastener be stout enough to transmit load without bending. A long, slender nail crushes the wood near the surface and bends, and then becomes loaded in withdrawal and "snakes" out of the hole.

The head design of a nail is also critical. The broad heads of common nails are usually large enough to carry full withdrawal load without pulling through the top member. However, the small heads of finishing nails limit the effectiveness of deep penetration since they readily pull through the top board. Thus, while a joint can be made stronger by using longer common nails, beyond a critical point a joint with finishing nails might best be improved by increasing the diameter or the number of nails, if space permits.

Appearance is a strong influence in the bias against using nails in woodworking joints. However, there are many places where nails could be the most efficient and effective fastener, for example, half-blind dovetail joints commonly used for drawer sides. It would be interesting to compare a well-nailed half-blind tongue-and-rabbet joint for overall strength. Dare I even suggest nailing a dovetail joint together?

Woodscrews

Most woodworkers recognize the superiority of correctly installed woodscrews *(2)* over other fasteners. Their great holding power is understandable in terms of the positive engagement of the threads into relatively undamaged wood structure. The key to maximum holding power is preboring pilot holes for the threaded portions of the screw and for the shank. Slightly undersized shank holes should be prebored to provide a snug fit and firm bearing without developing enough stress to cause splitting. Pilot holes for the threaded portion should be from 70% of root diameter in low-density woods to 90% in high-density woods. (In the densest woods, it may be best to bore the pilot hole the same diameter as the root, especially for brass screws.) Lubricating the threads with wax facilitates driving and minimizes screw breakage without loss of holding power.

When woodscrews are correctly installed into the side grain of seasoned wood, maximum withdrawal loads can be estimated by the empirical formula:

$$p = 15{,}700\ G^2DL,$$

where

- p = maximum withdrawal load, in pounds
- G = the specific gravity of the wood (*Table 3,* p. 8)
- D = the screw shank diameter, in inches
- L = the depth of penetration, in inches, of the threaded portion of the screw into the member receiving the point.

As with nails, the holding power of screws increases directly with diameter and length, and exponentially with the density of the wood. Similarly, holding power diminishes over time, and withdrawal resistance for loading conditions of long duration might be as little as 20% of the values estimated by the formula given above. Screws driven into end-grain surfaces average only about 75% as much holding power as those driven into side grain, and holding power will be more erratic. As with nails, it is preferable to design joints to load screws laterally rather than in direct withdrawal.

Besides nails and screws, a host of various other fasteners are available for use with wood, such as clamp nails, corrugated fasteners and staples. In addition, various hardware items serve as fasteners in the role of a "third party," which is fastened to the two or more wood components by nails and screws. These include mending plates, flat corner irons, angle braces, T-plates and hinges. In evaluating the use and effectiveness of each, the integrity usually depends on the holding power of the attachment fasteners in terms of stress application relative to grain direction. In construction joints, for example, special framing anchors have been developed which transfer loads from one member to the other by loading fasteners laterally rather than in withdrawal. These anchor plates also provide for the proper number and placement of fasteners (usually nails) for maximum strength.

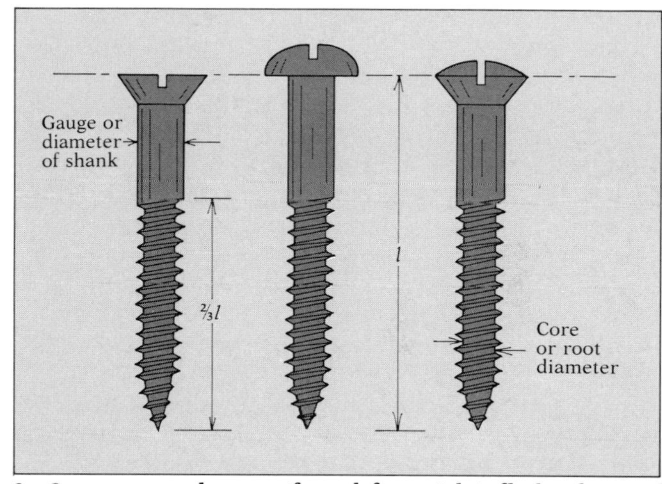

2—Common woodscrews (from left to right): flathead, roundhead, ovalhead.

Adhesive joints

Laminated items assembled with glue have been discovered in the tombs of early Egyptian pharaohs, and it is probable that the use of adhesive substances for holding wood parts together predates recorded history. Through the ages, most glues were made from fish, animals and vegetable starch and showed little change or improvement. In this century, however, development of the plywood industry initiated drastic changes in the types and properties of adhesives. Further stimulated by the demands of World War II and the scientific plunge into the space age, a dynamic adhesive technology has given us numerous multipurpose and specialized adhesives, with the promise of a continued parade of new ones well into the future.

Today's woodworkers use adhesives in a number of ways—to make large pieces out of smaller ones (such as carving blocks and laminated beams), to create combinations for strength or aesthetic improvement (such as plywood, veneers and marquetry) and to join parts to create a final product, as in furniture, sporting goods and structures. A complete discussion of gluing technology is impossible here. However, certain basic considerations that may be overlooked or misunderstood often cause serious gluing problems and are worth a systematic review.

The general term **adhesive** includes any substance having the ability to hold two materials together by surface attachment. Those most commonly used for wood are called **glues** although materials described as **resins, cements** and **mastics** are equally important in the assembly of wood products.

No truly all-purpose adhesive has yet been manufactured and probably never will be. A general-purpose adhesive cannot hope to attain all the individual capabilities and attributes of closely designed ones. Although any of the standard commercial glues will do a satisfactory job if the moisture content of the wood is controlled and the temperature remains within the human-comfort range, there is an increasing trend toward development of special adhesives. Adhesive selection must therefore take into account factors such as species, type of joint, working properties as required by anticipated gluing conditions, performance and strength, and, of course, cost.

One interesting adhesive is water. It is easily spread, wets wood well and solidifies to form a remarkably strong joint. It is delightfully inexpensive. However, it is thermoplastic and its critical maximum working temperature is 32°F. At temperatures at which it will set it has a very short assembly time. But due to its temperature limits, water will never capture a very important position among woodworking adhesives.

A wide and confusing array of adhesive products confronts the woodworker. A common pitfall is the belief that some glues are better than others; the notion that simply acquiring "the best" will ensure success is careless and may give disastrous results. With certain qualifications, all commercially available adhesives will perform satisfactorily if chosen and used within their specified limitations. An important corollary is that no adhesive will perform satisfactorily if not used properly. Within the specified limitations, most woodworking adhesives will develop joints equal in strength to the woods being joined. Thus, the wood, rather than the glue or its bond, is the weak link in a well-made joint.

Glues made from natural materials have been used from earliest times and even today, hide glue (made from the hides, tendons and hooves of horses, cattle and sheep) and casein (primarily a milk derivative) are still in use. However, the bulk of modern wood glues are synthetic compounds. Perhaps the most versatile are the polyvinyl acetate emulsions (PVA), commonly called white glues. More recently the yellow glues (modified PVA) have emerged, which have greater rigidity, improved heat resistance and better "grabbing" ability. These yellow glues are satisfactory for the bonding jobs of most craftsmen. They are easy to use and are more tolerant to unfavorable conditions than are white glues. Yellow glues also give less trouble in clogging abrasive paper.

Urea-formaldehyde, or plastic-resin glues, are water-resistant but not heat-resistant, as are the resorcinol-formaldehyde adhesives. A number of other adhesives with a variety of special uses and properties are also on the market, including epoxies, contact cements, mastics and acrylic adhesives.

A logical starting point is to wonder why glue sticks at all. It is sometimes assumed that adhesion results from the interlocking of minute tentacles of hardened adhesive into the fine porous cell structure of the wood surface. However, scientific research has shown that such **mechanical adhesion** is insignificant compared to the chemical attachment due to molecular forces between the adhesive and the wood surface, or **specific adhesion**. The assembled joint, or bond, is often discussed in terms of five intergrading phases *(1)*, each of which can be

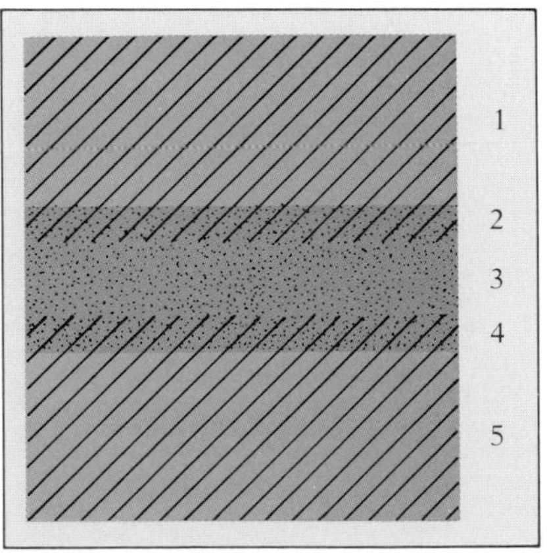

1—The five phases of a glue joint.

thought of as a link in a chain. The weakest phase determines the success of the joint. Phases 1 and 5 are the pieces of wood, or **adherends**, being joined. Phases 2 and 4 are the interpenetrating areas of wood and adhesive, where the glue must "wet" the wood to establish molecular closeness for specific adhesion. Phase 3 is the adhesive itself, which holds together by **cohesion**.

Fundamentally, then, gluing involves machining the two mating surfaces, applying an adhesive in a form that can flow onto and into the wood surface and wet the cell structure, and then applying pressure to spread the adhesive uniformly thin and hold the assembly undisturbed while the adhesive solidifies. The typical adhesive is obtained or mixed as a liquid but sets to form a strong glue layer, either by loss of solvent, which brings the adhesive molecules together and allows them to attach to one another, or by a chemical reaction that develops a rigid structure of more complex molecules.

Different woods have different gluing properties. In general, less dense, more permeable woods are easier to glue, for example, chestnut, poplar, alder, basswood, butternut, sweetgum and elm. Moderately dense woods such as ash, cherry, soft maple, oak, pecan and walnut glue well under good conditions. Hard and dense woods including beech, birch, hickory, maple, Osage-orange and persimmon require close control of glue and gluing conditions to obtain a satisfactory bond. Most softwoods glue well, although in uneven-grained species, earlywood bonds more easily than denser latewood. Extractives, resins or oils may introduce gluing problems by inhibiting bonding, as with teak and rosewood, or by causing stain with certain glues, as with oaks and mahogany.

Since most adhesives will not form satisfactory bonds with wood that is green or of high moisture content, wood should at least be well air-dried. Ideally wood should be conditioned to a moisture content slightly below that desired for the finished product, to allow for the adsorption of whatever moisture might come from the adhesive. For furniture, a moisture content of 5% to 7% is about right. However, when using urea-formaldehyde glues, the moisture content should not be below 7%. For thin veneers, which take up a proportionately greater amount of moisture, an initial moisture content below 5% might be appropriate.

Machining is especially critical. In some cases, especially for multiple laminations, uniform thickness is necessary for uniform pressure. Flatness is required to allow surfaces to be brought into close proximity. The surfaces to be glued should have cleanly severed cells, free of loose fibers. Accurate hand-planing is excellent if the entire surface, such as board edges, can be surfaced in one pass. On wide surfaces, peripheral milling (planing, jointing) produces adequate surfaces. Twelve to twenty-five knife marks per inch produce an optimum surface. Fewer may give an irregular or chipped surface; too many may glaze the surface excessively.

Dull knives that pound, heat and glaze the surfaces can render the wood physically and chemically unsuited for proper adhesion even though it is smooth and flat. Planing saws are capable of producing surfaces acceptable for gluing, but in general sawn surfaces are not as good as planed or jointed ones.

Surface cleanliness must not be overlooked. Oil, grease, dirt, dust and even polluted air can contaminate a wood surface and prevent proper adhesion. Industry production standards usually call for "same-day" machining and gluing. Freshly machining surfaces just before gluing is especially important for species high in resinous or oily extractives. Where this is not possible, washing surfaces with acetone is sometimes recommended. One should not expect a board machined months or years ago to have surfaces of suitable chemical purity. If lumber is flat and smooth but obviously dirty, a careful light sanding with 240-grit or finer abrasive backed with a flat block, followed by thorough dusting, can restore a chemically reactive surface without seriously changing flatness. Coarse sanding, sometimes thought to be helpful by "roughening" the surface, is actually harmful because it leaves loose bits. Tests have shown that intentionally roughening a surface, as in "toothed planing," does not improve adhesive-bond quality. In summary, wood should be surfaced immediately prior to gluing, for cleanliness and to minimize warp, and should be kept free of contamination to ensure an acceptable gluing surface.

Shelf life is the period of time an adhesive remains usable after distribution by the manufacturer. Unlike photographic films, adhesives are not expiration-dated. Beware the container that has been on the dealer's shelf too long. Outdated package styles are an obvious tip-off. It is wise to mark a bottle or can with your date of purchase. It is amazing how fast time can pass while glue sits idle in your workshop. If possible, refrigerate glues in tight containers to prolong shelf life. In general, if the glue is spreadable when mixed according to instructions, it is suitable for use. Adding water to restore spreadability is not a good practice.

The adage "when all else fails, read the instructions," all too often applies to glue. It is unfortunate that instructions are so incomplete on retail glue containers. Manufacturers usually have fairly elaborate technical specification sheets but supply them only to quantity consumers. Too often, many critical factors are left to the user's guesswork or judgment. Mixing proportions and sequence usually are given clearly; obviously they should be followed carefully.

Glues with a pH above 7 (alkaline), notably caseins, will absorb iron from a container and react with certain woods such as oak, walnut, cherry and mahogany to form a dark stain. Coffee cans or other ferrous containers can contribute to this contamination. Nonmetallic mixing containers such as plastic cups or the bottoms of clean plastic bleach jugs work nicely.

Once glue is mixed, the **pot life**, or **working life**, must be considered. Most adhesives have ample working life to handle routine jobs. The period between the beginning of spreading the glue and placing the surfaces together is called **open assembly time**; **closed assembly time** indicates the interval between joint closure and the development of full clamping pressure. Allowable closed assembly time is usually two or three times open assembly time. With many ready-to-use adhesives, there is no minimum open assembly time; spreading and closure as soon as possible is recommended, especially in single spreading, to ensure transfer and wetting of the other surface. If the joint is open too long, the glue may precure before adequate pressure is applied. The result is called a **dried joint**. In general, assembly time must be shorter if the wood is porous, the mixture viscous, the wood at a low moisture content, or the temperature above normal. (As a rule, it is a good policy to avoid gluing where temperature of the room or of the wood is below 70°F.) With some adhesives, such as resorcinol, a minimum assembly time and double spreading (that is, applying adhesive to each of the mating surfaces) may be specified for dense woods and surfaces of low porosity, to allow wetting of the wood and to permit thickening of the adhesive to prevent excessive squeeze-out.

Proper spread is difficult to control. Too little glue results in a starved joint and a poor bond. A little overage can be tolerated, but too much results in wasteful and messy squeeze-out. Though some squeeze-out is assurance that sufficient adhesive has been applied, squeeze-out may cause problems in machining or finishing. With experience the spread can be eyeballed. It is useful to obtain some commercial specifications and conduct an experiment to see just what they mean. Spreads are usually given in terms of pounds of glue per thousand square feet of single glueline, or MSGL. A cabinetmaker will find it more convenient to convert to grams per square foot, by dividing lb./MSGL by 2.2. Thus a recommended spread of 50 lb./MSGL, typical of a resorcinol glue, is about 23 grams per square foot. Spread it evenly onto a square foot of veneer for a fair visual estimate of the minimum that should be used. Usually, the recommended spread appears rather meager.

Double spreading is recommended where feasible. This ensures full wetting of both surfaces, without rely-

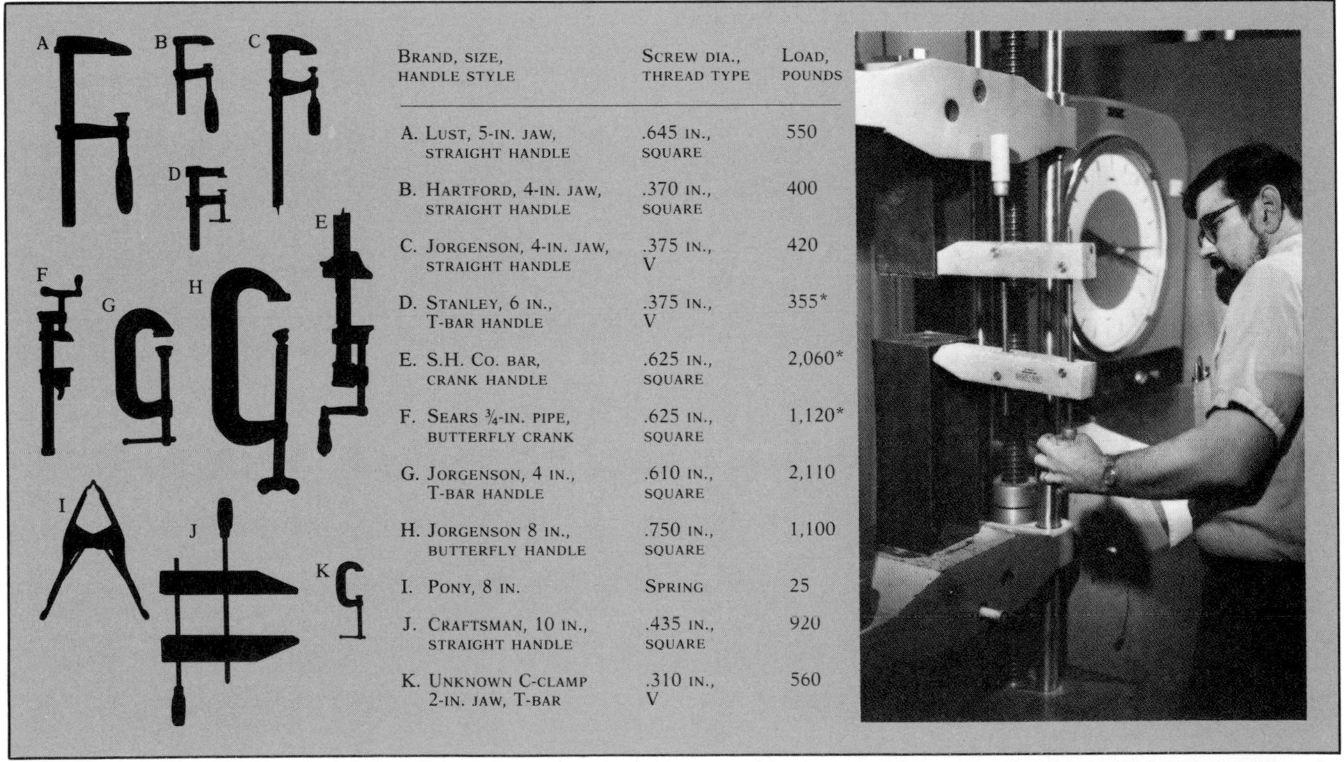

Brand, size, handle style	Screw dia., thread type	Load, pounds
A. Lust, 5-in. jaw, straight handle	.645 in., square	550
B. Hartford, 4-in. jaw, straight handle	.370 in., square	400
C. Jorgenson, 4-in. jaw, straight handle	.375 in., V	420
D. Stanley, 6 in., T-bar handle	.375 in., V	355*
E. S.H. Co. bar, crank handle	.625 in., square	2,060*
F. Sears ¾-in. pipe, butterfly crank	.625 in., square	1,120*
G. Jorgenson, 4 in., T-bar handle	.610 in., square	2,110
H. Jorgenson 8 in., butterfly handle	.750 in., square	1,100
I. Pony, 8 in.	Spring	25
J. Craftsman, 10 in., straight handle	.435 in., square	920
K. Unknown C-clamp 2-in. jaw, T-bar	.310 in., V	560

Table 15—Average clamping pressure of typical woodworking clamps. To find out just how much load typical clamps could apply, Hoadley attached open steel frames to the crossheads of a universal timber-testing machine. With a clamp positioned to draw the frames together, the load applied was indicated directly. The clamps are described in the table, with the last column giving the average of three trials by average-sized Hoadley, tightening as if he were trying to get maximum pressure in a gluing job. The quick-set clamp listed first in the table was used to calibrate the setup: A secretary squeezed 330 lb., a hockey player squeezed 640 lb., and Hoadley squeezed 550 lb. Repeated trials by each person yielded readings that agreed to within 10%. An asterisk indicates that the clamp began to bend, and the test was stopped at the value listed.

ing on pressure and flatness to transfer the glue and wet the opposite surface. With double spreading, however, a greater amount of glue per glueline is necessary, perhaps a third more.

Glue should be spread as evenly as possible, even though some degree of self-distribution will of course result when pressure is applied. Toothed scrapers, rollers or stiff brushes are best for this purpose. Some speed is necessary; when many pieces are to be assembled, it pays to have them in the order of assembly.

The object of clamping a joint is to press the glueline into a continuous, uniformly thin film, and to bring the wood surfaces into intimate contact with the glue and hold them undisturbed until setting or cure is complete. Since loss of solvent causes some glue shrinkage, an internal stress often develops in the glueline during setting. This stress becomes intolerably high if gluelines are too thick. Gluelines should be not more than a few thousandths of an inch thick.

If mating surfaces were perfect in terms of machining and spread, pressure wouldn't be necessary. The "rubbed joint," skillfully done, attests to this. But unevenness of spread and irregularity of surface usually require considerable external forces to press properly. The novice commonly blunders on pressure, both in magnitude and uniformity.

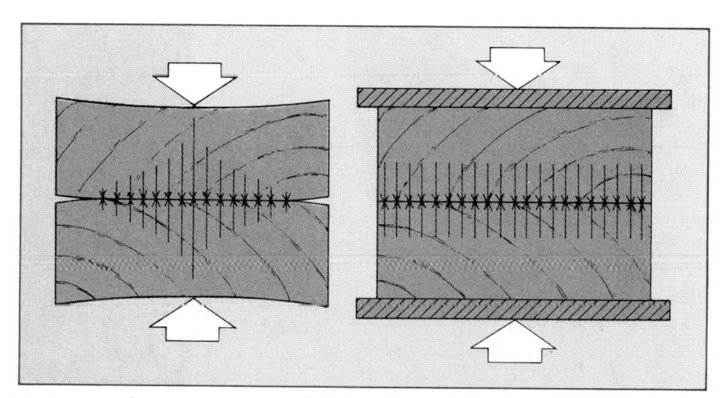

1—Cover boards (cauls) distribute clamping pressure evenly.

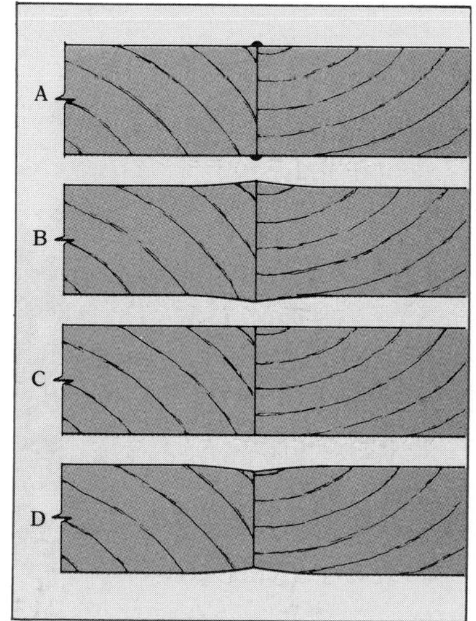

2—When an edge-glued panel (A) is surfaced while the glueline is still swollen with moisture (B, C), a sunken joint (D) is the result.

Clamping pressure should be adjusted according to the density of the wood. For domestic species with a specific gravity of 0.3 to 0.7, pressures should range from 100 psi to 250 psi. Dense tropical species may require up to 300 psi. In bonding composites, the required pressure should be determined by the lowest-density layer. In gluing woods with a specific gravity of about 0.6, such as maple or birch, 200 psi is appropriate. Thus gluing up one square foot of maple requires pressure of (12 in. x 12 in. x 200 psi) 28,800 pounds. Over 14 tons! This would require, for an optimal glueline, 15 or 20 *C*-clamps, or about 50 quick-set clamps. Conversely, the most powerful *C*-clamps can press only 10 to 11 square inches of glueline in maple. Jackscrews and hydraulic presses can apply loads measured in tons. But since clamping pressure in the small shop is commonly on the low side, one can see the importance of good machining and uniform spread. *Table 15* gives an indication of how much gluing pressure can be delivered from various clamps.

Another troublesome aspect of clamping is uniformity, usually a version of what I call "the sponge effect." Lay a sponge on a table and press it down in the center; note how the edges lift up. Similarly, the force of one clamp located in the middle of a flat board will not be evenly transmitted to its edges. It is therefore essential to use heavy wooden cover boards or rigid metal cauls to ensure proper distribution of pressure *(1)*.

Clamp time must be long enough to allow the glue to set well enough so that the joint will not be disturbed by clamp removal. Full cure time, that is, for development of full bond strength, is considerably longer. If the joint will be under immediate stress, the clamp time should be extended. Consider a dry run to check for tightness of joints and to rehearse the process.

Finally, cured joints need conditioning periods to allow moisture added at the glueline to be distributed evenly through the wood. Ignoring this can result in sunken joints *(2)*. When edge-gluing pieces to make panels, moisture is added to the gluelines, especially at the panel surfaces where squeeze-out contributes extra moisture *(A)*. If the panel is surfaced while the glueline is still swollen *(B, C)*, when the moisture is finally distributed the glueline will shrink *(D)*, leaving a joint that is sunken.

Finishing Wood

The word finish in woodworking usually describes some final treatment to the wood that protects and enhances its appearance. Most craftsmen agree that some form of protection is usually needed. The matter of appearance, however, is more controversial, depending on individual taste and preference.

Let's first consider protection. It is usually desirable to protect wood surfaces from accumulating dirt, or to create a surface that can be cleaned easily. Finishes may also protect against abrasion or indentation and prevent changes in color due to light or atmospheric pollutants. But their most important function is to impede the exchange of moisture with the atmosphere, thus helping to avoid the consequences of wood movement.

It is impossible to generalize on the subject of surface appearance because of the variation in circumstance and personal preference as to what looks best. Some woodworkers want to preserve wood in its natural state as much as possible, while others wish to change the wood in both color and appearance. Some prefer to retain any visible surface irregularity due to cell structure, others desire a surface that is perfectly smooth. Some want a matte finish, others a high gloss. Some try to retain or even accentuate variation in figure and color, others attempt to achieve uniformity. In this chapter I will concentrate on basic points about protection and appearance, without regard to functional requirements or aesthetic preferences. I shall not be giving recipes for stains.

Achieving a finish on wood involves a combination of two elements, the **surface condition** of the wood and the **finishing treatment** applied to it. Although done separately, they are interrelated and must be planned with respect to one another. Certain surface conditions will call for particular treatments and vice versa, but there is no such thing as the single best combination for all projects. I have the most fun experimenting, and I have trouble remembering ever finishing two items in exactly the same way.

1—The smoothness and gloss of a finish are indicated by how it reflects a black-lined target card. Surface roughness is indicated by distortion or break-up of the lines. Glueline creep will show as an abrupt break-up in diagonal vertical-line reflections. Top, grid reflection of Formica over plywood; at center, the reflection of a pearwood panel; bottom, the reflection of a marquetry tabletop.

Surface condition

Most finishing instructions begin with surface preparation, emphasizing such things as proper sanding and dusting just prior to treatment. But the concern must begin long before, because surface condition is influenced by every step of woodworking, from sawing the log and drying the lumber to machining the surfaces and gluing the joints. It is appropriate to evaluate surface condition using four criteria: trueness, evenness, smoothness and quality.

Trueness compares the actual to the intended geometry of the surface. Planed surfaces are expected to be flat, turnings are expected to be round, edges are expected to be straight, and so forth. Residual stresses due to improper drying of lumber and warp resulting from change in moisture content are the most common causes of cup, bow and twist in flat surfaces. But crowning over of surfaces near edges can result from careless sanding or planing.

The **evenness** of surfaces can be spoiled in several ways. Raised grain is a common cause, traceable to machining and moisture problems. The unevenness of elevated latewood can result from careless handsanding that scours more deeply into earlywood than latewood in uneven-grained woods, especially on flat-grained surfaces. Raised, sunken or mismatched joints can produce an uneven surface as a complication of poor gluing procedures. When these problems develop in a core material they can telegraph through face veneer. An otherwise attractive and successful finishing job can be overshadowed by lack of trueness or evenness of the surface.

Surfaces may, of course, be intentionally made uneven with satisfying results. Sandblasting and scorching out earlywood to provide a textured surface are examples of novel techniques used successfully in both sculptured and paneled surfaces.

Smoothness is the absence of surface irregularity, such as the undulating knife marks left after machine-planing or the chatter marks left by careless scraping. Corrugations in veneer, especially those associated with knife checks, are further examples. Minute tear-outs, which may occur when planing against the grain, destroy surface smoothness. (I do not include the surface voids traceable to cell cavities as departures from smoothness, because they are an inherent feature of wood, not of its condition.) We generally measure smoothness by the depth and uniformity of the scratch pattern left from sanding. The smoothest surfaces result from hand-planing with the grain, scraping and fine sanding (*1* through *4*).

Of equal importance is the **quality** of the surface cell structure and the cell damage that results in forming the surface. The ideal surface for finishing would be produced by light skim cuts with a razor blade, which would cleanly sever exposed cell walls with no damage to the remaining structure. Such an ideal surface, however, can hardly be expected in common woodworking practice. Try to think of any surface in terms of cellular damage. One illustration of this point would be knife marks on a longitudinal surface. The surface may be true and even, and the knife marks may leave the surface amazingly smooth. With a well-sharpened planer, with lumber fed at a rate to produce 20 knife marks per inch, knife marks would be imperceptible to the touch. We would certainly consider the surface smooth, yet the variation in cell damage along the surface would cause each knife mark to stand out as visually distinct. Microscopic examination reveals that variable light reflection

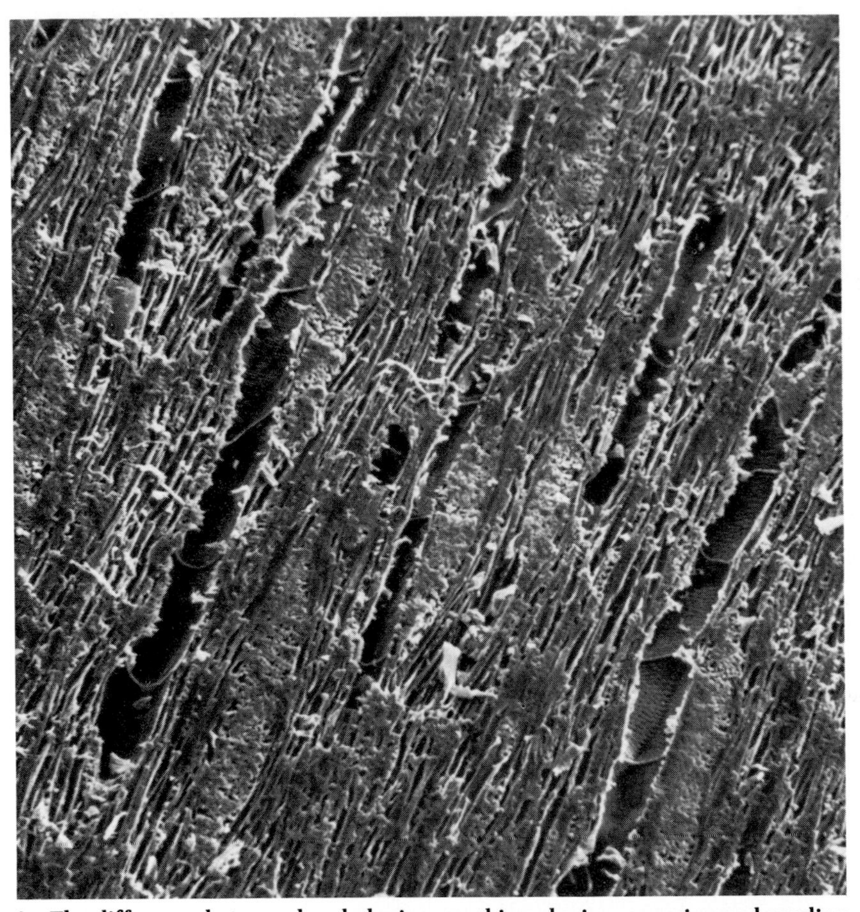

1—The difference between hand-planing, machine-planing, scraping and sanding is clear in these photomicrographs of rock maple. This one is at ×100 magnification, the others on this page at ×50. The long openings in this hand-planed specimen are various vessel elements, some divided by cross-walls. The lighter, more densely structured areas are ray cells in cross section. Minute details of the wood's fiber structure are visible, and most of the fibers are cleanly severed.

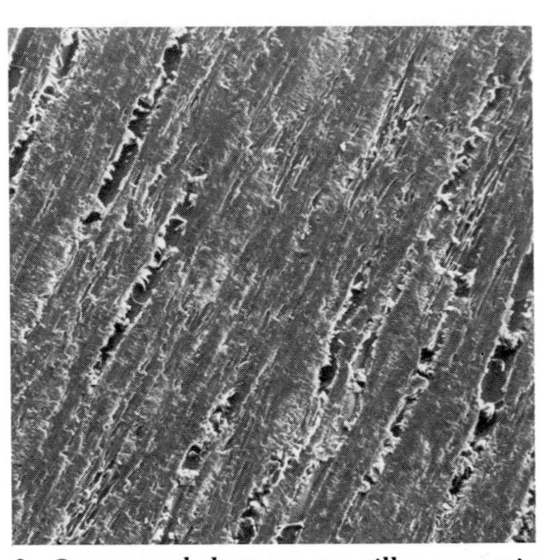

2—Open vessel elements are still apparent in machine-planed sample, but smaller features are obscured by torn and pounded fibers. The knife has moved across the surface from lower left to upper right, burnishing the fibers onto one another.

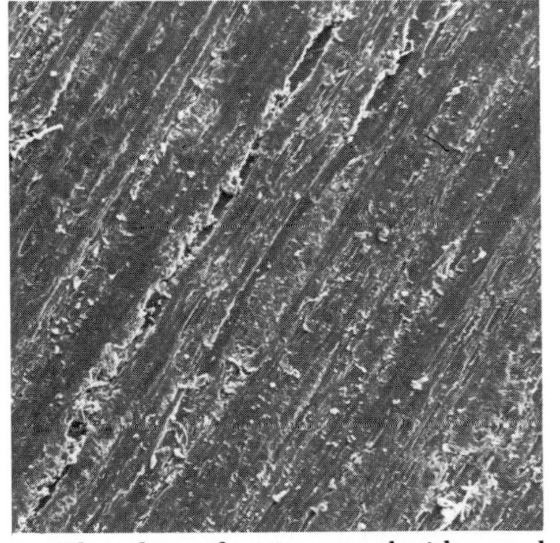

3—When the surface is scraped with a steel scraper blade, torn and rolled wood tissue fills most of the wood vessels, and the surface becomes scratched by the minute raggedness of the scraper's edge.

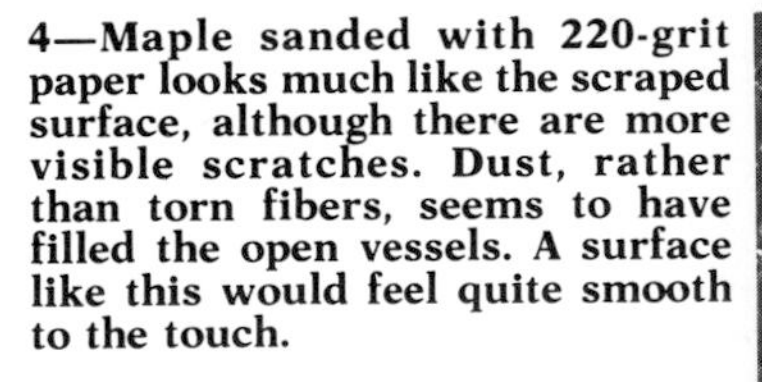

4—Maple sanded with 220-grit paper looks much like the scraped surface, although there are more visible scratches. Dust, rather than torn fibers, seems to have filled the open vessels. A surface like this would feel quite smooth to the touch.

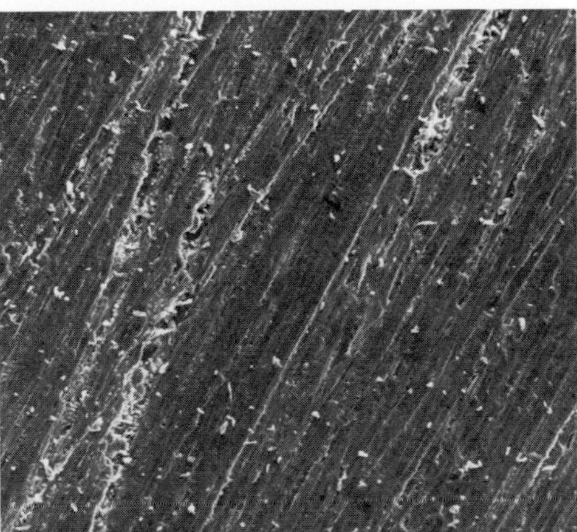

from damaged cells, more than physical surface irregularity, is responsible for the visibility of the knife marks. This damage can be obscured by the more uniform pattern of damage that is created by fine sanding or scraping along the grain.

Another example relates to sanding. If you sand with the grain using 180-grit paper, the surface will feel quite smooth. Sand the same wood across the grain with 240-grit paper, and it will also feel just as smooth; yet when this piece is stained, the scratches will show up because of the very different manner in which the surface cell structure was broken up and absorbed stain.

No point needs greater emphasis than sanding *parallel to* rather than *across* the grain. On the abrasive paper, each granule of abrasive is a tiny cutter *(5, 6)*. Since most of these granule faces have negative cutting angles, a scraping type of chip forms *(7)*. This cutting action carves out cell-wall material from the surface parallel to the grain, but when directed across the longitudinal cells, frayed and broken-out cell walls result. As in plan-

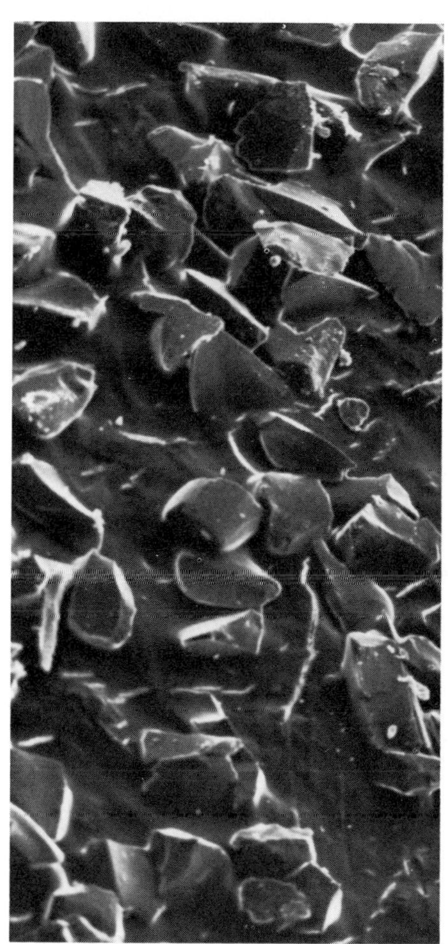

5—Photomicrograph at ×100 magnification shows surface of new 200-grit opencoat garnet sandpaper. Each granule on the paper acts as a tiny cutter that produces a scraping type of chip.

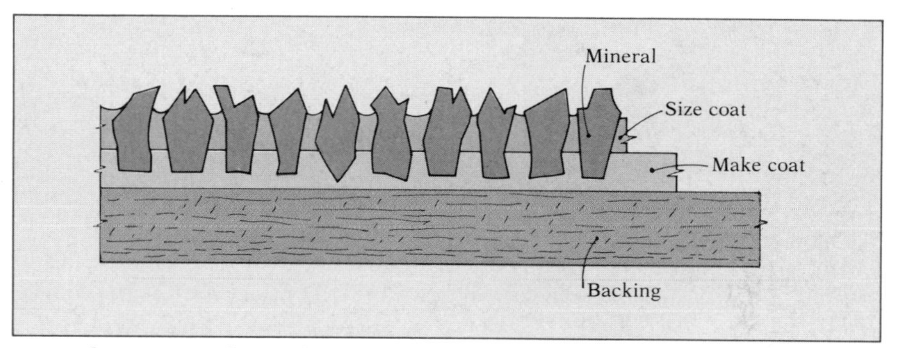

6—Sandpaper is made up of mineral particles attached to a backing. The minerals adhere to the make coat (glue coat) and are locked in place with a size coat.

7—The cutting action of a sandpaper particle yields a scraping type of chip.

ing, sanding with the grain where crossgrain occurs is preferable. Sanding end grain leaves some broomed-over cell material, so sanding in one direction will produce the most uniform surface damage.

Developing surface smoothness by sanding is best done using a progression of grit sizes, each of which produces a scratch pattern at least to the depth of the previous one *(1)*. The consequences of skipping grit sizes are well known. Nothing is more frustrating than to come down to an apparently smooth surface after arduous sanding with a series of grit sizes only to discover a few deep scratches left by the coarsest grit.

Damaged cell structure, though finished smoothly, may later show variable light reflection or uneven stain or finish retention. Common problems are minute seasoning checks or compression failures that have gone unnoticed, and hammer indentations or cell structure "bruises" below the surface from the action of rasp teeth. The glazed, pounded and scorched surfaces produced by a dull cutterhead can hardly be considered as having quality even if they are smooth. Such surfaces may show later problems of grain raising, uneven stain retention or poor adhesion of coatings. In hardwoods having tension wood, surfaces may be sanded to apparent smoothness. However, the microscopic woolliness of the severed cell walls will result in blotchy staining.

In any machining process, some fragile projections of damaged cell-wall material remain on the wood surface. Eventual adsorption and desorption of moisture will cause these fibers to distort and to project out from the surface. Where a surface coating will bury and lock them in place, they may be of no consequence. Otherwise, the raising of surface debris may detract from smoothness. It may therefore be desirable to remove loose cell-wall material as a final step in surface preparation. In a warm, dry atmosphere, the wood surface can be wiped quickly with a slightly damp (not moist or wet) cloth. The moisture from the cloth will be adsorbed quickly by the damaged cell-wall fragments, causing them to raise from the surface. The surface will soon reestablish moisture equilibrium with the environment without significant increase in overall moisture content. The projecting "whiskers" can then be removed by very light sanding with very fine (600-grit) abrasive paper. The trick is to remove the whiskers without further abrading the surface, which will produce only more whiskers. An extremely smooth and high-quality surface can be produced in this manner.

Surfaces should regularly be wiped or blown free of dust during and after sanding. Accumulated dust may cause "corns" on the abrasive paper, which can mar the surface. In addition, excess dust packed into the cell structure may later mar the finish. The final cleaning should be thorough. An air hose or vacuum cleaner may help if you have one, and it's a good idea to get in the routine of completing the cleaning job with a tack rag. Commercially available tack rags seem well worth the money, but a fairly good one can be made easily from a lint-free cloth, such as an old handkerchief. Dampen the cloth slightly with turpentine, sprinkle on a teaspoonful of varnish or lightly paint meager streaks of varnish across the cloth with a brush. Then thoroughly wring

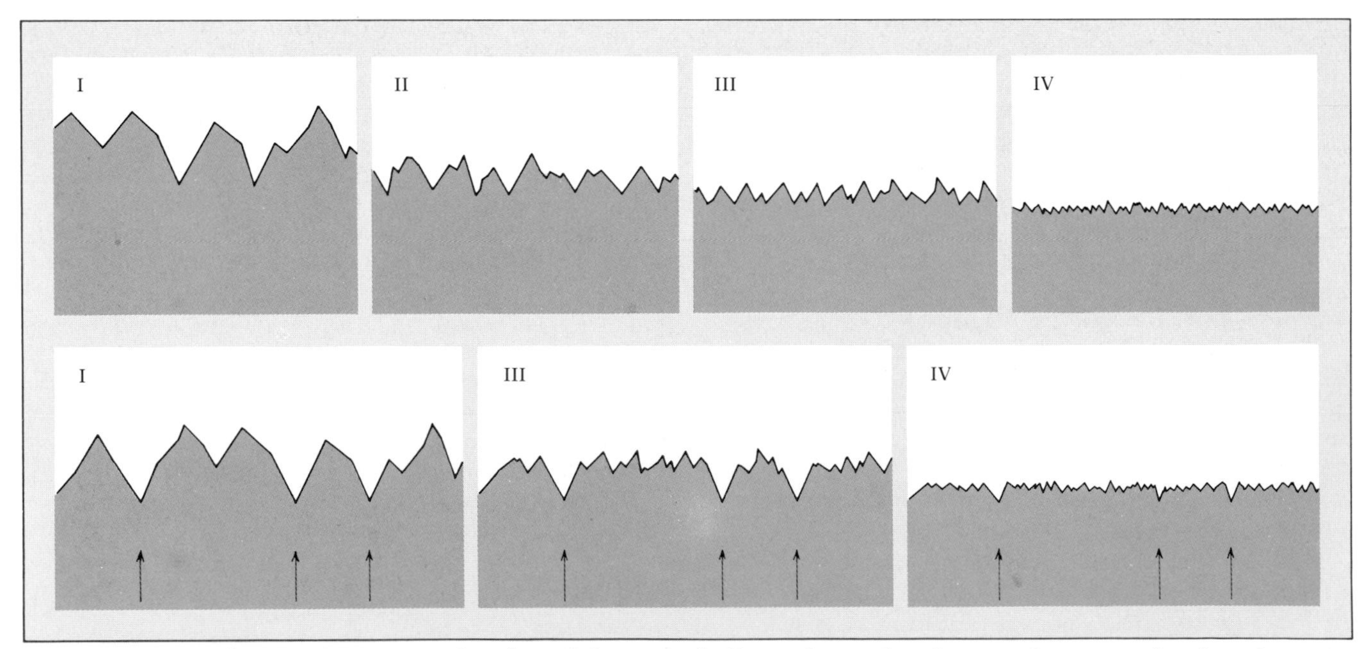

1—Top, proper sanding requires progressing through increasingly finer grits, so that the scratch pattern of each replaces the coarser pattern of the previous one. Skipping a grit, above, will leave deep scratches in the finished surface.

the cloth to distribute the varnish. It should feel tacky, not wet. Store in a glass jar. Whisk the surface lightly to pick up dust, repeatedly folding the cloth. Discard and make a new one when it has lost its effectiveness.

Commercial spray products (such as Endust) for treating household dust rags work quite well for me.

Surface quality also should be considered from the chemical standpoint. Chemical discoloration resulting from such things as sticker stain in drying or fungal activity may cause visual defects in the finish. Traces of previous finish, glue spills or other contaminants can interfere with the evenness of stain retention or the adhesion of finish coats. As with glues, bonding of finishes depends in large measure upon molecular adhesion. If there is any doubt as to possible contamination of the surface, a final sanding prior to finishing will promote good adherence.

The four criteria of surface condition must be considered separately. For example, a tabletop that is machined to true and even flatness may have poor quality if it has been sanded across the grain. On a carved surface, the trueness must be judged in relation to the desired shape. If the surface is produced by a sharp gouge properly used (with the grain), the surface may be of high quality but intentionally uneven. If the unevenness of a high-quality carved surface were undesirable, sanding might make the surface more even, but at the same time might reduce its smoothness and quality. In a sense, the moisture content of the wood also should be considered a factor in surface condition, for if it changes after finishing, the trueness, evenness or surface quality may be belatedly altered.

In considering finishing treatments for wood there are no "right" answers, only countless alternatives. Function, aesthetics, time and cost ultimately are the deciding factors. As with the drying of wood, a great deal of lore and tradition influences our modern practices, yet few areas of woodworking are so touched by modern advances. Although no subject as complex as finishing can be generalized or simplified, I have come to recognize three basic categories of surface treatment: coatings, that is, treatment **on** the surface; penetrating finish, that is, treatment **in** the surface; and **no** treatment at all.

No treatment

Usually, some sort of surface application is required for protection and appearance, and the instances where no finishing treatment at all makes sense are apt to be few and far between. Yet too often tradition seems to force the assumption that something must be brushed, swabbed, wiped or sprayed onto the surface of a completed work, while leaving the wood untreated is not even considered.

The more one works with wood and the more deeply one comes to understand wood, the more sensitive one becomes to the value of natural tactile surfaces, and the more appreciation one has for the appearance of wood in the raw. Here more than ever, however, the surface condition, especially smoothness and quality, must hold its own. I always try to recognize those few special cases where the absence of finish can be the most gratifying treatment. To most, the term natural finish means a clear application; to me it suggests no finish at all.

Many items, if kept indoors, really need no finish. These are often decorative, such as carvings and sculpture, but may also be functional, such as trays, bowls or utensils. They often will be made of a single piece of wood, which can change dimension without affecting function or appearance. For example, I have a small abstract carving of eastern white pine *(2)*. Its smooth, dry surface is light in color, with only a subtle growth-ring figure displayed at the surface. Any treatment of the sur-

2—Carving in eastern white pine, left, was made 12 years ago and finished with nothing. Periodic sanding with 400-grit paper keeps its color bright and fresh. Catalpa statue, right, standing 13 in. tall, was sanded with coarse sandpaper and left unfinished.

face would bring out this figure too strongly. About as often as you might oil or polish a coated item, I simply resand the surface lightly with 400-grit paper to remove any accumulated dirt, dust and discoloration from handling. After 12 years, it still looks fresh and clean.

For such items as utensils and tool handles, the normal dirt accumulation and surface abrading from handling create a finish that is both unique and appropriate. Many years back I needed a net-maker's needle, so I whittled one out of black cherry and put it to work immediately without coating it with anything. The years of use have given it a finish I would never trade for anything that comes in a can. I also marvel at the natural finish that develops on well-worn hammer and wheelbarrow handles, railings and chair arms, once the original coating of paint or varnish has finally worn off. Unfinished wood usually darkens or "ages" more rapidly than wood protected with coatings, especially coatings that contain ultraviolet filters. However, the anticipation of color change can be an integral part of the design of any wooden object, and the eventual patina developed in a wood surface can be regarded as a valuable asset.

The no-treatment finish also has fantastic potential for outdoor wood objects as well. But the effects of the elements will be far more drastic and complicated, and the changes that will take place must be understood and anticipated. We somehow seem obsessed with the idea that everything must be made to last forever. Consequently, we often fail to take advantage of nature's own progression. Why not consider a finite life for an object and allow gradual deterioration to take place, especially where the effect is beautiful?

In nature we see examples of fallen trees and weathered driftwood where silvery-grey sculptured surfaces surpass all human creativity. In building design and architecture, the natural aging of materials has long been used to both decorative and functional advantage. Likewise, sculpture can become more and more attractive as the ravages of time erode the surface. By sensible selection of wood species and intelligent sculptural design, this deterioration can be programmed into the life of the piece. If a decay-resistant species is chosen and the design permits water to run off, deterioration can be restricted to surface weathering. I have a carved ruffed grouse on a post next to our driveway *(1)*. It was carved out of catalpa and set out, unfinished, 14 years ago. The weathered surface of greys and browns is more appropriate to the subject of the carving than any finish I could have applied. Because it is mounted "high and dry," and because catalpa is highly resistant to decay, I'm convinced it will outlast me.

The weathering of wood is a combination of physical, mechanical and chemical effects. The wetting and drying of the surfaces cause expansion and compression set followed by shrinkage, resulting in surface checking. Water that freezes and expands in the surface leads to further breakdown. Ultraviolet radiation also causes the surface structure to deteriorate. Windborne particles abrade the surface. Despite all this, weathering alone will remove only about 1/4 in. of wood per century from exposed surfaces.

Normally the breakdown of lignin leaves a cellulosic residue on the surface which, along with water staining, produces a predominantly grey color. Dark woods tend to lighten as they weather, and light woods tend to darken. Some species develop a silvery-grey color, others a dark grey or a brownish tinge. However, the moisture condition of the wood can complicate the process, especially when it remains high enough to allow fungi to grow. In such cases, uneven surface discoloration and darkening may result before normal weathering develops. Commercial "bleaching oils" that contain water repellents and fungicides are used as an initial treatment for exposed shingles and boards to give temporary, superficial protection until natural weathering eventually takes over. Understanding and using natural weathering to advantage seems to be among the lost arts. But it frequently is far more gratifying to understand and work with nature than to strive for results in defiance of natural forces.

1—The 'unfinished finish' of this carved catalpa grouse, obtained from weathering for about 14 years, is more appropriate to the nature of the carving than any finish that comes out of a can.

Coating treatments

The most universally used finishes are the transparent coating treatments applied to the surface. The word varnish is sometimes used loosely to include any or all such treatments. Usually, however, it refers more specifically to those clear finishes consisting of tough resins dissolved in oil-based solvents. When the solvent, or vehicle, evaporates, the resin hardens, or polymerizes, and remains firmly adhered to the wood surface. Modern varnishes are specified according to their resins. The newer synthetic varnishes, especially the urethanes, are applied by hand easily and are extremely tough. Various chemical additives can produce a full range of surfaces from high gloss to dull satin. A varnished surface is highly resistant to water and alcohol.

Another traditional favorite is shellac varnish, usually called simply shellac. It is quick-drying, easily applied, adheres well, and although not as water-resistant as other varnishes, is generally appropriate for interior surfaces. Shellac is a natural gum secreted by the lac bug, an insect found in southern Asia. The finish is prepared by dissolving this gum in denatured alcohol. When applied, the alcohol quickly evaporates, leaving a film of shellac. The shellac can be resoftened by alcohol, however, so the finish is not effective on surfaces where alcoholic beverages might be spilled.

The third major coating finish is lacquer. The principal variety has a nitrocellulose resin in a vehicle such as amyl acetate. Lacquers are crystal-clear and available in formulations suited to either spraying or brushing. They harden by loss of solvent but do not build layers as thick as most varnishes.

Before applying varnish-type coatings, the surface should be freshly sanded to avoid raised whiskers, then cleaned well with a tack rag. Woods with open grain—that is, which drink up finishing material, as redwood does—are often sealed with a dilute coat of shellac, with a lacquer sealer or with a dilute coat of the final finish itself. When a perfectly smooth surface is wanted, woods with large open pores, such as oak or walnut, should be given a coat of paste wood filler. Like much advice in finishing, fillers are a matter of taste, not an obligatory step. If you like the surface open pores impart, there is no rule requiring you to fill them.

From this point, a major problem in finishing is failure to study the label on the can. The label will specify suitable staining and sealing materials, and usually will warn against incompatible materials. Many modern resin varnishes must be recoated within a specified time, else the second and subsequent coats will not bond with the first.

Another great difficulty encountered in applying varnish-type finishes in the home shop or small commercial shop is dust. The surface tension around a dust particle landing in a film of wet finish causes a noticeable blemish, which must later be worked out. For those who must do finishing in the same location as woodworking, it is impossible to produce even a reasonably dust-free surface. The faster-drying lacquer and shellac finishes have an advantage in these situations.

Bubbles are another problem. They sometimes result from striking the brush off on the side of the can. Then the bubbly varnish drips back onto the liquid surface. Bubbles can largely be avoided by striking off the brush into an empty coffee can. A few bubbles are to be expected, but if the varnish is thinned properly they will break, and the film will settle without a blemish.

Temperature change also can cause serious bubble problems. I stumbled onto this fact one time when I decided to avoid my dusty cellar shop and varnish a yellow birch candlestand in the most dust-free room in the house—the dining room. I spread my drop cloth, set everything up, dusted the room and returned to the cellar to let any remaining dust settle. Meanwhile I strained the varnish and got the brush worked in. Returning to the dining room, I gave the stand a last whisk with the tack rag and started by varnishing the underside of the top. Everything appeared to be going well, but as I finished the second leg I noticed the first leg was speckled with bubbles. As I brushed out the bubbles on the first leg, I could see more developing on the second leg. I was baffled. The brush was in perfect condition, and the varnish can was virtually free of bubbles. After long puzzling moments of watching bubbles appear before my eyes, I noticed that each bubble developed at the end of a vessel opening. The cellar was considerably cooler than the dining room. When I brought the work into the warmer room, the air inside the wood gradually began to expand. Each vessel had become a minute bubble pipe! I've since verified my observation by controlled experiments in the laboratory. Since then I always make certain that a piece to be varnished is kept at an even temperature or moved from a slightly warmer to a slightly cooler location just before finishing. No more bubbles.

Since everything I varnish seems to wind up with dust specks, I sand lightly between coats with 280-grit paper on a flat block just enough to knock the tops off the dust spots, then go over the whole surface lightly with 5/0 steel wool followed by the tack rag. After the final coat, I use 600-grit paper on a good flat block and work carefully to level every high spot flush with the surroundings. Here is where corns on the paper cause trouble. Then I rub with pumice and oil and finally rottenstone and oil. Last of all is a rub with lemon oil or sometimes paste wax. No question about it, it makes an attractive finish. But during all these stages of work you really become aware that you are working on the finish coating, not upon the wood.

Penetrating finish

The third general type of finish is *in* the wood, not on the surface. Oil finishes, or the penetrating resin-oil finishes such as Watco and Minwax, are in this category. In application the finish is simply flooded onto the surface and as much as possible is allowed to soak in. Additional finish is applied to any dry spots that develop. After 15 to 30 minutes, any remaining liquid is removed from the wood surface, and the surface is buffed dry in the process. Most of the finish remains in the cell cavities or is absorbed by the cell walls. Only an imperceptible amount covers the exposed wood surfaces. Repeated coats give more complete and deeper treatment and result in a very slight build on the surface. Enough finish remains to accent the figure of the wood, but there is the illusion that none really covers the surface *(1)*. This finish is a delightful compromise where the natural wood surface is preferred but some protection is also needed. A penetrating oil finish also can fill the open pores of the wood if it is sanded with fine-grit wet-dry paper while it is soaking in. This makes a fine dust of wood mixed with finishing material, and subsequent buffing pushes this mixture into the pores and levels the surface.

Linseed oil is a traditional favorite, but since it does not harden completely, it may later bleed out on the surface. It also attracts dirt, yellows in color and darkens the wood. Commercial penetrating finishes have resins that polymerize in time and become permanently set in the wood, consolidating and hardening the surface.

A real advantage of oil finishes in the small shop is that there is no trouble from dust, because any remaining liquid is wiped free. They are truly quick and easy to apply. However, experience soon reveals that the time saved in finishing with oils might well be invested in preparing the surface. Penetrating finishes are the acid test of surface condition, especially smoothness and quality, for every imperfection is not only exposed by lack of surface build, but in fact accented even more than if the wood were left unfinished. It really pays to "de-whisker" the surface, because the real quality of an oil finish is determined by the surface quality of the wood itself. This is in contrast to a varnish finish, which masks many slight imperfections, scratches and tear-outs in the wood and where the final surface belongs to the varnish, not to the wood.

1—Mushroom carving stands 5 in. tall and is finished with several coats of Watco oil and a coat of paste wax.

Combinations and compromise

I love to experiment and I always wind up trying something I've never tried before *(2, 3)*. I especially like to try to amalgamate varnish and oil finishes. A good starting point is a mixture of one part boiled linseed oil, one part alkyd varnish and two parts turpentine. Go heavy on the turpentine for better penetration, go heavy on the varnish for more build. Don't go heavy on the linseed oil, but you might substitute something else, such as tung oil. The result is somewhere in between a varnish finish and a commercial penetrating finish. It wipes on dust-free but gives more build, depending on proportions.

In recent years I have become intrigued with tung, or chinawood, oil. It is about as close to the one-shot all-purpose finish as I can imagine. Tung oil is an aromatic natural drying oil that is obtained from the nut of the tung tree (*Aleurites* spp.), originally from China but now grown extensively in the southern United States. Commercial preparations contain a drying agent and can be used as obtained. Tung oil can be applied directly to the wood surface much as other oil finishes, but it's a good idea not to allow it to remain more than about 15 minutes before wiping clean. This is because it sets up more quickly than most oil finishes. After a couple of hours drying, the surface can be recoated. It gives a better build than the usual penetrating oil finishes, and it holds up well outdoors. I have found it to be the most satisfactory treatment for outdoor thresholds. I have also used it for everything from kitchen furniture to woodcarvings and wooden jewelry.

2—Black walnut carving, measuring 8 in. tall, is finished with an oil/varnish mixture. Hoadley rough-sanded the hair, fine-sanded the face and left the marks of the chisel on the neck.

3—Black shoe polish with a top-coat of paste wax makes an attractive finish for this 11-in. tall eastern white pine carving.

Slowing moisture exchange

Although a primary objective of finishing treatments is to prevent moisture exchange, no finish is totally effective. Given enough time, moisture will be adsorbed into wood from a humid atmosphere, or will escape to a dry atmosphere, through any finish. But as discussed earlier, the important role of the finish is to retard the rate of exchange enough to buffer the temporary extremes of high and low humidity.

Obviously, some finishes are better than others in this respect. But despite the importance of the subject, the information regarding the relative effectiveness of different finishing materials has been inconsistent and incomplete. Only some generalizations can be made. In my limited experience, for single-coat applications, marine spar varnish and urethane varnish give best protection, brushing lacquer and shellac somewhat less. Linseed oil gives virtually no protection, penetrating oil finish is better but still only slightly retards moisture pick-up. However, with any finish, multiple coats make a tremendous improvement. In this regard, finishes are rated on a scale from 100% effective (no moisture pick-up as the humidity increases) to 0% effective, which describes unfinished wood. The U.S. Forest Products Laboratory rates single coats of linseed oil as having only 1% effectiveness, but three coats are 20% effective. I have heard of linseed oil being used to "waterproof" gunstocks—by the application of a mere 20 coats, each one hand-rubbed. FPL figures also indicate that whereas one coat of varnish gives only 8% effectiveness, three coats are 73% effective.

It is easy to imagine that the first coat of finish, while penetrating, may disperse itself into the cell structure. After curing, however, it provides a barrier that concentrates subsequent layers at the surface to form a more complete barrier. It would seem then that in the case of penetrating finishes, multiple coats are especially crucial to developing moisture retardance.

If there is anything worse than no moisture barrier at all, it's an uneven moisture barrier, which would allow moisture to be adsorbed or desorbed unequally in different areas of the wood. In carcase pieces, for example, it is tempting to work conscientiously on the exposed surfaces and forget the insides. It is crucial that all sides of every board receive equal finish. The concept of balanced construction also applies to finishes. Forgetting this is a major cause of surface cupping. For this reason, many experienced cabinetmakers finish all the wood in a carcase before final assembly, taking care not to drip finishing material onto gluing surfaces, which may even be protected with masking tape. In frame-and-panel construction, this is the only way to be sure that an unfinished line will not appear along the edge of a raised panel. It is also an effective way to avoid having to rub down finish in tight corners.

Evaluation of surfaces

One of the most effective ways to evaluate the quality of finished surfaces is by observing line patterns reflected at low angles across a surface. You need only a target card with boldly ruled horizontal, vertical and diagonal lines (Figure *1*, page 180). Hold this card perpendicular to the surface, and examine the lines reflected on the surface. The clarity of the reflection will reveal the relative uniformity of gloss developed in the finish. Waviness or discontinuities in the lines will indicate the lack of surface trueness, evenness and smoothness. Generally, such defects as sunken joints, raised grain or lathe checks can be pinpointed.

Modifying Wood

CHAPTER 11

For most uses, the woodworker finds that wood is satisfactory in its natural condition, requiring only that the moisture or sap be dried down to a level consistent with the environment. In special cases, however, it may be desirable or even necessary to bend the wood to an unnatural shape or to impart special properties to the wood by chemical treatments. Various special chemical treatments are important commercially, but too numerous to be included here. The chemical processes that can be used conveniently by the average woodworker fall into two categories: treatment to prevent the development of wood-inhabiting fungi or insects, and treatment to reduce or eliminate moisture-related dimensional change.

Preservative treatment of wood

When wood is used in a location where its moisture content can range above 20%, wood-inhabiting fungi will probably take up residence. Termites and carpenter-ant infestations also are encouraged by high moisture content, and some insects are troublesome even in dry wood. Certain wood species have heartwood extractives that resist the attack of fungi and are termed decay-resistant or durable woods (listed in *Table 4*, page 36), and certain woods have selective resistance to insect attack. In many cases, however, where conditions favorable to biological deterioration cannot be avoided, and where resistant species are not available, the best alternative may be to treat the wood with a substance that will give it the desired durability. Such chemicals are called **wood preservatives.** (This term sometimes includes treatments to make the wood nonflammable, although the term **fire-retardant** is preferred for such materials.)

The ideal preservative would readily penetrate the wood and would be permanent, toxic to fungi and insects, safe to handle, colorless, compatible with coatings and finishes, and of course, inexpensive. No one chemical has yet been developed that has all of these attributes, but a wide array of chemicals with various advantages have emerged for specific purposes.

Coal-tar creosote has been used commercially to preserve such things as railroad ties and utility poles. Oil-borne preservatives such as pentachlorophenol and copper naphthenate, and some water-borne preservatives, mostly salts of copper, zinc, chromium and arsenic, also have been employed—each of these has specific advantages and disadvantages. Various proprietary brands are sold as building materials—it's not safe to try mixing your own.

The key to using preservatives is penetration. Only areas of the wood that are penetrated by preservative chemicals will be protected. A first consideration, then, is choosing the most penetrable wood. Generally, sapwood or species with low extractive content (e.g., spruce)—ironically those that usually have the least natural decay resistance—are the best choice for preservative treatment. Except for very thin pieces, the only way to attain any worthwhile degree of penetration is under pressure. Commercially, this is done using cylinders that produce pressures up to about 150 psi, and sometimes also with vacuum treatment or elevated temperatures. Since such conditions are beyond the capability of the average woodworker, it is usually most logical to buy commercially treated lumber for use where constant moisture problems prevail.

1—PEG-treated tulipwood bowl made by Ed Moulthrop is 36 in. in diameter.

Non-pressure treatments include soaking, dipping and brush application. For any use involving contact with the soil or constantly wet or moist conditions, such as fence posts or sills lying on bare ground, nothing less than immersion in preservative for several days will be worth the expense and effort. The wood should be at least air-dried to facilitate penetration and to ensure that no further drying occurs after penetration, which might open checks and thus expose untreated wood.

Where possible to do so safely, heating the treating solution will improve penetration. Heating the wood expands and drives out air from the cell structure; when allowed to cool, the remaining air contracts, drawing the preservative solution into the cell structure. Cutting open test pieces indicates the degree of penetration. Chemical indicators are available for determining the penetration of colorless materials.

Brush and dip methods give only superficial treatment and should be relied upon only where the wood needs surface protection, as with above-ground parts of a structure exposed to intermittent rainfall. Dip treatments (total immersion for a few minutes) will do a far better job than brush treatment of reaching vulnerable voids such as bolt holes, deep end checks, splits and loose knots. Dipping or flooding the surface may give fairly good end penetration, but side-grain penetration by either method may be as little as 1/32 in., varying somewhat according to species.

The most common mistake in using surface treatments is application after rather than before construction. Consider an outdoor structure such as a deck, porch, bench, boardwalk, railing or flower trellis. During a rain, water seeps and settles into joints and crevices and is absorbed by the wood, especially into concealed end-grain surfaces, such as the bottom ends of vertical posts resting on horizontal surfaces. After the rain, most exposed surfaces, especially side-grain surfaces, dry quickly enough that fungal activity does not make significant progress. In hidden joints, water is held longer, absorption is prolonged and drying is delayed. The hidden surfaces of joints are therefore the most vulnerable places. These places are seldom reached by preservative brushed on after construction. For this reason, every effort should be made to apply preservative to bolt holes, joint surfaces and inside mortises before assembly. In nailing exposed horizontal surfaces, such as deck boards or stair treads, nail heads should be driven in flush. Setting nails below the surface exposes end grain and creates a water pocket.

Preservative treatment, especially superficial brush treatment, can never compensate for poor design. For exterior application, promoting runoff and preventing entrapment of water should be primary considerations. Many modern fungicidal preservatives are both water-repellent and fungicidal. They are called water-repellent preservatives and are marketed as such. In combination with good design, brush application of these preservatives can be quite effective. Remember, however, that no brushed-on preservative will last forever. The chemical itself eventually leaches out of the wood, becomes diluted or simply degrades after prolonged exposure to the weather. This deterioration takes place from the exposed surfaces inward, another reason why depth of penetration is so important.

In some instances, such as on millwork, trim and window sash, a superficial treatment does give enough protection. A brief dip of window sash, for example, will keep wooden interior surfaces bright that might otherwise discolor when periodic condensation increases the moisture content.

Stabilization treatment

Another category of chemical treatment includes the methods designed to prevent moisture-caused dimensional change in wood. Although many stabilization processes have been developed experimentally or are used commercially, few are readily appropriate for use in small-scale workshops.

Wood-plastic composite (WPC)—One such process is the impregnation of wood with chemicals that are then transformed into rigid plastic. The resulting product is termed wood-plastic composite (WPC). The chemical most practical and logical for small-shop use is the monomer (single molecule) form of methyl methacrylate. A special apparatus forces the chemical into the wood by vacuum and pressure, and the plastic is then cured with heat and a catalyst. This curing, or **polymerization,** links together monomers into multiple molecules, called polymers, in which form the plastic is hard and stable. (Commercially formed polymethyl methacrylate is familiar under the trade names Plexiglas and Lucite.) A composite structure of wood and plastic is thereby formed. In addition to being quite stable, WPC has superior hardness, toughness and abrasion-resistance in comparison to untreated wood. Natural wood figure can be accented by adding dye to the monomer. WPC can be polished to a high luster, and further finishing is not necessary. WPC is available commercially in small pieces from some suppliers of woodworking materials, and is popular for novelties such as chess pieces, jewelry, bowls and paperweights.

PEG-1000—The stabilization treatment that is probably the most widely used by woodworkers is polyethylene glycol (PEG), a polymer of ethylene glycol, the basic ingredient of most automobile antifreeze (although antifreeze cannot be used to stabilize wood). The most appropriate polymer for stabilization has an average molecular weight of 1000, hence the designation PEG-1000. At room temperature, this chemical is a whitish solid with much the same appearance as paraffin wax. It melts to a syrupy liquid at 104°F and is very soluble in water.

PEG-1000 stabilizes wood by preventing shrinkage. When green or fully swollen wood is soaked in a solution of PEG-1000, the molecules of PEG-1000 replace the water molecules in the cell walls. They remain in the cell wall when the wood dries and "bulk" the cell wall, that is, keep it in its fully swollen dimension.

PEG will not prevent already dried and shrunken wood from swelling. To be effective, therefore, PEG must penetrate thoroughly—it depends on diffusion through free water in the cell cavities to reach the cell walls. So not only must the wood be green and fully swollen, but its moisture content must be above the fiber saturation point (about 30% moisture content). When wood at 100% moisture content is treated, shrinkage is reduced by nearly 90%. The first key to success with PEG is to keep the wood at its original green moisture content. For woods whose green moisture content is low, such as white ash or the heartwood of many conifers, expect less than maximum stabilization.

The second key to success is grain direction. Just as liquid movement and diffusion are many times greater along the grain than across, PEG penetrates end grain much better than side grain. In practice, you cannot expect effective penetration into boards (of average density) much more than an inch in thickness, or into cross-sectional pieces more than 3 in. to 4 in. along the grain. These limitations suggest the most appropriate applications of PEG: stabilizing cross-sectional discs of wood, and small turnings or carvings.

Penetrability varies among species. In general, the

greater the density, the more difficult the wood is to treat. Woods with high extractive or resin content may also resist PEG penetration. Sapwood and heartwood within a species may behave differently—sugar maple sapwood treats well for its density, but the heartwood is considered untreatable.

Strength of solution, temperature and soaking time are critical factors. Solutions of 30% to 50% PEG work best. Stronger solutions penetrate better but require more chemical to prepare. Equal parts (by weight) of PEG and water yield a 50% solution: 10 lb. of PEG in an equal weight (4.8 qt.) of water yield about 8.5 qt. of 50% solution. To mix a 30% solution, dissolve 3 parts of PEG in 7 parts water, that is, 4.46 lb. of PEG in 5 qt. of water, for 7 qt. of solution. Melting the PEG (in a double boiler) or fragmenting it speeds the process. Hot water also accelerates dissolving.

It is best to check the solution density with a hydrometer, as well as the temperature. At 70°F (21°C) a 30% solution will have a specific gravity of about 1.05; a 50% solution about 1.083. As the wood absorbs the PEG, the solution may become diluted, or if water can evaporate from the soaking vat, the solution may become more concentrated. A hydrometer, along with the graph below *(1)*, will indicate when it's time to add more PEG or more water. Leftover solution can be reused by restoring it to its original strength.

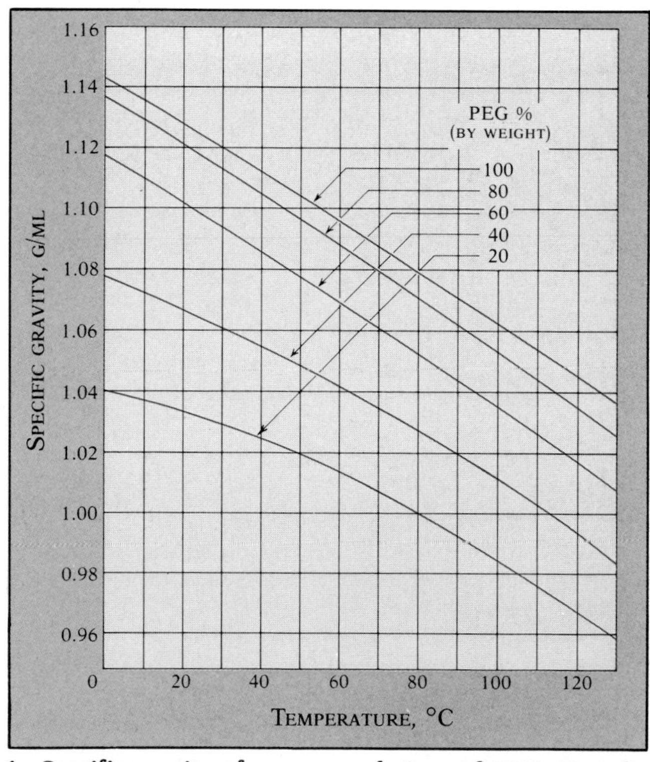

1—Specific gravity of aqueous solutions of PEG. Periodic monitoring with a hydrometer allows one to keep the solution up to original strength.

Soaking procedure and equipment can be quite simple. The objective is to find a container about the same size and shape as the wood being treated, to minimize the solution volume needed for immersion. Any ceramic, enamel or plastic containers will work. Because metals (other than stainless steel) discolor many woods, line metal containers with a puncture-resistant rubber sheeting or 6-mil polyethylene. For large, long or irregular objects, you could make a wooden or plywood tub and line it with polyethylene. Where a great deal of treating will be done, line the box with fiberglass.

In open tanks or vats, some means of weighting or holding down pieces has to be improvised, along with separating strips to ensure that the solution circulates around all surfaces. You may get layering of high or low concentration during soaking, so stir at least daily with a stick or paddle.

The solution penetrates much faster at temperatures of up to 160°F, so where the project warrants it, it is worth heating the treating vat. A thermostatically controlled heat source would be ideal (my wife's Crock-Pot would make a perfect soak vat for 6-in. discs). Even an improvised tank can be fitted with a thermostatically controlled immersion heater, and the walls insulated with fiberglass wool. A simple way to get heat is to put the tank over or near a radiator. Placing the vat outdoors in the summertime to be warmed by the sun will work, except that it will be difficult to assess the treatment in terms of reproducing it closely another time. I know of one PEG supplier who sells insulated vats with thermostatic heaters.

How much PEG retention, how much stabilization and how much soaking time are obviously the critical questions; unfortunately, they are hard to answer and are best considered in terms of specific categories of treatment.

The first category involves pieces of wood with the pith included, where the greater tangential than radial shrinkage would result in radial cracks. Cross-sectional discs, intended for clock faces, coasters, lamp bases or even tabletops, are typical examples. Although such pieces usually cannot be dried without defect when untreated, they are ideal subjects for PEG stabilization as long as their thickness (i.e., dimension along the grain) is not greater than 3 in. to 4 in. It is impossible to specify precise soaking schedules, but start with the guidelines suggested by the Forest Products Laboratory *(2)*. These treating times have been successful in preventing cracks in black walnut discs treated green.

The advantage of elevated temperature and increased solution concentration is obvious. For lower-density woods such as white pine, cottonwood or yellow poplar, try half to two-thirds the times listed; for higher-density woods such as beech or hickory, double or triple the times. I have soaked cross sections of red oak, 6 in. in diameter and about 1 in. thick, in a 50% solution for a week at 120°F. No defects developed whether they were then dried rapidly or slowly. But red oak is quite permeable; I would not expect the same results with white oak. Experience will help identify the easy and difficult woods to stabilize: I have given up trying to treat sugar maple, but I have had excellent results with oak, black walnut, butternut and elm. Others report good success with a wide variety of species including pines, spruce, redwood, cottonwood, willow, soft maples, beech and apple. Cherry may tend to develop internal checks (honeycomb) if treated at temperatures above 110°F.

A second major category of PEG application is where any dimensional change due to moisture variation is undesirable in the final product, and maximum stabilization is the objective. A good example is gunstocks for target rifles, where small changes in the stock's dimension may affect accuracy. Another application involves the maple strips used in core laminations for archery bows, where humidity changes can cause twisting. Other PEG-stabilized products are musical instrument parts, bases and framing, large engraving blocks and patterns. PEG-treated wood can be glued with resorcinol or epoxy, but polyvinyl (white) and aliphatic (yellow) glues do not work well. Treated lumber should be dressed with a jointer or surface planer as in routine gluing and laminating. Removing surface traces of PEG by first scrubbing with toluol and washing with methanol (methyl or wood alcohol) produces the best glue joints. If joints will not be subjected to high stress, washing with alcohol alone will suffice.

PEG is also used by bowl turners (see Figure *1*, page 190). It is difficult to dry blocks of considerable thickness (say 5 in. to 6 in.) without degrade. Thus turners often rough-turn the bowls in the green condition. But when an untreated rough-turning is dried, unequal dimensional changes may distort the bowl so much that final turning is impossible. With hard-to-season species or with irregularly grained pieces such as burl or crotch, where drying defects are common, PEG treatment of the rough turning will usually overcome these problems. If bowls are green-turned to wall thickness of ½ in. to ⅝ in., an average species like black walnut can be treated with a moderate schedule (e.g., three weeks at 70°F in a 30% PEG solution or one week at 140°F in a 50% solution) to achieve a high degree of stabilization.

PEG is described by the Forest Products Laboratory

2—Suggested period of soak for walnut discs. Below, discs properly treated with polyethylene glycol will dry without the usual radial cracking.

Solution concentration and temperature	Up to 9 in. in diameter and 1 in. to 1½ in. thick	More than 9 in. in diameter and 2 in. to 3 in. thick
30%, 70°F	20 days	60 days
50%, 70°F	15 days	45 days
30%, 140°F	7 days	30 days
50%, 140°F	3 days	14 days

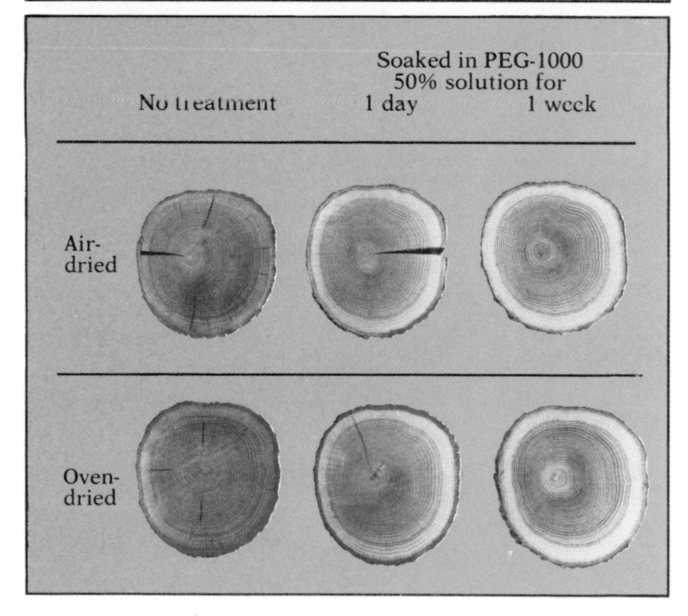

as "non-toxic, noncorrosive, odorless." It has been declared safe as an ingredient in cosmetics, ointments and lotions, and as a binder for pharmaceutical tablets. Results of tests involving laboratory animals also support this classification. Where PEG-treated wood is used for utensils or containers such as bowls, PEG contamination of food does not appear to present any problems. Possible toxicity of finishes should, of course, always be considered with any food containers.

PEG treatment is also applicable to green woodcarvings *(1)*. The green wood is carved to within ¼ in. or ⅛ in. of the final surface. Between carving sessions the piece is kept wrapped to prevent drying. Then the wood is treated enough to stabilize the surface to a depth beyond that which will finally be carved, thus avoiding surface checks. Minor changes in overall dimensions are usually acceptable. For a rough carving of approximately 8 in. by 6 in. by 24 in. in yellow poplar, I found that a four-week soak in 30% PEG solution at room temperature was sufficient. For me as a hobby carver, the drawback is that I seldom finish a sizable carving quickly, so storing the green wood without stain or molding becomes a problem. Sometimes I put the carvings in my freezer. Remember, PEG treatment will offer only surface control of dimension; pieces containing the pith will still develop large radial cracks due to the tangential/radial shrinkage differential.

PEG has also proven useful in stabilizing waterlogged artifacts recovered by archaeologists. Long immersion in water causes cell walls to break down by hydrolysis, and exaggerated shrinkage results when the wood is dried. PEG can stabilize such material. Woodcarvings produced in tropical climates will suffer disastrous checking when imported to drier climates. Soaking and then treating with PEG can prevent troublesome checks.

A final category of application is superficial treatment to eliminate drying degrade in thick planks or irregularly grained stock, used perhaps for furniture parts, carvings or bowls. The objective is to get just enough penetration to control surfaces, where many drying de-

1—Head, above, was rough-carved in a piece of cork tree *(Phellodendron sachalinensis)*, then soaked for three weeks in a 40% solution of PEG-1000 (center). The treated piece was then dried in an oven at 212°F, without checking, and finish-carved. The carving, 6 in. high (right), was finished with a penetrating oil finish.

fects begin; the treated surface material will later be totally removed from the completed item. Thick planks, soaked for a week to 10 days in 50% solution at 140°F, can be air or kiln-dried without major defects even under drastic conditions. Surface checks, even end checks, are virtually eliminated. The retarded outer shrinkage apparently relieves internal compression that might result in collapse during early stages of drying. Since PEG is highly hygroscopic, it probably also reduces the severity of the moisture gradient by holding moisture near the surface. Elimination of surface checks, which can become internal checks, helps avoid honeycombing.

Treatment will leave a heavy concentration of PEG at the surface. In precarved sculpture, rough-turned bowls or thick lumber, this excess is routinely removed as surface material is machined away. In large cross sections for tabletops, where the original surface may have been cut with a chain saw, the surface might best be leveled using a router with a guide rack fixed above the disc surface. Abrasive belts will load with PEG-saturated dust, though they can be cleaned with a bristle brush and warm water. Other than sanding, PEG-treated wood is easier to carve and machine than untreated wood. The PEG lubricates the tool, and there is less splintering.

Small discs cut smooth in the green condition and requiring only fine sanding should be leached by flooding with hot water, scrubbing and sponging to remove excess PEG. Bark edges especially will retain excess PEG, which should be carefully rinsed away.

Before finishing, surfaces should be dried, which can be accomplished in a variety of ways. For thorough stability, drastic means, including ovens and direct sunlight, can be employed without harm. Wood dried in an oven, however, may darken noticeably. Pieces can be placed over radiators or in improvised drying boxes heated with light bulbs or simply left to come to equilibrium at room conditions.

It is not a good idea to leave treated objects unfinished. PEG is very hygroscopic, and treated wood will pick up enough atmospheric moisture in humid weather to feel damp on the surface. Surfaces that are left unfinished, especially when exposed to the light, may also develop a sooty and dirty discoloration.

After final sanding of dried pieces, apply a finish right away. Many finishes commonly used for untreated wood (such as shellac, alkyd varnish, lacquer) cannot be used. The two best finishes are oil and moisture-cure polyurethane-resin varnishes, and PEG suppliers have formulated special varnishes.

I use regular Watco Danish oil, as it is easy to apply and gives the finish I prefer. Three or four well-rubbed coats are enough if the surfaces have been sanded smooth. It is the best finish for discs with the bark on, because bark retains heavy concentrations of PEG and varnish will not adhere.

If you prefer a built-up finish, try four or five coats of moisture-cure polyurethane, sanded between coats with 200-grit paper. The heaviness of this type of varnish makes it difficult to apply without leaving brush marks. A couple of finish coats of conventional urethane varnish, rubbed with pumice and oil, will leave a flat luster.

In considering whether to use PEG, the cost of the chemical may be important. At this writing it retails for about $2 per pound. Ten pounds will make enough solution to treat a stack of discs in a small bucket. But a large cross-sectional disc, say 30 in. in diameter and 3 in. thick, will require several gallons of solution and some extra PEG to maintain the concentration. One final suggestion: Whenever you treat wood with PEG, always process a few similar pieces *without* treatment. This will give you a clearer indication of what effect the PEG really has. Without the control material for comparison you will not be able to assess the improvement. It will help you decide whether the treatment was worth it.

Bending solid wood

Bending wood is among the ancient arts of woodworking. That is not surprising, since bending wood into a curved form has two distinct advantages over machining. First, it wastes much less wood and requires narrower pieces, and second, it reorients the grain to the curving axis of the part and results in much greater strength by avoiding cross-grain weaknesses.

The question is how to bend a piece of wood to a prescribed shape—often well beyond its normal breaking point—and keep it there.

In some cases the weaker strength properties of green wood can be taken advantage of. Green wood can be bent beyond its proportional limit, and allowing for a wee bit of elastic springback, can be dried in the bent position. Back spindles for chairs are traditionally formed this way, and the weaving of ash splints in basketmaking is another traditional application *(1)*. But since green wood must then be dried, the resulting shrinkage and drying defects may present a severe problem. What's more, bending to even greater extremes than can be accomplished with green wood is sometimes desired. Some means of additional plasticizing of the wood must therefore be found.

The age-old method for plasticizing is with heat and moisture, which in combination are capable of markedly extending the plasticity of wood. Let's review our earlier discussion of strength properties, and in particular, beam mechanics. A bent beam is in compression on the concave side and in tension on the convex side. When a beam is bent beyond its elastic limit, greater deformation occurs in compression than in tension. In fact, about 1% elongation in tension causes failure. Steaming the wood increases the compressibility to as much as 30% or more, but only slightly increases the elongation ability in tension, perhaps no more than 2%. Yet by bending thin pieces to moderate curvature, many woods can be successfully shaped without failure.

In the trout-net bow *(2)* for example, the ½-in. thick strip of air-dried walnut bent to form like wet leather after 15 minutes of steaming. However, for sharper bends and for thicker stock, the 1.5% to maybe 2.0% tensile strain limit would be exceeded in a "free" bend. To

1—Two traditional applications of green bending: Post-and-rung chair by John D. Alexander, Jr., has bent oak posts and bent hickory back slats. Martha Wetherbee's Shaker-style basket is made of green-bent brown ash splints and handle.

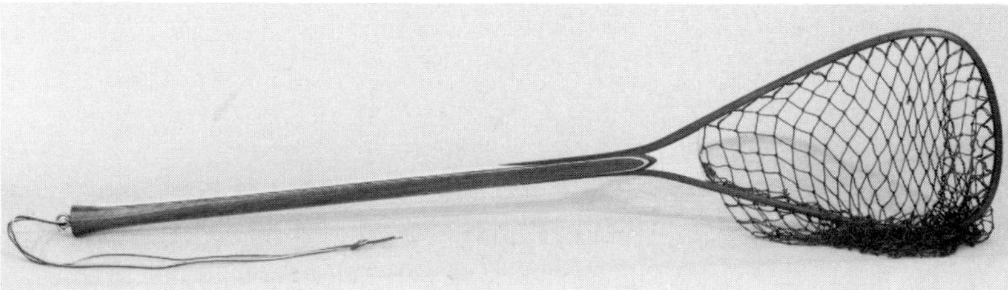

2—The bow of this trout net was made from black walnut free-bent after steaming, without a tension strap.

restrict elongation of the convex face, a tension strap is used to restrain that surface of the wood *(3)*. The strap is made of steel or other metal to which end blocks are attached. Because the tension strap acts as part of the beam, it carries the stress on the tension side of the bend. As bending takes place, the neutral axis is effectively shifted to or very near to the tension strap, so most of the bending strain in the wood must occur in compression. But since steaming will greatly increase the wood's plasticity in compression, the extreme compressive strain, if uniformly distributed, can be absorbed without undesirable results. A typical bending jig with tension strap, for making a reverse curve in one plane, is shown below *(4)*. The critical interrelated factors that enter into bending are species, moisture content, steaming time and geometry. The ratio of the radius of the bend to the thickness of the stock usually determines whether the wood can be bent free or whether a tension strap will be needed.

3—When wood is bent without restraint, tension and compression stresses are balanced on opposite sides of the centrally located neutral axis. A steel strap with end-block restraints, added to the convex side of the bend, carries most of the tension stress, shifting the neutral axis toward the strap so the wood undergoes mainly compression strain.

Softwood species are not well suited to steam-bending. Among the hardwoods, there is extreme variation in bending success. The species that bend well include ash, hickory, beech, walnut, white oak, birch, elm and hackberry. Judging by its ability to take a bend in the green condition, I would guess catalpa is another species worth trying.

Discussions are endless about what might be the best moisture content for bending stock, and how long the piece should be steamed. Such considerations are interrelated, since the drier the wood is, the stronger it is to begin with and the longer it has to be steamed to become sufficiently plasticized. Wood bends best at a moisture content near its fiber saturation point, but then the wood must be dried down after bending and drying defects may occur. Low moisture content would eliminate drying defects, but the bending would be riskier. A fair compromise is air-dry levels (12% to 20% moisture), wet enough to bend fairly well, but dry enough to minimize drying defects. Because surface discoloration is usual and resurfacing generally is needed, some feel that drying distortion is not a serious problem and prefer to begin with wetter stock. In some species having extremely

4—Bill Keyser's bending setup (above) for a reverse curve in one plane, with tension strap and end blocks clamped in place.

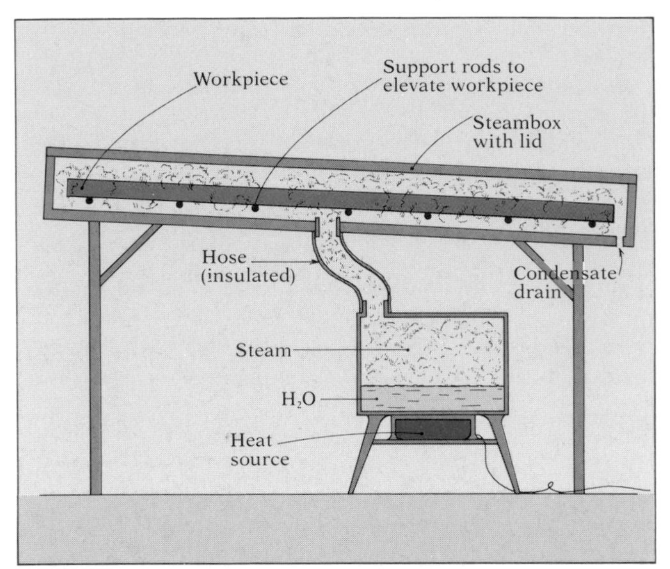

1—The basic elements of a steaming apparatus are the steam generator and the box; the rest depends on ingenuity. The main considerations are insulating wherever possible, allowing condensate to drain from the chamber, introducing steam into the chamber at a point along the workpiece where the bend is most critical, and supporting the workpiece to ensure steam circulation and to avoid immersion in the condensed water.

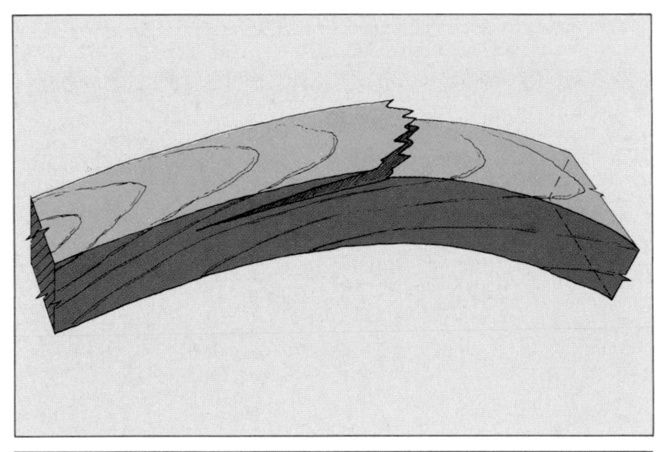

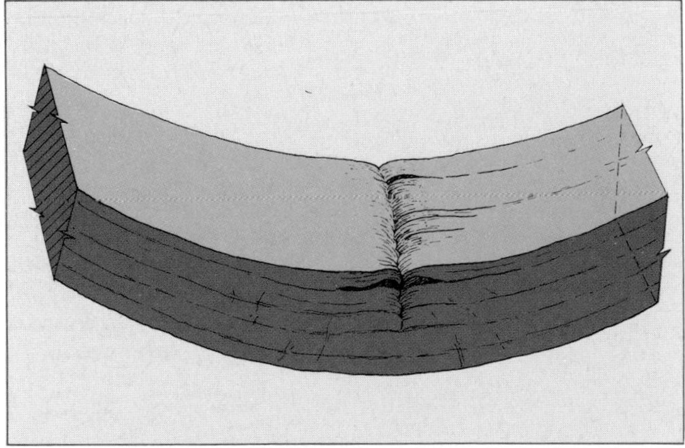

2—Top, failure in free-bending wood with cross grain. Above, a typical compression failure.

high green moisture content, the cell cavities are virtually filled with free water. If bending stress is applied, they cannot deform normally, and the internal hydrostatic pressure developed may cause the cells to burst.

As a general starting point, the rule of thumb is, "one hour of steaming for an inch of thickness of air-dried wood, one half-hour per inch of green wood." Everyone must adjust that rule to the circumstances at hand. As an old man once told me, "Whatever works best for you, you use *that.*" (The advice was for something else, but it would apply to steam-bending, too.) It is also important to remember the effects of heat on wood properties (as briefly summarized in Chapter 7) to recall that permanent reduction in strength can result from prolonged heating. Excessive steaming should be avoided. When the steaming time is held to the minimum necessary for successful bending, strength losses should be no more than 10% to 20%.

The curvature that can be achieved using steamed 1-in. air-dried stock varies. The better bending species can be free-bent to a radius of 12 in. to 15 in.; using a restraining strap, bends to a 2-in. radius or less can be achieved. It always amazes me that bent stock stays in place as well as it does, especially if it can stay restrained on the form while its moisture content equalizes. Upon drying, the curve will often bend even more sharply, due to the greater longitudinal shrinkage of the over-compressed concave side.

Intuition, rather than instruction books, has probably designed most steaming arrangements. Any vessel capable of boiling water and directing it into a steam chamber will do the job *(1).* A wallpaper steamer makes a nice steam source. For a steam chamber I have used everything from aluminum irrigation pipes to aluminum downspouts. I avoid ferrous metals because of discoloration problems. Although I've never tried it, I like the idea of building a steam chest out of plywood, because its insulating value should minimize condensation compared to a thin-walled metal enclosure.

My limited experience has left me with one very strong impression: Don't waste time trying very severe bending with wood you don't consider to be nearly perfect. Any irregularity is apt to concentrate the strains and develop a compression buckle, the usual catastrophe once you have tension restraint worked out *(2).* Any knots, other irregularities or reaction wood should be expected to cause trouble. Straight grain is a must. Decay, shakes, surface checks, pith and similar defects should be eliminated. Care should be taken to condition the stock to a uniform moisture content before bending. If you are choosy about the quality of the bending stock, steam-bending will probably be easier and more fun than you imagined.

Plasticizing with ammonia

In recent years, much interest has focused on treating wood with ammonia to plasticize it for bending. Ammonia treatment is more effective than steaming, apparently because ammonia interacts with the lignin as well as with the cellulosic portion of the cell-wall structure. Species that bend well with steam are most readily bent with ammonia plasticizing, and even most species considered unsuited to steam-bending can be successfully plasticized with ammonia. Experiments in bending thin ammonia-treated hardwood strips (1/16 in. to 1/2 in. thick) have shown that the technique can be applied to creating art forms and sculpture *(3)*, functional bentwood objects such as room dividers and lamp bases, novelties, picture frames and furniture parts, as well as sporting equipment.

Two basic systems have been developed: immersion in liquid anhydrous ammonia at atmospheric pressure, and treatment with gaseous anhydrous ammonia in closed chambers at 145 psi pressure *(4)*. (Household ammonia, commonly used for cleaning, is a water solution of ammonia and useless in plasticizing wood.) Depending on the permeability of the species, treatment takes from a half-hour for 1/16-in. veneers to many hours for 3/4-in. lumber. Upon removal from the liquid or gas treatment, the pieces can be bent to shape with little springback. Within minutes to an hour, the ammonia dissipates from the wood sufficiently to set the bend in place, and eventually stiffness and rigidity are restored virtually to original levels.

3—Sculpture by Anni Berman of ammonia-plasticized wood.

In spite of the striking results possible with ammonia treatment, it is hardly a panacea. Although the scientific principle is straightforward enough, the process involves expensive equipment in addition to working conditions that are disagreeable and potentially hazardous. With the liquid immersion process, heavy-duty refrigeration equipment is required to keep the ammonia below its boiling point of −28°F (−33°C). With gaseous treatment, a pressure chamber with connected holding tanks and pumping equipment is necessary. Therefore, only people experienced in the handling of hazardous chemicals should experiment with anhydrous ammonia. Because the ammonia fumes emitted while treated wood is being bent are extremely offensive, work must be done in the open air, with powerful exhaust facilities, or while wearing a gas mask and protective equipment. Although plasticizing with ammonia may have important future potential for commercial woodworking, it is probably not likely to replace simple steam-bending in the small workshop.

4—Pilot ammonia plasticizing plant at the State University College of Forestry at Syracuse (N.Y.) University treats wood with either liquid or gaseous anhydrous ammonia.

The Woodworker's Raw Materials

CHAPTER 12

With what form of wood does woodworking begin? There is no single or simple answer. Only a few crafts, such as sculpture, post-and-rung chairmaking and hand-hewn cabin construction, actually start with the tree itself. Most depend on some degree of conversion into lumber having already been completed, perhaps only a rough milling to nominal thickness of random length and width, as is common with hardwoods. The wood may be additionally dressed to uniform thickness, standard width and length, as is usual for softwoods. It may be milled to pattern, as in cove molding or picture-frame stock. Or the wood may be manufactured to a considerable extent, as when logs are peeled into veneer and then processed into plywood.

All of these wood forms are considered primary products, for it is assumed that they will be further worked and assembled into finished secondary products such as furniture, toys or cabinets. The woodworker's raw material is usually a primary product, and locating it is an integral part of the challenge, success and enjoyment of woodworking. But the task is not always easy. Because of the vast array of primary products, the veteran woodworker as well as the beginner is often baffled and dismayed in the search for the appropriate form of wood.

In this chapter I will explore some of the alternatives among local sources for wood, as well as the more traditional, established markets. I'll also survey the major categories of standard commercial primary products available to the woodworker.

The search for wood

In beginning the search for wood, the woodworker should carefully size up all the alternatives first, from lumberyards and craft supply houses to sawmills. Sometimes, however, the prices are beyond the budget or the desired items are not readily available. A little individual reconnaissance is then in order, for "wood is where you find it," as Bob Butler so aptly states in his book, *Wood for Wood Carvers and Craftsmen.*

The tree — It is ironic that we are surrounded by trees, and yet wood for woodworking is so hard to come by. Trees are a source of wood only for those willing to expend much time and physical effort, because extracting workable wood from the tree is no casual pastime. But if you love wood and woodworking, the psychological reward is worth the effort.

The basic equipment required is a chain saw and accessories (or a good two-man crosscut saw if you can find one and know how to file it) and safety gear, including gloves, hard hat, safety shoes and eye protection. A peavey or log jack will be a valuable addition. If you are a woodcarver, a splitting maul or sledge hammer and wedges, an ax and a drawknife will harvest all the various blocks, chucks, slabs and flitches you will ever require. A froe is handy for splitting out slats. If you want boards or a number of fairly uniform slabs, a chain-saw mill may best suit your needs.

Be sure to think through the whole job and be prepared to end-coat, stack and store your entire harvest properly (Chapter 5). If you wait to end-coat until checking begins, your stock will be good only for firewood.

Where does one find trees? If you live in or near the country, farmers and landowners often will be willing to give or sell you a tree or two, or will arrange to give you the choice portions of the tree trunk if you buck up the remainder into firewood for their use. Anyone who cuts a fair amount of firewood will probably swap some choice trunk wood for your labor. Local loggers may sell you a few logs at prevailing prices. Keep track of construction work in your area. Construction crews on large jobs are seldom concerned about trees destroyed in clearing land, and are likely to burn or bury the wood. You may get it for the asking. Orchard managers often replace unproductive, overmature trees with young ones, and the wood of nut trees, apple, pear and other fruit trees might be obtained in such instances.

Residential and roadside tree work can also yield some invaluable material. Whenever I hear a chain saw, I stop and investigate. Often it is just one more elm that has succumbed to Dutch elm disease, but sometimes it turns out to be a species I am eager to have or a new one I've never worked with. In addition to oak, butternut,

1—Rotary-cut veneer 'unrolls' from the lathe.

basswood, maple and pine, I have picked up beautiful carving blocks of black walnut, honeylocust, yellowwood, cork tree and catalpa, species not native to our area but grown as ornamentals. In fact, I have never been refused my "pick-of-the-litter" when I have stopped and asked a tree crew for a chunk of wood to use for carving. It is helpful to establish acquaintance with crews of local highway departments, firewood cutters, utility companies and tree-service companies. They may be able to let you know when a particularly exotic tree is being cut down. In short, spread the word. Let people know what you want.

Some years ago I developed a fondness for catalpa for carving, so I began to make my wishes known. One morning a short time later I found a dozen short logs dumped at the foot of my driveway—I later learned that a friend had happened upon them and remembered my remarks. Soon afterward another friend brought a 12-ft. catalpa log 20 in. in diameter. I've since received three more offers of catalpa trees in people's yards, on a come-and-get-it basis. When a farmer, landowner or neighbor offers a tree, be prepared to do a neat and total job of removing all the branchwood, or of bucking it into firewood.

Two words of caution: First, residential and roadside trees often contain foreign material—fence wire, staples, clothesline pulleys, gate hinges—particularly in the lower six feet. This is why tree crews can't sell the logs to sawmills, and why mills aren't usually willing to custom-saw a log for you. Second, residential trees often are available precisely because they are defective. Unless a construction clearing or road widening is involved, perfectly healthy trees usually aren't cut down. Likewise, defective trees are the most likely to be blown down in windstorms. Be wary of limbwood, which usually has pronounced reaction wood.

Over the years, I've learned to keep my eyes and ears open for those choice logs and chunks that are free for the asking. At the same time, I have learned to decline material that is not in excellent form and condition, material I can't deal with promptly and properly, or material in excess of what I really want or need.

Recycling used wood—Used lumber and timbers are an excellent potential source of wood, although somewhat unpredictable and sporadic. But old frame houses, barns and factories can yield fantastic material. By our modern standards, the superb grade of lumber used for building years back is unbelievable. Old timbers were often straight-grained and free of defects. Although nails can be a problem, they were used sparingly as a rule—worked joints were most often employed. The most hardware-free material is often found in roof and barn framing. Attics and lofts often have wide floor planks laid in place with little or no fastening. Occasionally, beautiful lumber is found squirreled away in a barn or attic.

Old furniture such as church pews and government surplus should also be examined, because the grade of wood is often quite good, even if the furniture is poorly constructed. Turn-of-the-century furniture was often made of veneered panels on cores of clear chestnut or yellow poplar. The crossbands and face veneers of walnut, oak or mahogany, glued with non-moisture-resistant adhesives, may easily peel free. The furniture is a total loss but the wood is in perfect condition. Dining-room tables may have massive pedestal bases of oak or mahogany that yield sizable carving blocks.

Local sawmills — Buying direct from local mills is one of the best ways to obtain lumber economically, but usually it will be rough green stock and you will have to do your own drying and surfacing.

Large mills are geared to high-volume wholesale trade and normally handle no retail sales whatsoever. Among smaller mills, however, the policy is variable. A woodworker often can build a working relationship by using the right combination of courtesy and diplomacy. When approaching a mill, keep one thing in mind: The mill is in operation to make money, not to do favors for woodworkers. You should therefore expect to pay at least the going rate for the logs or lumber you buy, plus perhaps a little something extra for the miller's time and effort.

Small sawmills dot the country, although they rarely advertise. Sometimes you can locate them through the telephone book, or through inquiries to the local chamber of commerce. But sometimes you come upon them only by diligent exploration of side roads in your own county, or by following a truck carrying logs. Small mills may supply hardwood to the furniture industry or make quantities of wooden parts to order. Many of them saw only rough stock for fencing, pallets and shipping crates. Even here, a log containing mineral-stained wood that is worthless in factory production might be perfect for the careful cabinetmaker. Some small mills will be happy to sell to an individual, some will refuse.

Here are a few tips that may help. First, order as much as possible at one time by pooling your needs and resources with other woodworkers. The larger the quantity you can buy, the larger the mill that will deal with you. Don't expect a mill operator to stop and rip apart a pile so you can pick a board or two. Second, indicate in advance what you are looking for in terms of species, quality, lengths, widths, grades and approximate quantity, leaving reasonable flexibility in your specifications. Make it clear you are willing to pay a fair price and be reasonable about accepting a portion of the material that does not make the grade if the main portion is well up to your specs. In time your fairness will be more than rewarded and once you become a "special" customer, a small-mill operator usually will outdo himself in putting up your order.

The lumberyard — Because lumber is perhaps the most common of the woodworker's primary materials, the lumberyard would seem a most likely place to shop. But the typical modern lumberyard is primarily a builders' supply center, vending everything from hardware, paint, electrical equipment and plumbing and masonry supplies to roofing, flooring, fencing and garden supplies. The line of wood products is mainly softwood structural and yard lumber, millwork and specialty items, plus an array of sheet materials—plywood, prefinished hardwood paneling, hardboard and particle board. While the lumberyard will provide some routine materials, the woodworker will want to have other sources for specialized stock.

A thorough get-acquainted trip to local yards is always worthwhile. The stock in each yard is largely influenced by the species available in that geographic region, the local building traditions and the prevalent construction in the area.

Structural lumber will be mostly the stronger, uneven-grained conifers such as Douglas fir, southern yellow pine, hemlock and larch, or the moderately even-grained species such as spruces and firs. Structural grades commonly used for residential framing are seldom appropriate for shop woodworking. Occasionally, however, large timbers and posts of cedar or Douglas fir contain clear portions suited to carving or sculpture. The major drawback of structural lumber is that the moisture content may be drastically higher than is suitable for most interior cabinet work. Measures must therefore be taken to dry the stock for general shop use.

The average yard stocks an assortment of lengths and widths of 1-in. (actually about ¾ in. thick) pine yard lumber, but possibly only one or two grades, usually No. 1 or No. 2 common. (Lumber types and grades are discussed later in this chapter.) In addition, matched boards of fir or hard pine (used for subflooring and sheathing) might be in stock. Western redcedar or redwood may be available in the higher grades (for exterior trim) or the common grades (for exposed siding or decking).

Be sure to check out the specialty and millwork items, also. Door jambs, panel strips, lattice stock, hand rails and rounds, stair treads and thresholds may offer useful material and usually are dried to levels suitable for interior use. Certain items may be in demand locally and therefore commonly stocked. In coastal towns, where boatbuilding and repair are popular, yards often carry mahogany, teak or Sitka spruce. Architectural demands for weathered exteriors may result in cedar, redwood or cypress being carried in stock.

Although the inventory of solid wood items is shrinking, the typical lumberyard now stocks a greater and greater variety of modified sheet products including plywood, particle board, fiberboard and hardboard, most of which are construction types and grades. Many are suited to do-it-yourself crafts, such as interior remodeling, and some are useful in cabinetmaking. For example, birch-veneered lumber-core panels in ¾-in. thicknesses are commonly sold for kitchen-cabinet work. Unfinished hardwood plywood paneling is useful for drawer bottoms, dust panels and carcase backs. For exterior painted surfaces as in door panels or signs, plywood with an overlay of resin-impregnated paper is excellent. Except for special panel edge banding, veneer is not commonly sold at lumberyards.

Other commercial sources — A diverse range of suppliers who deal in various species, forms and quantities of cabinet woods exists across the country. Because of their diversity, they cannot be described easily or put into neat categories. At one extreme are large commercial wholesalers who deal mostly in hardwood shop lumber, which is sold by grade and which is sometimes dried and dressed, sometimes rough and green. These wholesalers normally sell to volume users such as furniture manufacturers, contractors and architectural woodwork companies, and their minimum order requirements are usually more than the average woodworker wants or can afford at one time.

At the other extreme, wood can sometimes be purchased by the board foot or by the measured piece at woodworking supply houses that stock a limited inventory of choice or exotic woods, high-grade cabinet lumber, turning squares, carving blocks and veneers. As expected, however, the price is relatively high. Such firms commonly operate a combination retail store/mail-order business.

In response to the demands of woodworkers for more moderately priced lumber in limited quantities, numerous small retail outlets or wood stores that usually specialize in local species are springing up all across the country. In many cases these enterprises are affiliated with local sawmills and their lumber is often "mill run," or ungraded. Boards are sold by the square foot of surface measure, carving blocks by the board foot or by the pound. Such operations may offer additional services such as surface-planing or milling-to-pattern.

The woodworker probably can realize the most economical buys on small quantities from such dealers. A frequent complaint, however, seems to concern the lack of moisture-content control, and especially the loose claims made about stock being dry that actually is not. It therefore behooves the buyer to have the ability to check the moisture content of each purchase. I cannot imagine operating such a business without humidity control for storage, and without a meter for measuring the moisture in the lumber. I can think of no more important quality-control feature upon which to build customer confidence than to provide reliable moisture-content information.

The average woodworker is probably interested in something between the extremes of wholesale quantities and the single board. The serious woodworker might typically be looking for moderately small quantities (say between 30 and 300 board feet) of the higher grades of hardwood lumber that already has been kiln-dried to the appropriate interior EMC conditions for the area. Since surfacing equipment is not common in many small shops, the woodworker might prefer the material surfaced on two sides to uniform thickness. Certain restrictions of length or width also may need to be imposed for a particular project. Whether your needs are large, medium or small, a major problem is often locating the appropriate firm.

Finding commercial sources

For locating sources of supply at all levels, but especially at the intermediate level, local high school or college industrial-arts teachers are excellent first contacts. They routinely face the problem of locating sources and also must pay attention to economics on behalf of their students. You will usually find them quite helpful.

Woodworking magazines (such as *Fine Woodworking*, *Workbench* and the *National Carvers Review*) feature periodically updated sources of supply and carry classified ads from dealers. Certain organizations (such as the National Woodcarvers Association and the International Wood Collectors Society) issue a journal or bulletin that includes announcements or classified listings of sources.

Specialty forms of wood may be found in association with specific fields of woodworking. For example, high-grade spruce spar stock and thin hardwood plywood are available through boating and aircraft supply houses. They can be located through the advertisements and classified sections in yachting, boating and sport-flying magazines. Similarly, musical-instrument makers rely on premium-quality wood in dozens of species, both domestic and imported. Miniature makers and dollhouse builders have sources of thin stock in small pieces. Patternmakers use great quantities of mahogany, clear pine and cherry, and might be willing to sell you some.

The Yellow Pages in the phone book will carry listings under at least the following headings: Cabinetmakers; Hobby and Model Construction Supplies-Retail; Hardwoods; Hardboard; Lumber-Drying; Lumber-Retail; Lumber-Treated; Lumber-Used; Lumber-Wholesale; Millwork; Patternmakers; Plywood and Veneer; Sawmills; Wood Products; Wood Turning; Woodworkers. Local chambers of commerce have listings of all the firms in the general area that produce various products. Newspaper classified ads contain an often overlooked building-materials section. At the public library, the *Thomas Register of American Manufacturers* lists the names of firms geographically, under appropriate material or product headings. Trade magazines also carry advertisements and classified listings. For example, *Panel World* and *Plywood and Panel* magazines list veneer and plywood firms all across the country. Most forest industry trade magazines publish an annual directory issue, and most trade associations publish various shoppers' guides. The agricultural extension service or forestry department at most state universities can help locate local sawmills and logging companies. Nobody knows the lumber business better than lumbermen—ask at the sawmill, lumberyard or millwork shop. Once you get started, one good supplier usually will lead to several more.

Lumber

The word **lumber** needs little introduction, and in fact, when the word wood is mentioned, lumber is probably what most people think of. Lumber has been defined as the product of the saw and planing mill, not further manufactured other than by sawing, resawing, passing lengthwise through a standard planing machine, crosscutting to length and matching. The word lumber further carries the connotation of thickness (say, ½ in. or more), in contrast to thin sheets of **veneer**. Lumber is also thought of as single-piece items of wood, in contrast to pieces glued together as in **plywood** or **laminated** beams, or fastened together as in **trusses**. Another feature of lumber is that the cell structure is undisturbed from its original state in the tree, in contrast to the reoriented cell arrangement in **hardboard**, **fiberboard** and **particle board** or the distorted cell structure in **compreg** (compressed, impregnated wood). Abbreviations of common lumber terms are listed in *Table 16*.

Lumber, then, is simply a solid, rectangular chunk of wood that has been separated from the log, usually by

Table 16—Abbreviations of common lumber terms.

AD—air-dried
AW&L—all widths and lengths
B&BTR, also B&B—grade B and better
BD.—board
BD. FT.—board foot
BDL—bundle
BM—board measure
B&S—beams and stringers
BTR—better
CLR—clear
COM—common
DET—double end trimmed
DF—Douglas fir
DIM—dimension
D&M—dressed and matched
DOUG FIR L—Douglas fir or larch
E—edge
EE—eased edges
EG—edge grain
EM—end-matched
EMC—equilibrium moisture content
ES-AF—Engelmann spruce, alpine fir
FAS—firsts and seconds
FG—flat or slash grain
FLG—flooring
FT BM—foot board measure
HDO—high-density overlay
HEM—hemlock
IC—incense cedar
IN.—inch or inches
IWP—Idaho white pine
J&P—joists and planks
JTD—jointed
KD—kiln-dried
L—length
L—larch
LBR—lumber
LFT.—lineal feet
LGR—longer
LGTH—length
LL—longleaf
LP—lodgepole pine
M—thousand
MBM—thousand (feet) board measure
MC—moisture content
MC 15—moisture content not exceeding 15%
MDO—medium-density overlay
MG—mixed grain
MH—mountain hemlock
MSR—machine stress-rated
NSR—not stress-rated
P—load, concentrated
PAD—partially air-dried
PP—ponderosa pine
P&T—posts and timbers
R—radial
RC—redcedar
RDM—random
RES—resawn
RGH—rough
RL—random lengths
RW—random widths
S—shrinkage percentage
S-DRY—surfaced at a moisture content not exceeding 19%
SEL—select
SG—slash or flat grain
S-GRN—surfaced at a moisture content above 19%
SIT SPR—Sitka spruce
SL—shiplap
SM—surface measure
SP—sugar pine
SP—specialty grade
SQ—square
SR—stress-rated
SS—select structural
STD—standard
STK—stock
STR, also STRUCT—structural
S4S—surfaced four sides
S2S—surfaced two sides
SYP—southern yellow pine
T—tangential
TBR—timber
T&G—tongue and grooved
VG—vertical (edge) grain
WCH—West Coast hemlock
WDR—wider
WF—white fir
WRC—western redcedar
WTH—width
WW—white woods, any combination of Engelmann spruce, true fir, western hemlock and white pine
YP—yellow pine

sawing. Lumber is produced in a lumber mill or sawmill, where logs go through a characteristic sequence of operations in being reduced to lumber *(1)*.

At the lumber mill, logs are stored in log ponds or in log piles. Ponding allows one man to handle large logs, and water storage inhibits fungal growth and end-checking. In large mills the logs are first debarked, for two important reasons. Debarking removes the embedded dirt and stones that would dull the milling equipment, and it allows the log slabs and edgings to be converted into high-grade (bark-free) pulp chips or particles for composite products. Bark is removed by grinding with mechanical cutterheads or by blasting with high-pressure water jets. The debarked logs are then rolled onto a sloping ramp called the **log deck** to await sawing.

A log to be sawn is mounted on a **log carriage**, a low, heavy trolley mounted on tracks. A mechanism at the log deck loads the log onto the carriage and turns it to the desired position. "Dogs" on the carriage firmly clasp the log in place. By moving the position of the dogging system, the side of the log can overhang the edge of the carriage so that as the edge of the carriage passes the headsaw, a slab is cut off. By advancing the log on successive passes of the carriage, board thicknesses are removed from the log.

Headsaws or headrigs are principally of two types, **band saws** and **circular saws**. Sawmills commonly are referred to in terms of their headsaws as well as the type of lumber cut, so one might hear of a **hardwood circular mill**, or a **softwood band mill**. Band mills can be built for larger-sized logs and are therefore more prevalent among the large softwood timber mills in the western United States. Band saws have the added advantage of taking a narrower kerf. Circular saws usually have smaller capacity, but they are less expensive and can be portable. They are generally suited to the small hardwood operations typical of the eastern and southern United States.

It is important to realize that logs usually contain their clearest material just under the bark. Because branches arise from the pith and commonly persist for a number of years, the core of the log near the pith is generally defective. Therefore, when opening a log, the best-looking face is sawn first. After taking a **slab cut**, successive boards are removed as long as they are clear. When defects are encountered, the log is turned to the next clearest face and sawn until more defects appear *(2A)*. This removes the maximum amount of material in clearer grades and leaves the knot defects mostly boxed into the central portion. The center can then be sawn into low-grade lumber for pallet stock or crating, or can be used intact as a timber or railroad tie. This process of sawing "around the log" produces mostly flatsawn or tangential lumber.

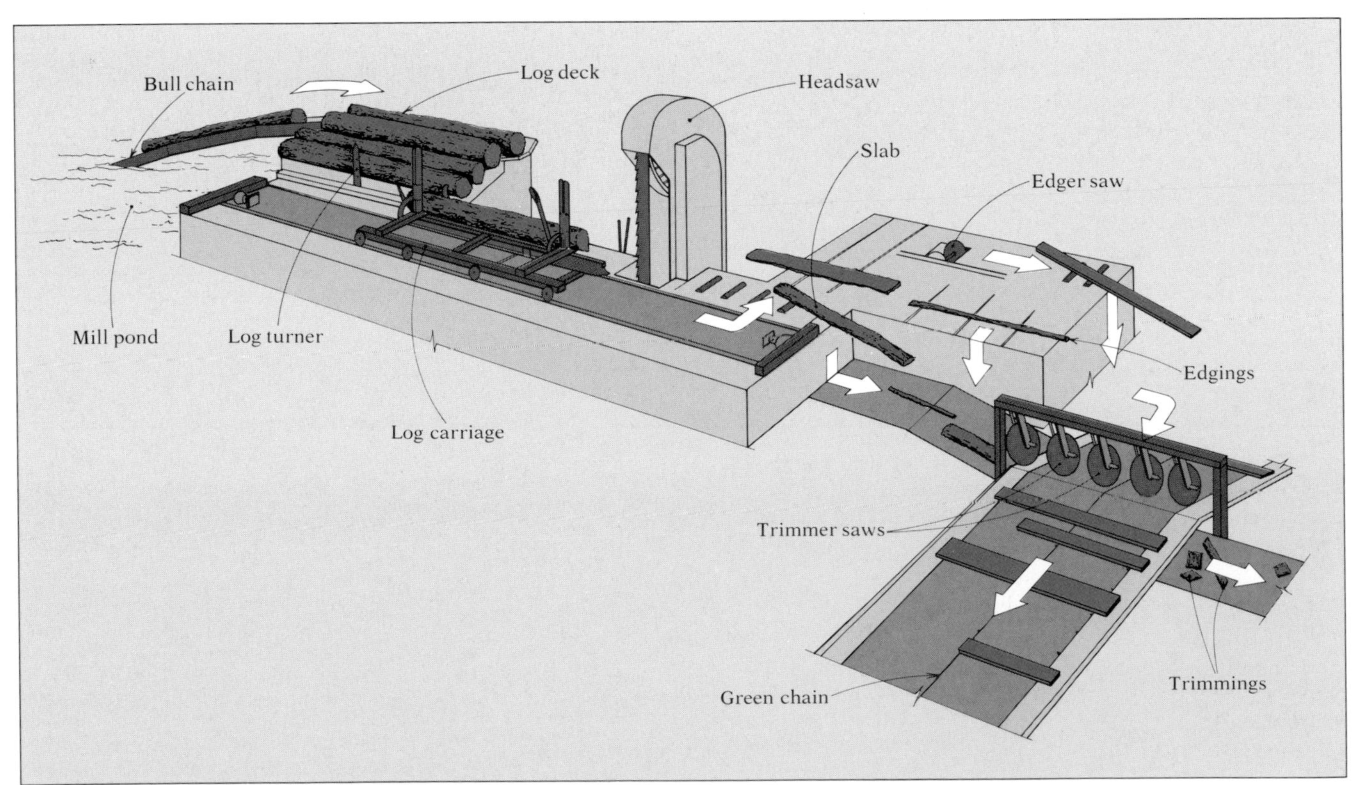

1—From log to finished board: the typical sequence of manufacturing rough, green lumber at a sawmill.

When a log has numerous defects, as is common with small logs, little may be gained by turning. Such logs may therefore be sawn "through and through" *(2B)*. Sawing low-grade logs this way is done most efficiently with a **gang saw**, a set of equally spaced sash saws or circular saws that makes an entire series of cuts during one pass of the log. A gang saw may serve as a headrig for low-grade logs, or a gang resaw is sometimes used to reduce large, clear cants or timbers into uniform boards for millwork, molding or sash stock.

Quartersawing, on the other hand, produces edge-grain stock. As the term suggests, a large log is halved and then each half is redogged onto the carriage and quartered, the quarters in turn being resawn either a board at a time on the headsaw or by being passed through a gang resaw. By this method, the widest boards will be most truly edge-grained *(2C)*. In many applications, edge-grained stock is preferred because the radial surfaces have more uniform wearing and finishing properties and radially cut boards have greater dimensional stability across the width. There are drawbacks to quartersawing, however. It is certainly more time-consuming and often more wasteful. Also, central knot defects are likely to occur along the wider boards, especially in smaller logs, and much of the outer clear material winds up in the narrowest boards. Quartersawing is therefore most appropriate for fairly large or reasonably clear logs, or where the products need not be very wide (as in strip flooring). The above considerations suggest why quartersawn lumber is less common and more expensive then flatsawn lumber. The woodworker often can take good advantage of softwood lumber that has been sawn through-and-through. Since many such boards contain the pith and lots of knots, they are usually sold in relatively wide pieces for utility shelving. Ripsawing down the center to remove the pith and crosscutting to remove the worst knots yields short lengths of clear, quartersawn stock at the utility price.

A modern innovation for optimum recovery from the log is the chipping headrig. Best suited for small logs of uniform size, this machine first trims them to a rectangular shape by profile cutterheads, thereby reducing to chips or flakes (for pulp or composite board manufacture) the portions of the log that would normally be removed as slabs, edgings and sawdust. The shaped log then proceeds through a gang saw where specific lumber sizes are produced *(2D)*.

After boards are separated, they are usually passed through an edger, a machine having twin saws of adjustable spacing that removes bark edges or edge defects and leaves parallel edges on the board. The board then may be crosscut to length, and defects cut out in the trimmer. The board then moves out of the mill along the **green chain**, to be sorted by size and species. At this stage the lumber has sawtooth marks on its surfaces and is termed **rough lumber**. Hardwood is then graded for shipping from the mill.

Softwood lumber is usually dried and dressed at the mill where it is produced. The lumber may be stacked outdoors in piles to air-dry, or dried to prescribed levels in kilns. The dry, rough lumber then goes to the **planing mill**, where it is dressed to uniform thickness. Usually both sides and both edges are dressed (surfaced four sides, abbreviated S4S), or at least both surfaces are dressed (S2S), leaving the edges roughsawn. In the planing mill the lumber may be **matched** (dressed on both surfaces and tongue-and-groove edged), **shiplapped** (edges rabbeted), or **patterned** (molded to a special shape, such as casings, coves and panel moldings).

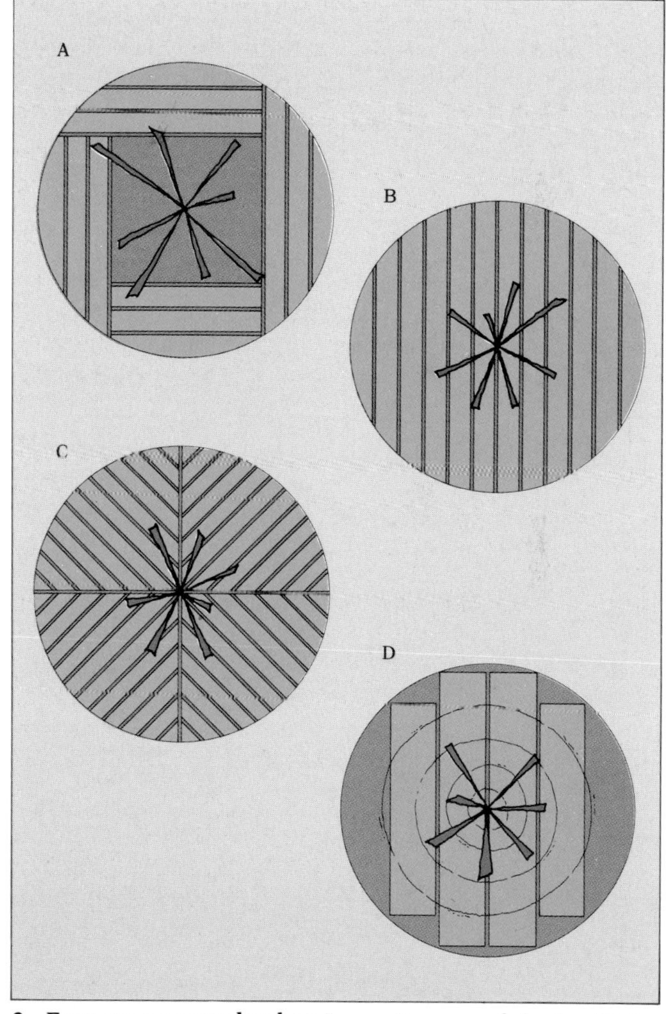

2—Four ways to saw lumber: In sawing around the log (A) the best face is sawn first. When defects are encountered, the log is turned to its next face, and so on. Defects are concentrated in the boxed heart. Sawing through and through (B), or gang-sawing, is used for low-grade logs. Quartersawing (C) yields edge-grain stock. First the log is quartered, and then each log in turn is gang-sawn or returned to the carriage and sawn. A chipping headrig (D), first removes the outer residue, then boards of predetermined size are cut by a gang resaw.

Lumber measure

Lumber usually is sold by volume according to its nominal size. Nominal size refers to the dimensions when roughsawn. The unit of volume is the **board foot** (bd. ft.), equivalent to a piece of lumber one foot long, one foot wide and one inch thick, or 144 cubic inches *(1)*. Softwood lumber usually is produced in rough size multiples, in which thickness and width are given in inches and length in feet. A piece of framing lumber, for example, might be sawn 2 in. x 10 in. rough (called a two-by-ten, simply written 2x10), although it would be dressed to standard dimensions of 1½ in. x 9¼ in. as shown in *Table 17*. Similarly a 1x12 board (one-by-twelve, roughsawn 1 in. x 12 in.) would be dressed dry to ¾ in. x 11¼. (Lengths are expressed in actual measurement.) A 1x12 contains 1 board foot per lineal (running) foot. A 2x10 contains 1⅔ board feet per lineal foot. The price then is calculated by the board foot volume the piece had *before* surfacing. This system is based on the logic that the discrepancy in volume between nominal and actual dimensions reflects the material lost in shrinkage during drying and the necessary removal of material to improve the surface—losses that the buyer would suffer later if these services were not provided as part of the manufacturing process. Millwork, such as moldings and sash stock, is usually sold by the lineal foot.

In hardwood lumber, rough nominal thickness is commonly expressed to the nearest quarter-inch as a non-reduced fraction, so that 1-in. lumber is referred to as 4/4 (four quarter) lumber, 1½-in. lumber is 6/4, 3-in. lumber is 12/4, and so on. Since hardwood lumber usually is sold on a random width and length basis, board footage is based on actual surface area and nominal thickness.

The way to calculate the board footage in a given piece of lumber is to multiply its width in feet (or a fraction thereof) by its length in feet by its thickness in inches. Lumber prices usually are calculated from the cost per thousand board feet (MBM). Thus red oak at $750/MBM costs 75¢ a board foot. However, MBM prices almost always increase as the quantity decreases because of the extra handling small quantities require.

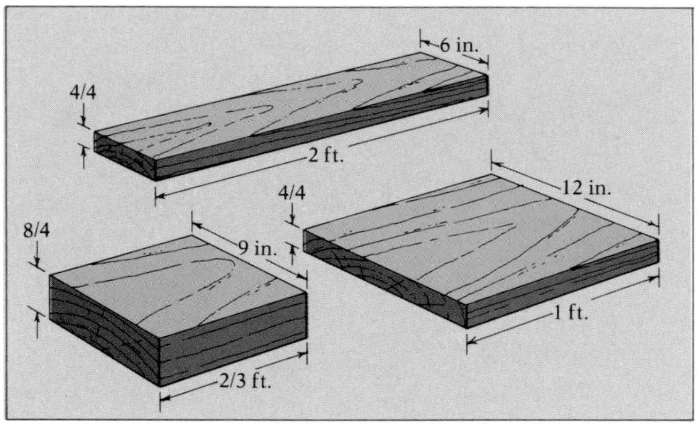

1—One board foot is 144 cubic inches of wood.

Table 17—Standard American sizes for construction lumber.

	Thickness (in.)			Face width (in.)		
	Nominal	Minimum dressed		Nominal	Minimum dressed	
Item		Dry	Green		Dry	Green
Boards	1	¾	25/32	2	1½	1 9/16
	1¼	1	1 1/32	3	2½	2 9/16
	1½	1¼	1 9/32	4	3½	3 9/16
				5	4½	4⅝
				6	5½	5⅝
				7	6½	6⅝
				8	7¼	7½
				9	8¼	8½
				10	9¼	9½
				11	10¼	10½
				12	11¼	11½
				14	13¼	13½
				16	15¼	15½
Dimension	2	1½	1 9/16	2	1½	1 9/16
	2½	2	2 1/16	3	2½	2 9/16
	3	2½	2 9/16	4	3½	3 9/16
	3½	3	3 1/16	5	4½	4⅝
	4	3½	3 9/16	6	5½	5⅝
	4½	4	4 1/16	8	7¼	7½
				10	9¼	9½
				12	11¼	11½
				14	13¼	13½
				16	15¼	15½
Timbers	5 and greater		½ less than nominal	5 and greater		½ less than nominal

Lumber classification and grading

Because of the many and varied uses and requirements for lumber, no single system of classification would be useful in every type of situation. It is therefore appropriate to classify lumber initially according to its use and further according to its degree of manufacture or on the basis of the size of the pieces *(Table 18)*. Under various classification levels, **grades** are established which attempt to group pieces according to quality. There are broad general differences between softwoods and hardwoods, both with respect to the end uses and the manner of working. Softwoods generally are used in large sizes for building and construction. Much of the final working of softwood lumber is completed during its manufacture, so that pieces can be used full-size with little or no additional surfacing. The bulk of hardwood lumber, on the other hand, is usually only roughly manufactured initially, because it is cut to smaller pieces and further worked in the final processing of the finished product.

Hardwood lumber — Hardwood lumber falls into three basic marketing categories: **factory lumber, dimension parts** and **finished market products**. The last category includes such items as flooring, millwork parts and moldings, in which little or no further manufacture is needed. Grade designations for finished parts indicate appearance and general freedom from defect.

Factory lumber and dimension parts are both expected to receive further working in assembling the finished product. Dimension parts are processed so each piece can be completely used virtually in the size provided. Pieces are produced to specific sizes (rough or dressed) as either flat stock or squares. Within shape categories, grades reflect freedom from defect and end-use suitability.

Factory lumber grades — Most hardwood is manufactured as factory lumber, which is primarily intended to serve the industrial customer. Because it is assumed that the lumber will be cut up into smaller useful pieces, factory lumber is handled in random lengths and widths within thickness sortings. The grade is based on the proportion of a board that can be cut into a certain number of clear-faced cuttings not smaller than a specified size.

The hardwood factory grade rules generally used

Table 18—Lumber classification based on use, degree of manufacture and size.

USE CLASSIFICATION

1. *Yard lumber*—Lumber of those grades, sizes and patterns generally intended for ordinary construction and building.
2. *Structural lumber*—Lumber 2 in. or more in nominal thickness and width for use where working stresses are required.
3. *Factory and shop lumber*—Lumber produced or selected primarily for remanufacturing.

MANUFACTURING CLASSIFICATION

1. *Rough lumber*—Lumber that has not been dressed (surfaced) but has been sawn, edged and trimmed at least to the extent of showing saw marks in the wood on the four longitudinal surfaces of each piece for its overall length.
2. *Dressed (Surfaced) lumber*—Lumber dressed by a planing machine (for the purpose of attaining smoothness of surface and uniformity of size) on one side (S1S), two sides (S2S), one edge (S1E), or a combination of sides and edges (S1SE, S1S2E, S2S2E, S4S).
3. *Worked lumber*—Lumber which, in addition to being dressed, has been matched, shiplapped or patterned.
   a. MATCHED LUMBER—Lumber worked with a tongue on one edge of each piece and a groove on the opposite edge to provide a close tongue-and-groove joint; when end-matched, the tongue and groove are worked in the ends also.
   b. SHIPLAPPED LUMBER—Lumber rabbeted on both edges of each piece to provide a close-lapped joint.
   c. PATTERNED LUMBER—Lumber shaped to a pattern or to a molded form, in addition to being dressed, matched or shiplapped, or any combination of these.

SIZE CLASSIFICATION

1. *Nominal Size*
   a. BOARDS—Lumber less than 2 in. in nominal thickness and 2 in. or more in nominal width. Boards less than 6 in. in nominal width may be classified as strips.
   b. DIMENSION—Lumber from 2 in. to, but not including, 5 in. in nominal thickness, and 2 in. or more in nominal width. Dimension lumber may be classified as framing, joists, planks, rafters, studs, small timber, etc.
   c. TIMBERS—Lumber nominally 5 in. or more in least dimension. Timber may be classified as beams, stringers, posts, caps, sills, girders, purlins, etc.
2. *Rough-dry size*—The minimum rough-dry thickness of finish, common boards, and dimensions of size 1 in. or more in nominal thickness shall not be less than $\frac{1}{8}$ thicker than the corresponding minimum finished dry thickness, except that 20% of a shipment may not be less than $\frac{3}{32}$ in. thicker than the corresponding minimum-finished dry thickness. The minimum rough-dry widths of finish, common strip, boards and dimension shall not be less than $\frac{1}{8}$ in. wider than the corresponding minimum-finished dry width.
3. *Dressed-sizes*—Dressed sizes of lumber shall equal or exceed the minimum American standard sizes *(Table 17, page 210)*.

throughout the country are those established by the National Hardwood Lumber Association (NHLA). These rules list the following grades in descending order of quality: Firsts, Seconds, Selects, No. 1 Common, No. 2 Common, No. 3A Common, No. 3B Common, Sound Wormy. The grade rules are summarized in *Table 19.*

Hardwood lumber is graded rough and usually in the green condition because there are no moisture-content specifications. Grading is determined from the poorer face of each board (except in Selects grade). Firsts and Seconds often are combined into one grade called Firsts and Seconds (FAS). Sound Wormy is similar to No. 1 Common, but worm holes are allowed.

Hardwood grades are specified according to the percentage of the total board surface ("surface measure," SM, in square feet) which could be cut out in the form of rectangular, clear-faced pieces (called "cuttings"). To determine this percentage, a unit of measure called the **cutting unit** is used. A **cutting unit** is equivalent to an area 1 inch wide and 1 foot long (or 12 square inches). The number of cutting units in a cutting can be calculated quickly by multiplying the width in inches by the length in feet. There are therefore 12 cutting units in a square foot. The total number of cutting units in any board is simply its surface measure times 12 (SM x 12).

Each hardwood grade specifies the percentage of the surface area that must be in clear cutting units. This percentage is stated as some fraction with the denominator 12. Therefore, to determine the number of clear units a particular grade board must have, the total units in the board (SM x 12) is multiplied by this fraction. For example, Firsts requires 11/12 of the board to be available

Table 19—Minimum requirements for grading factory hardwood lumber.

Grade	Board Size Minimum Width	 Length	Conversion Factor (% Clear Face)	Minimum Size of Cuttings	Maximum Number of Cuttings for Board SM
Firsts	6″	8′-16′	11xSM (91⅔%)	4″x 5′ or 3″x 7′	1 for SM 4′x9′ 2 for SM 10′-14′ 3 for SM 15′ or more
Seconds	6″	8′-16′	10xSM (81⅔%)	4″x 5′ or 3″x 7′	1 for SM 4′-7′ 2 for SM 8′-11′ 3 for SM 12′-15′ 4 for SM 16′ or more
Selects	4″	6′-16′	11xSM (91⅔%) 10xSM (83⅓%)	4″x 5′ or 3″x 7′	1 for SM 2′-3′ 1 for SM 4′-7′ 2 for SM 8′-11′ 3 for SM 12′-15′ 4 for SM 16′ or more
No. 1 Common	3″	4′-16′	9xSM (75%) 8xSM (66⅔%)	4″x 2′ or 3″x 3′	1 for SM 2′ 1 for SM 3′-4′ 2 for SM 5′-7′ 3 for SM 8′-10′ 4 for SM 11′-13′ 5 for SM 14′ or more
No. 2 Common	3″	4′-16′	6xSM (50%)	3″x 2′	1 for SM 2′-3′ 2 for SM 4′-5′ 3 for SM 6′-7′ 4 for SM 8′-9′ 5 for SM 10′-11′ 6 for SM 12′-13′ 7 for SM 14′ or more
No. 3A Common	3″	4′-16′	4xSM (33⅓%)	3″x 2′	Unlimited
No. 3B Common	3″	4′-16′	3xSM (25%)	1½″x 2′	Unlimited

This chart gives the minimum requirements a board must meet to merit a particular grade. In general, a high-grade board is relatively long and wide, and a high percentage of its area is free of defect. The clear lumber in a high-grade board must be obtainable in relatively few large cuttings.

in clear face cuttings. The total clear units in a board to make the grade of Firsts must therefore be

$$\left(\tfrac{11}{12}\right) \times (\text{SM} \times 12) = 11(\text{SM}).$$

The required units to make a particular grade can be calculated by multiplying the numerator of the fraction by the surface measure. Thus, if a board has a surface measure of 9 square feet, it must have 11 x 9 = 99 cutting units in clear face cuttings to be graded Firsts. This logic explains the fractional percentages (e.g., $91\tfrac{2}{3}\%$ = $\tfrac{11}{12}$) specified in the grade rules.

A hardwood grader goes through several steps to evaluate a typical board *(1)*. In practice, an experienced grader can assign grades to most boards with rather brief inspection, often only a few seconds. However, boards at the borderline of a grade category must be measured individually by going through every step of the procedure. Because the hardwood rules are specific, the grade of each board can be assigned by precise numerical measurement rather than by depending on subjective judgment alone.

For the average woodworker, No. 1 Common might be the best all-around grade when price is considered against its yield of about 65% clear material. Where larger cuttings are needed, the FAS grade with its 80% to 90% yield may justify the much higher price. The best way to assess the overall quality of each grade, however, is to visit a dealer in order to look over material in the various grades.

Being able to grade hardwood lumber rapidly takes years of experience. However, with a basic knowledge of fundamentals and a grade book in hand to check the many details (which the grader has committed to memory), the woodworker can check-grade a shipment to determine whether it meets the specifications of the grade that was ordered.

Softwood lumber — The basis for most classification and grades of softwood lumber is the American Softwood Lumber Standard PS 20-70, established by the U.S. Department of Commerce. Although specific grade rules for various species have been developed by many different associations throughout the country, there is gen-

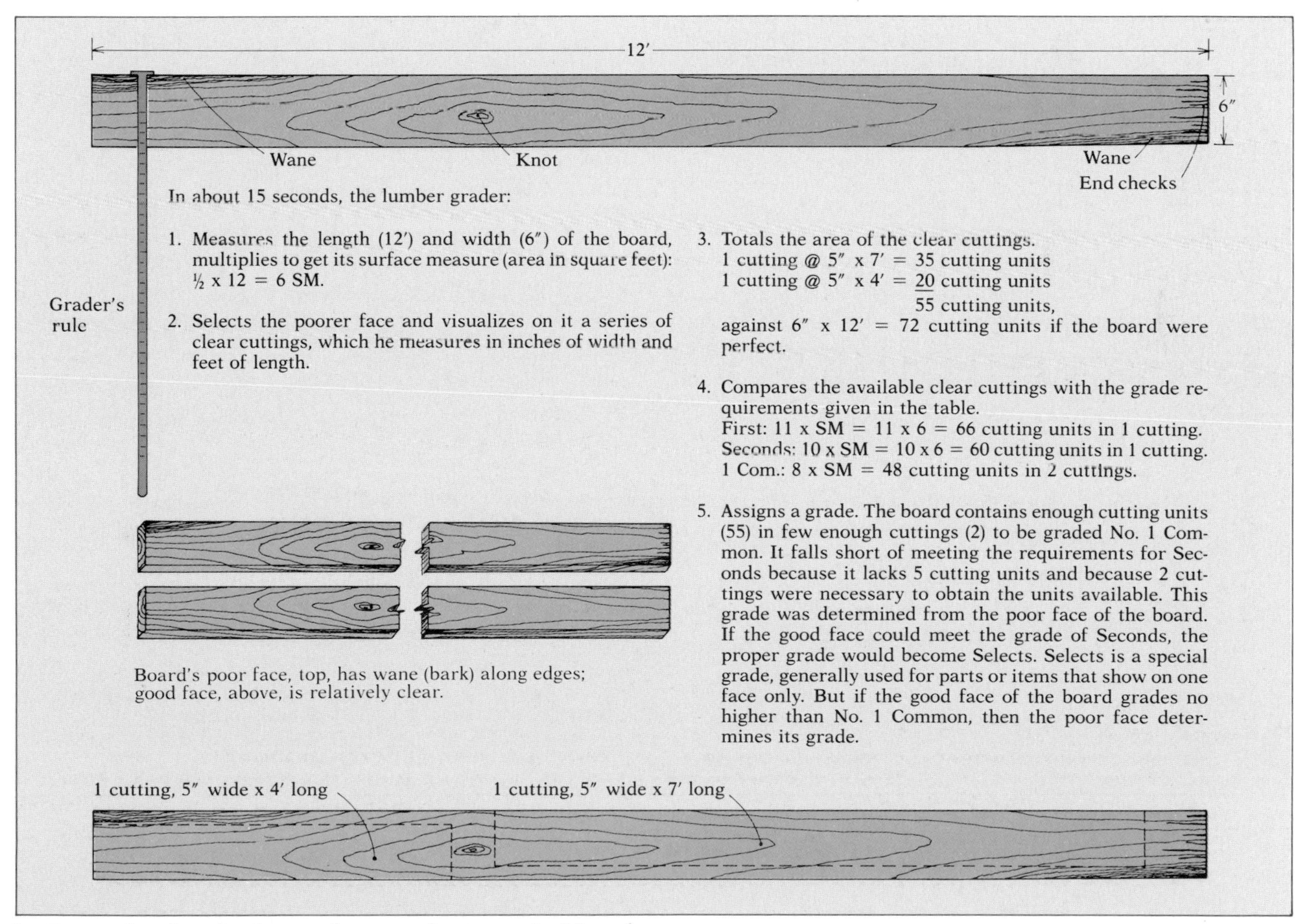

1—Steps taken by a lumber grader to evaluate a typical board.

Table 20—Some major lumber-grading associations and the species under their jurisdiction.

Association	Species
HARDWOOD GRADING ASSOCIATIONS	
NATIONAL HARDWOOD LUMBER ASSOCIATION 59 EAST VAN BUREN STREET CHICAGO, ILLINOIS 60605	HARDWOODS (FURNITURE CUTTINGS, CONSTRUCTION LUMBER, SIDING, PANELS)
HARDWOOD DIMENSION MANUFACTURERS ASSOCIATION 3813 HILLSBORO ROAD NASHVILLE, TENNESSEE 37215	HARDWOODS (HARDWOOD FURNITURE DIMENSION, SQUARES, LAMINATED STOCK, INTERIOR TRIM, STAIR TREADS AND RISERS)
MAPLE FLOORING MANUFACTURERS ASSOCIATION 424 WASHINGTON AVENUE, SUITE 104 OSHKOSH, WISCONSIN 54901	MAPLE, BEECH, BIRCH (FLOORING)
NATIONAL OAK FLOORING MANUFACTURERS ASSOCIATION 814 STERICK BUILDING MEMPHIS, TENNESSEE 38103	OAK, PECAN, BEECH, BIRCH AND HARD MAPLE (FLOORING)
NORTHERN HARDWOOD AND PINE MANUFACTURERS ASSOCIATION SUITE 207, NORTHERN BUILDING GREEN BAY, WISCONSIN 54301	ASPEN (CONSTRUCTION LUMBER)
SOFTWOOD GRADING ASSOCIATIONS	
NATIONAL HARDWOOD LUMBER ASSOCIATION 59 EAST VAN BUREN STREET CHICAGO, ILLINOIS 60605	BALDCYPRESS, EASTERN REDCEDAR
NORTHEASTERN LUMBER MANUFACTURERS ASSOCIATION, INC. 13 SOUTH STREET GLENS FALLS, NEW YORK 12801	BALSAM FIR, EASTERN WHITE PINE, RED PINE, EASTERN HEMLOCK, BLACK SPRUCE, WHITE SPRUCE, RED SPRUCE, PITCH PINE, TAMARACK, JACK PINE, NORTHERN WHITE-CEDAR
NORTHERN HARDWOOD AND PINE MANUFACTURERS ASSOCIATION SUITE 207, NORTHERN BUILDING GREEN BAY, WISCONSIN 54301	BIGTOOTH ASPEN, QUAKING ASPEN, EASTERN WHITE PINE, RED PINE, JACK PINE, BLACK SPRUCE, WHITE SPRUCE, RED SPRUCE, BALSAM FIR, EASTERN HEMLOCK, TAMARACK
RED CEDAR SHINGLE & HANDSPLIT SHAKE BUREAU 5510 WHITE BUILDING SEATTLE, WASHINGTON 98101	WESTERN REDCEDAR (SHINGLES AND SHAKES)
REDWOOD INSPECTION SERVICE 617 MONTGOMERY STREET SAN FRANCISCO, CALIFORNIA 94111	REDWOOD
SOUTHERN CYPRESS MANUFACTURERS ASSOCIATION P.O. BOX 5816 JACKSONVILLE, FLORIDA 32207	BALDCYPRESS
SOUTHERN PINE INSPECTION BUREAU BOX 846 PENSACOLA, FLORIDA 32502	LONGLEAF PINE, SLASH PINE, SHORTLEAF PINE, LOBLOLLY PINE, VIRGINIA PINE, POND PINE, PITCH PINE
WEST COAST LUMBER INSPECTION BUREAU BOX 25406 1750 SW SKYLINE BOULEVARD PORTLAND, OREGON 97225	DOUGLAS FIR, WESTERN HEMLOCK, WESTERN REDCEDAR, INCENSE CEDAR, PORT-ORFORD-CEDAR, ALASKA CEDAR, WESTERN TRUE FIRS, MOUNTAIN HEMLOCK, SITKA SPRUCE
WESTERN WOOD PRODUCTS ASSOCIATION 700 YEON BUILDING PORTLAND, OREGON 97204	PONDEROSA PINE, WESTERN WHITE PINE, DOUGLAS FIR, SUGAR PINE, WESTERN TRUE FIRS, WESTERN LARCH, ENGELMANN SPRUCE, INCENSE CEDAR, WESTERN HEMLOCK, LODGEPOLE PINE, WESTERN REDCEDAR, MOUNTAIN HEMLOCK, RED ALDER

eral conformance to Standard PS 20-70. *Table 20* lists some of the major associations and the species for which they establish rules.

Although some softwood lumber is sold for remanufacture through industrial suppliers and wholesalers, most softwood lumber is for construction and falls into three major categories: appearance lumber, non-stress-graded lumber and stress-graded lumber. The term "yard lumber" is loosely applied to the appearance lumber and non-stress-graded softwood lumber commonly carried by retail lumberyards. This lumber is manufactured to final dressed size and is assumed to be usable in its full size. Unlike hardwood lumber, softwood lumber is assumed to be dried to some extent, so dimension specifications are tied to moisture content. For example, a 2x10 is dressed dry to 1½ in. x 9¼ in.; if green, it is dressed to 1⁹⁄₁₆ in. x 9½ in. (see *Table 17*, page 210). Pieces are graded after final surfacing. The grading is done on the better face of the piece, and grades are designated by describing the allowable size and number of defects (rather than by defining the necessary clear cuttings as in hardwoods).

Appearance grades are the highest grades, commonly termed Select or Finish. Originally grades ran A through D, but now the highest grade is usually B and better (B&BTR). For western white or Idaho white pine (IWP) the B&BTR, C and D Select grades are called Supreme, Choice and Quality, respectively. It is important to note that although appearance grades are the highest grades, they are not the strongest grades, being limited as to knots but not as to grain deviations, density, rings per inch, etc. Other yard lumber is graded on the basis of its general integrity for building. Boards are usually separated into various common grades. Within dimension yard lumber, various grades are designated *(Table 21)*. For example, light framing 2x4s are graded as Construction, Standard or Utility. Structural joists and planks are graded as Select Structural No. 1, No. 2 and No. 3.

Structural lumber, that is, lumber two or more inches in nominal thickness for use where working stresses are encountered, is stress-graded. As indicated in Chapter 6, lumber can be either visually stress-graded or machine stress-rated to assign working stress values for bending stress (F_b), modulus of elasticity (E) and other properties. Lumber is grade-stamped at the mill. In addition to the grade, the stamp usually includes the emblem of the association, the mill number, the species and sometimes the moisture condition *(1)*.

Table 21—Grades and sizes of dimension yard lumber as designated by the American Lumber Standards Committee.

Category	Grades	Sizes
Light framing	Construction, Standard, Utility	2" to 4" thick 2" to 4" wide
Studs	Stud	2" to 4" thick 2" to 4" wide
Structural light framing	Select Structural No. 1, No. 2, No. 3	2" to 4" thick 2" to 4" wide
Appearance framing	Appearance	2" to 4" thick 2" and wider
Structural joists and planks	Select Structural No. 1, No. 2, No. 3	2" to 4" thick 6" and wider

1—Examples of softwood lumber grade stamps. A stamp usually designates the association whose grading rules are used and the manufacturing mill number, plus the grade, the species, the moisture condition and sometimes the stress rating.

Veneer

Veneer is wood in the form of a thin layer or sheet having the grain direction of the wood parallel to the surface. Typical thicknesses are $\frac{1}{40}$ in. or $\frac{1}{28}$ in. for hardwood veneers used in furniture and cabinetmaking; thicknesses of $\frac{1}{10}$ in. to $\frac{3}{16}$ in. are common for softwood veneers used to manufacture plywood. Veneer can be cut as thin as $\frac{1}{100}$ in., but veneer this thin is too fragile for most uses. Thicknesses of up to about $\frac{1}{4}$ in. are classed as veneer, while knife-cut sheets of wood of greater thickness are called **slicewood.**

Wood in the form of veneer has three important categories of application. The first is to decorate surfaces. Typically, veneer of high value or striking appearance is applied over a lower-quality substrate. Valuable or unusual wood can thereby be extended to cover far greater surface area than it could if used as boards. Veneer is also used to make decorative inlay, in the form of pictures for covering large surfaces (marquetry), as well as in the form of geometric border strips for ornamenting furniture and small wooden objects.

The second application of veneer is in crossply construction, to make plywood panels. The advantages of plywood include uniform strength across the panel in all directions and virtual elimination of both splitting and dimensional instability.

The third category is veneer bent to shapes, as in commercial basketmaking. Also, veneers can be formed to shape in a mold by gluing multiple layers. Contours and angles can be formed that otherwise cannot be attained with a single piece of wood of similar thickness without chemical modification. Bent laminations and molded plywoods are examples. A given application may embrace any or all three of the above categories.

Veneer dates back thousands of years. Its earliest uses were as surface decoration for art objects and furniture. Veneer was probably first made by splitting thin slats of wood and painstakingly scraping them down to desired smoothness and thinness. The development of saws allowed larger sheets of wood to be cut, even from logs with irregular grain direction. In post-Renaissance Europe, handsawn veneer was the material of choice for the most lavish furniture. It permitted intricate pictorial effects and abundant use of rare and precious woods, on both flat and curved surfaces. But it wasn't until the last half of the 19th century that knife-cutting was developed as an important production technique. Early in our century veneering flourished in commercial furniture production, even in applications where solid wood had always been serviceable. But because the adhesives then in use could not resist moisture, eventual delamination was

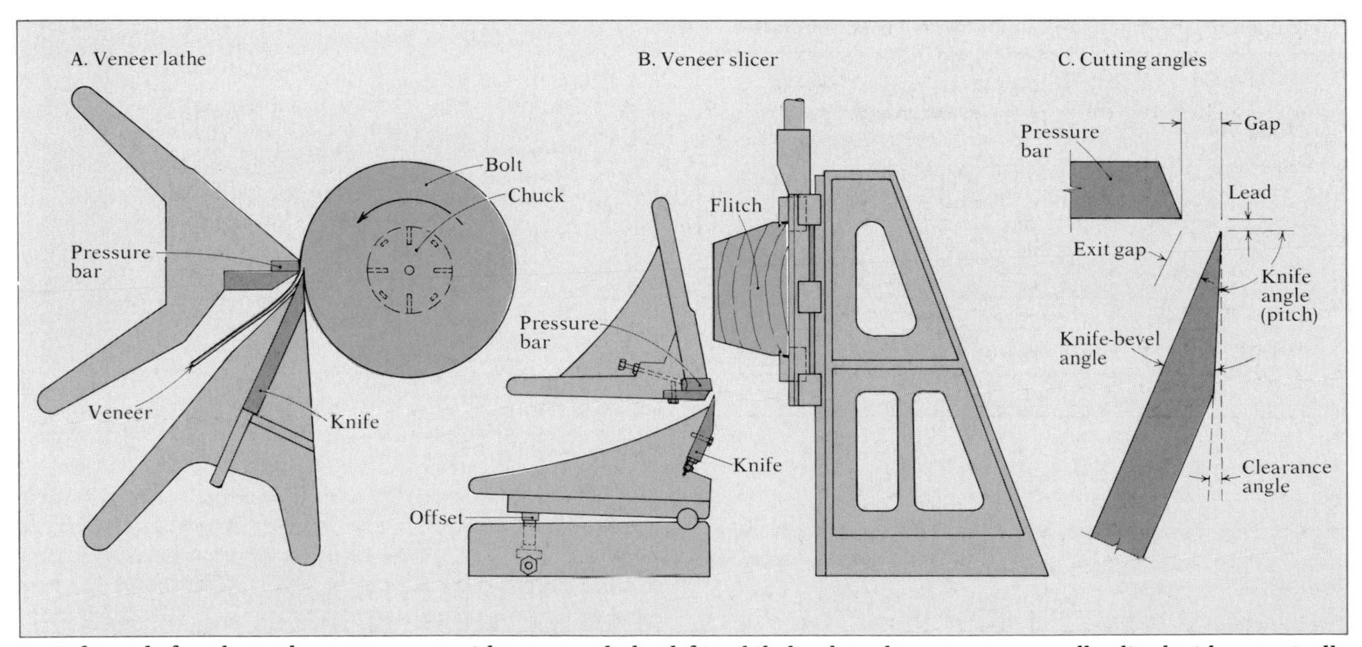

1—Softwoods for plywood are rotary-cut with a veneer lathe (left), while hardwood veneers are usually sliced, either vertically (center) or horizontally. The diagram at right shows the relationships between pressure bar (nosebar) and knife, which determine the quality of the veneer. Also see Figures 1 and 2, page 156.

commonplace. This earned veneered products a bad name that prevailed through mid-century. The assertion that furniture was "solid mahogany" or "solid cherry" or "solid" whatever implied superiority, despite the fact that some of the very finest furniture of the past several hundred years had been veneered. Even today, advertisers proudly proclaim that their rustic benches are made of "solid pine."

Since the development of moisture-resistant and fully waterproof adhesives, veneered products now can routinely be made with all the moisture-resisting integrity of solid wood. Veneer is regaining its rightful place as a respectable medium for woodworking of all types.

Today, commercial sawing produces only a limited amount of veneer, either for specialty uses or from very dense or brittle species. Some craftsmen do saw their own veneer on the band saw, to get the most from an unusual plank or to achieve a curved shape. Most commercial veneer, however, is knife-cut by one of two basic techniques: rotary cutting and slicing.

In **rotary cutting**, a log is center-chucked at its ends in a large lathe *(1)*. As the log revolves, a large knife peels off a continuous layer of veneer, in a manner analogous to unrolling a roll of paper towels (Figure *1*, page 202). For each revolution of the log, the knife automatically advances toward the center of the log by one increment of veneer thickness. Rotary cutting has the advantage of high production rate, and the continuous veneer can be clipped into sheets of desired width. It is thus well suited to high-volume production of softwood veneer for structural plywood. However, peeling produces a continuous tangential cut, whose figure is recognizable as rotary veneer (Figure *4*, page 21). It cannot masquerade as sawn lumber of greater thickness.

In **slicing** veneer, a log portion, called a **flitch**, is first prepared by sawing. The flitch is firmly dogged in a horizontal position against a rugged frame that can move up and down *(2)*. A veneer knife, also mounted horizontally, is positioned alongside the flitch. A slice of veneer is produced as each downward stroke of the frame forces the flitch against the knife, the knife being advanced toward the flitch by one thickness of veneer after each upstroke. As they are cut, the veneer slices are kept in sequence so that after drying they can be restacked in the original order of cutting. This reassembled package of all the veneers from the same piece is also called a flitch. For industrial use, the entire flitch is the customary sales unit. For smaller-quantity sales, a flitch of veneer may be divided into **books**, each consisting of successive **sheets** or leaves of veneer.

2—Flitch firmly dogged in rugged frame is sliced into veneer.

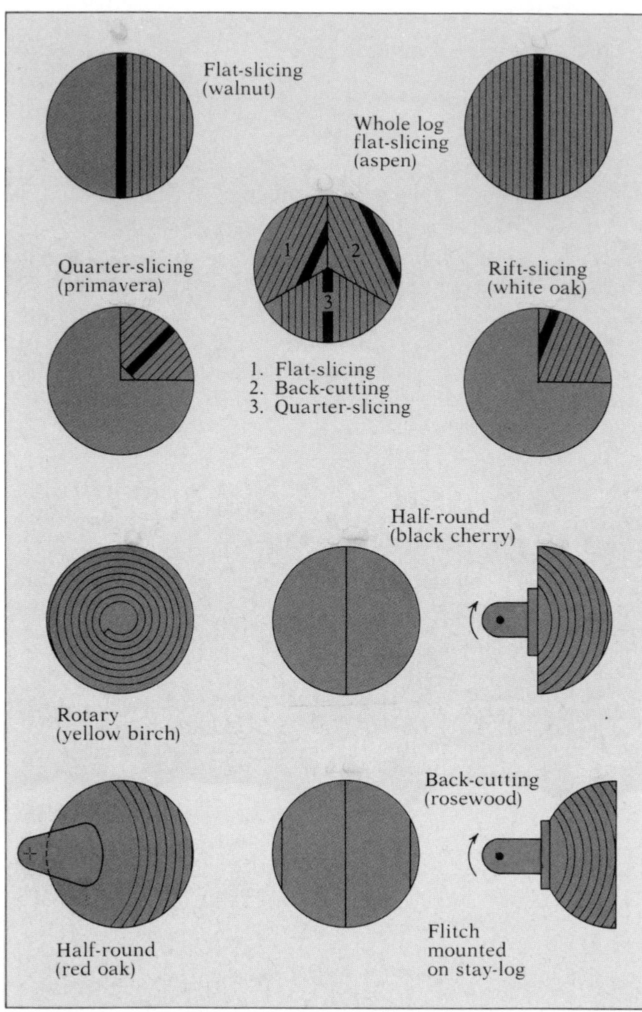

1—The basic cutting directions—flat-slicing, quarter-slicing, rift-slicing and rotary—can be modified by half-round cutting to vary the figure, through off-center chucking or stay-log cutting. Species in parentheses are commonly cut by each method; wide dark lines are backboards left after slicing.

In slicing, the production rate is much lower than in peeling, and the width of veneer sheets is governed by the dimensions of the flitch. However, the manner of sawing the flitch and the position in which it is mounted in the slicer control the growth-ring orientation to produce a desired figure. Because the veneers are sliced straight through the log, their figure is indistinguishable from that of board surfaces. Various methods of flitching logs are shown in the drawing at left *(1)*.

Rotary cutting can be modified by off-center chucking or by **stay-log** cutting to produce **half-round** *(2)*, **back-cut** or **rift-cut** veneers. Despite the obvious differences between lathes and slicers, the cutting action that separates the veneer layer is essentially similar in both. As the knife separates the veneer from the flitch, the separated layer of wood is severely bent, and stresses build up in the region near the knife edge. When the strength of the wood is exceeded, the stress is relieved by failure, and the plane of failure thus formed is called a **knife check** (see page 156). This bending and breaking cycle continues, and each sheet of veneer is liable to have regular checks across the side that was against the knife. This side is called the **loose side**, or **open face**. The side that was away from the knife is called the **tight side** or **closed face**.

The tendency to develop knife-checking varies according to the species of wood, the temperature of the wood and thickness of cut, and the accuracy maintained in the slicing machinery. Diffuse-porous hardwoods with fine, well-distributed rays, such as birch, are more likely to yield tight, uniform veneer. On the other hand, coarse ring-porous hardwoods such as red oak can be severely knife-checked, particularly in thicker sheets.

The tightness of a piece of veneer can be assessed by manually flexing it. The veneer will feel stiffer when flexed to close the checks, but more limp when the checks are flexed open *(3)*. Veneer with a rough or corrugated surface is probably loosely cut *(4)*. Veneer that flexes

2—Veneer being cut by the half-rotary method.

about as easily both ways is tight.

Knife checks are of more than academic interest. The most common consequence is parallel-to-grain cracks through the finish on the veneered surface *(5, 6)*. This problem is especially aggravating because the cracks may not appear for months or years after the work is done. Glue may also bleed through the checks, showing up as dark lines in light-colored woods. In plywood, knife checks may cause the type of failure known as **rolling shear** (see page 120). Thus in woodworking, veneer must routinely be inspected for tightness, and whenever possible the loose side should be the one spread with glue. With luck, the glue will penetrate the checks and keep them closed but care must then be taken not to sand through the tight surface into the glue. Bookmatched patterns are a predicament, because the veneer must be placed with alternate open and closed faces up. Thus it is especially important when bookmatching to avoid loose veneers.

On the whole, most hardwood cabinet veneer produced by reputable mills is cut with adequate quality control to ensure reasonable tightness. But it nevertheless pays to be on guard against loose veneer being sold at bargain prices. The best guideline is to buy veneer from reputable dealers and to know how to detect and cope with knife checks when they do occur. As the supply of larger trees diminishes and the cost of wood continues to increase, the beauty and functional versatility of veneer will ensure its increasing importance.

3—Veneer flexes more easily when checks are opened, left, than closed, right. This maneuver distinguishes the tight and loose faces.

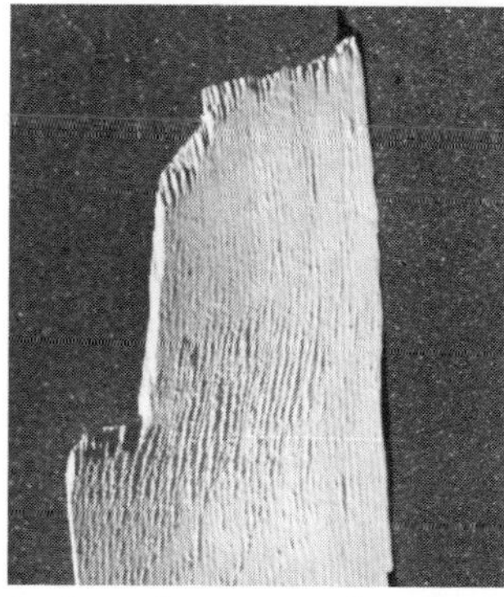

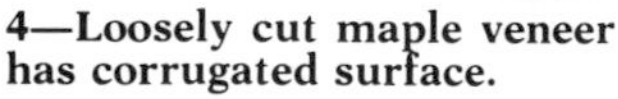

4—Loosely cut maple veneer has corrugated surface.

5—Left, checks broke through paint on fir plywood in a year. Baltic birch panel, right, was smooth when finished with Deft several years ago.

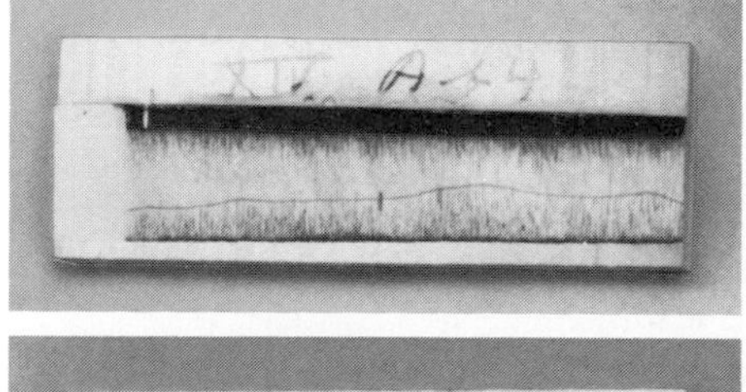

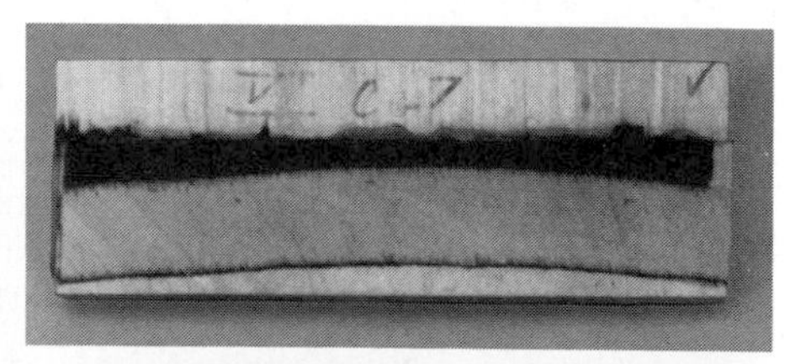

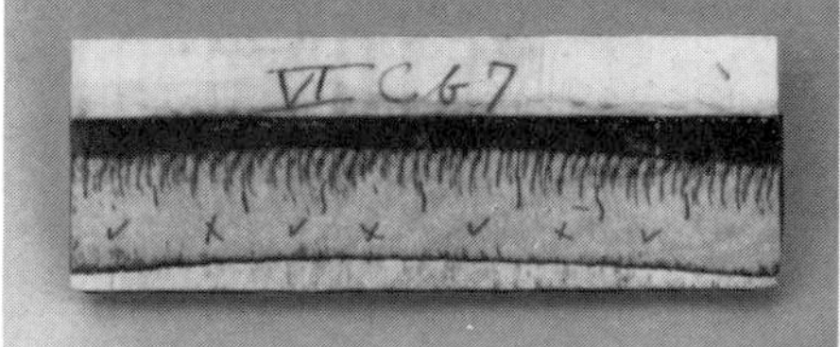

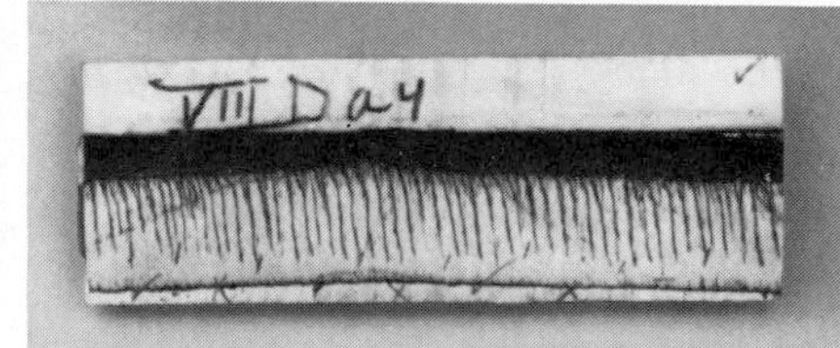

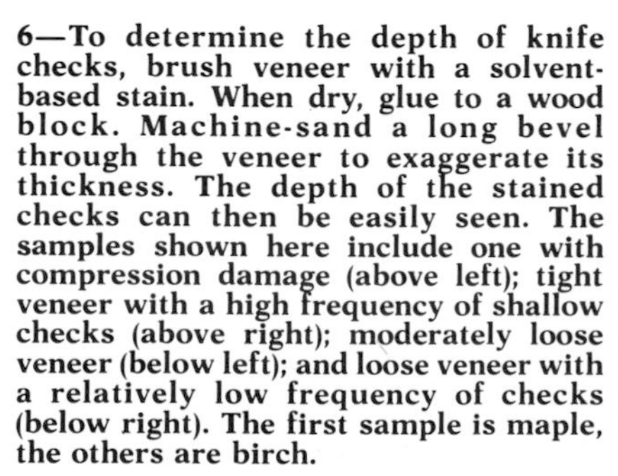

6—To determine the depth of knife checks, brush veneer with a solvent-based stain. When dry, glue to a wood block. Machine-sand a long bevel through the veneer to exaggerate its thickness. The depth of the stained checks can then be easily seen. The samples shown here include one with compression damage (above left); tight veneer with a high frequency of shallow checks (above right); moderately loose veneer (below left); and loose veneer with a relatively low frequency of checks (below right). The first sample is maple, the others are birch.

Plywood

Plywood typically is composed of an uneven number of thin layers of wood glued together with the grain direction of adjacent layers perpendicular to one another *(1)*. The thin layers, called plies, are usually wood veneers, but may be other materials, such as edge-glued lumber panels. Because of the crossply construction, most properties are approximately equalized across the surface of the panel, but are dominated by the greater strength and dimensional stability of wood parallel to the grain. However, bending properties are usually somewhat greater in the grain direction of the face ply. Thus, when it is used as sheathing, sub-flooring and shelving, there is a preferred direction. Crossply construction also virtually eliminates the parallel-to-grain splitting problems characteristic of solid wood. On the other hand, plywood splits easily in the plane of the panel, which must be remembered when attaching hinges or fasteners to panel edges. Another major advantage of plywood is its availability in panels vastly wider than any natural boards.

Although examples of crossply construction have been known since ancient times, plywood as we know it can be rightfully thought of as a modern material, for only within the present century has plywood become established among the principal primary wood products. Veneering techniques developed in the last century paved the way, but ultimately the success of today's plywood industry must be attributed to the development of economical moisture-resistant adhesives. Although plywood can be produced in the small shop, there are two common problems in making your own. The first is the failure to balance completely the panel plies on either side of the centerline with respect to thickness of plies, grain direction and moisture content. The second is inadequate pressure. For example, a piece 12 in. by 12 in. would require a minimum of 7 tons and an optimum of 14 tons to develop adequate pressure on maple veneer, especially if it is at all rough cut. Commercially produced plywood is readily available but the array is so broad that the woodworker should become familiar with the many combinations of species, veneer grades, adhesive types and specialty variations.

The plywood industry is broadly segregated along softwood and hardwood lines. Most softwood plywood is produced from rotary-cut veneer for structural uses. Only a small percentage is used for cabinet work and interior wall paneling. Hardwood plywood is manufactured from both rotary and sliced veneers for furniture, cabinet work, paneling and a host of specialty items.

In manufacturing plywood, veneers can be produced to a fairly high degree of accuracy in thickness, and also can be routinely dried to low moisture content. After gluing under pressure, the plywood panels are machine-sanded to final panel thickness and surface smoothness, and trimmed to final dimensions and squareness. The ability to control the moisture of veneers, the stability afforded by crossply construction and the greater dimensional accuracy in manufacturing usually result in quite consistent panel dimensions.

Plywood is manufactured in a variety of thicknesses up to an inch or more. In surface dimension, the 4-ft. by 8-ft. sheet is the common standard, although specialty types are produced in other sizes. Plywood is usually priced by the square foot of surface measure (or by the standard 4x8 sheet).

The outermost plies of plywood are called **faces** or **face plies.** If they are different in grade, value or appearance, the better of the two is designated as the **face**, the poorer as the **back**. When of equal quality or of unspecified quality, both are considered faces. The center ply (whether veneer, edge-glued lumber, particle board or other material) is termed the **core**. In plywood having more than three plies, the veneers immediately beneath the faces are usually termed **crossbands** or **crossbanding.** These terms are used especially to designate the layers of veneer running perpendicular to the faces and core in lumbercore plywood panels.

The terminology is somewhat inconsistent, however. For example, sometimes all the plies between the central core and the faces are termed crossbands; in other cases, all the plies within the faces are termed core plies, or collectively, simply the core. Materials other than wood used to form the surfaces of the panel, such as metal sheets, foil or resin-impregnated paper, are called **overlays**.

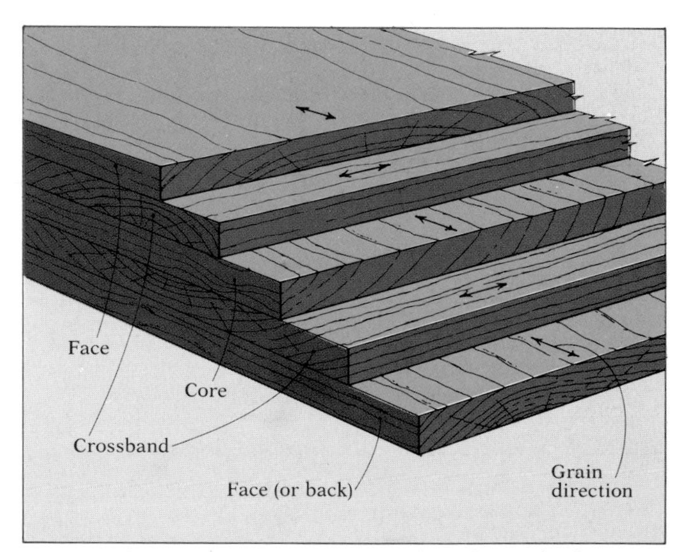

1—Typical plywood construction is five-ply, with the grain direction of adjacent plies running perpendicular to one another.

Softwood plywood

The softwood-plywood industry grew principally around the manufacture of Douglas fir plywood. Once established, however, other species gained prominence, including true firs, southern yellow pines, western larch, western hemlock, cedars and redwood. Most softwood plywood is now manufactured by member mills of the American Plywood Association (APA) whose grading rules and manufacturing standards have been developed in cooperation with the Federal Government. Species are segregated into five groups based on strength properties (*Table 22)*, with Group 1 representing the strongest and stiffest.

Softwood plywood is manufactured in either of two types: **exterior**, with 100% waterproof gluelines; and **interior,** with highly moisture-resistant gluelines. In some interior types, exterior glue is used to improve performance under prolonged wetting. Never use interior-type plywood for exterior applications. One sees too many delaminating garage doors and even front doors on houses.

After that, softwood plywoods are divided into two main categories, **engineered grades** and **appearance grades.** As the terms indicate, engineered grades are designed for applications where strength and serviceability are of primary concern. Appearance grades assume the plywood will be used where appearance is important. Face veneers are therefore graded as N (best), A, B, C plugged, C and D (poorest). Grade N veneers are special-order and intended to take a transparent finish, as in furniture and cabinet work. Other veneer grades are established on the basis of the size and number of repairs or defects such as knots, splits and insect damage. The grade designation generally incorporates the veneer grade of the face and back plies. For example, in A-A grade plywood, both faces have grade-A veneer; in C-D grade plywood, the better face has C-grade veneer, the poorer face has D-grade.

In certain engineered grades of unsanded plywood, such as C-D sheathing, an identification index of two numbers indicates the maximum roof-frame spacing if the panel is used as roof sheathing, and the maximum spacing of floor framing if the panel is used as subfloor. For example, an identificaton index of 32/16 would indicate a maximum 32-in. rafter spacing and a maximum 16-in. floor-joist spacing.

A number of specialty plywoods may be of particular interest to the woodworker. **Marine plywood** is exterior

Table 22—The American Plywood Association's classification of plywoods. The lower the group number, the greater the strength and stiffness of the plywood.

Group 1	Group 2		Group 3	Group 4	Group 5
Apitong	Cedar, Port Orford	Maple, black	Alder, red	Aspen	Basswood
Beech, American	Cypress	Mengkulang	Birch, paper	bigtooth	Fir, balsam
Birch	Douglas fir 2	Meranti, red	Cedar, Alaska	quaking	Poplar, balsam
sweet	Fir	Mersawa	Fir, subalpine	Cativo	
yellow	California red	Pine	Hemlock, eastern	Cedar	
Douglas fir 1	grand	pond	Maple, bigleaf	incense	
Kapur	noble	red	Pine	western red	
Keruing	pacific silver	Virginia	jack	Cottonwood	
Larch, western	white	western white	lodgepole	eastern	
Maple, sugar	Hemlock, western	Spruce	ponderosa	black (western poplar)	
Pine	Lauan	red	spruce	Pine	
Caribbean	almon	Sitka	Redwood	eastern white	
ocote	bagtikan	Sweetgum	Spruce	sugar	
Pine, southern	mayapis	Tamarack	black		
loblolly	red lauan	Yellow poplar	Engelmann		
longleaf	tangile		white		
shortleaf	white lauan				
slash					
Tanoak					

plywood that has A or B-grade faces and no voids in interior plies, so that the plies are supported and so that the edges will not have voids even if the panel is cut—all features especially important in boat hulls. **Medium-density overlay (MDO)** and **high-density overlay (HDO)** plywood is faced with resin-impregnated paper. This plywood is particularly good for exterior painted surfaces such as signs, where surface defects that commonly result from knife checks in face veneers are a problem. Special surface effects, particularly for use as building, siding or interior wall covering, are also available. Striated and brushed surfaces are especially effective in disguising the knife-check defects on panel surfaces. Kerfed and grooved surfaces will hide panel joints. Plywood with hardboard surfaces, called Plyron, is also manufactured.

Grade stamps *(1)* summarize the pertinent information: face-veneer grade, species group, type and sometimes also identification index number and glue.

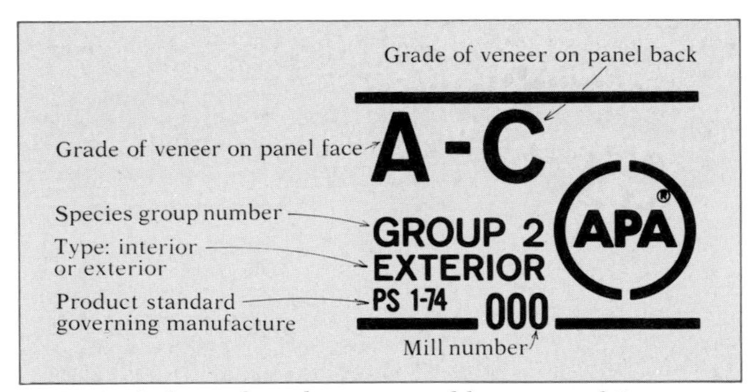

1—Typical plywood grade-stamp, and how to read it.

Table 23—Categories of commonly used species for hardwood plywood based on specific gravity ranges.*

CATEGORY A (0.56 OR MORE SPECIFIC GRAVITY)	CATEGORY B (0.43 THROUGH 0.55 SPECIFIC GRAVITY)	CATEGORY C (0.42 OR LESS SPECIFIC GRAVITY)
ASH, COMMERCIAL WHITE	ASH, BLACK	ALDER, RED
BEECH, AMERICAN	AVODIRE	ASPEN
BIRCH, YELLOW SWEET	BAY	BASSWOOD, AMERICAN
BUBINGA	CEDAR, EASTERN RED**	BOX ELDER
ELM, ROCK	CHERRY, BLACK	CATIVO
MADRONE, PACIFIC	CHESTNUT, AMERICAN	CEDAR, WESTERN RED**
MAPLE, BLACK (HARD)	CYPRESS**	CEIBA
MAPLE, SUGAR (HARD)	ELM, AMERICAN (WHITE, RED, OR GREY)	COTTONWOOD, BLACK EASTERN
OAK, COMMERCIAL RED	FIR, DOUGLAS**	PINE, WHITE AND PONDEROSA**
OAK, COMMERCIAL WHITE	GUM, BLACK	POPLAR, YELLOW
OAK, OREGON	GUM, SWEET	REDWOOD**
PALDAO	HACKBERRY	WILLOW, BLACK
PECAN, COMMERCIAL	LAUAN (PHILIPPINE MAHOGANY)	
ROSEWOOD	LIMBA	
SAPELE	MAGNOLIA	
TEAK	MAHOGANY, AFRICAN	
	MAHOGANY, HONDURAS	
	MAPLE, RED (SOFT)	
	MAPLE, SILVER (SOFT)	
	PRIMAVERA	
	SYCAMORE	
	TUPELO, WATER	
	WALNUT, AMERICAN	

***Based on oven-dry weight and volume at 12% moisture content**
****Softwood**

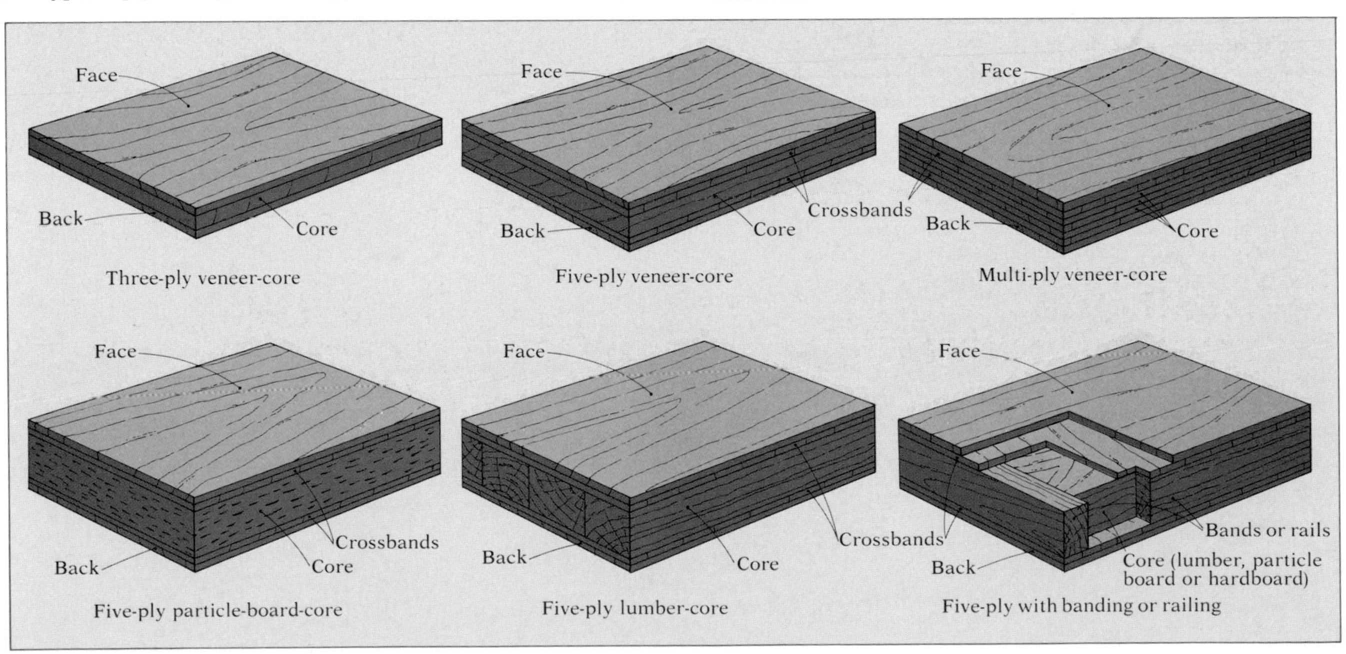

2—Typical hardwood-plywood constructions.

Hardwood plywood

Hardwood plywood is made especially to display the aesthetic qualities of its face veneers. Panels are, in fact, identified according to the species of the face ply; the backs and inner plies may be of another hardwood species, a softwood species or even another material such as particle board. On the basis of specific gravity, species are separated into three categories *(Table 23)*. Depending on the use, the veneers are incorporated into a wide range of panel constructions, ranging from special three-ply, 3/64-in. thick aircraft plywood to 2¼-in. thick flush doors. Common constructions are shown on the previous page *(2)*. Hardwood plywood is used in countless products, from architecture and interior design, cabinetry and furniture to sporting goods, musical instruments and jewelry boxes.

In addition to species, the grade of the face ply is identified in hardwood plywood. Hardwood face-veneer grade standards are listed in *Table 24*. (Certain softwood veneers commonly used in faces because of their decorative features are included with the hardwood standards.) A requirement of **Premium Grade** (A or #1) is that multiple veneer pieces used to make a panel face must be **matched**, that is, arranged in special sequence according to their figure display. Some common arrangements are bookmatching, slip-matching and random-matching *(3)*. Matching of veneers and of successive sheets of plywood can produce beautiful results. Special combinations of figure and matching combinations can be stipulated by specifications within the **Specialty Grade** (SP), and often can be produced to order.

Four types of hardwood plywood are recognized on the basis of adhesive specifications. **Technical Type** and **Type I** both use waterproof adhesives and will therefore withstand all degrees of moisture exposure without delamination. **Type II** is made with moisture-resistant adhesives but is not intended for use where repeated or prolonged wetting might occur. **Type III** is made with adhesives of even lower moisture resistance, adequate for uses such as packing cases or crates.

Most hardwood plywood is available in standard 4x8 sheets. Sheets of greater widths and lengths are sometimes produced to fulfill special needs in building and manufacturing.

When hardwood plywood is manufactured to U.S. Federal standards, certification may accompany the shipment or may be stamped directly on each panel, indicating the species and grade of the face veneer, the type of plywood, and identification of the producer and of the inspection or testing agencies involved.

Table 24—Face grades of hardwood plywood.

SPECIALTY GRADE (SP): This is a plywood made to order to meet the specific requirements of a particular buyer. Plywood of this grade usually entails special matching of the face veneers.

PREMIUM GRADE (#1): The veneer on the face is fabricated for matched joints, and contrast in color and grain is avoided.

GOOD GRADE (#1): The veneer on the face is fabricated to avoid sharp contrasts in color and grain.

SOUND GRADE (#2): The veneer on the face is not matched for color or grain. Some defects are permissible, but face is free of open defects and is sanded and smooth. It is usually used for surfaces to be painted.

UTILITY GRADE (#3): Tight knots, discoloration, stain, wormholes, mineral streaks, and some slight splits are permitted in this grade. Decay is not permitted.

BACKING GRADE (#4): This grade permits larger defects. Grain and color are not matched, and the veneer is used primarily as the concealed face. Defects must not affect strength or serviceability of the panel made from it. At the manufacturer's option, this face can be of some species other than the exposed face.

Hardwood plywood is manufactured in several specific grades. As in softwood plywood, each face must be specified.

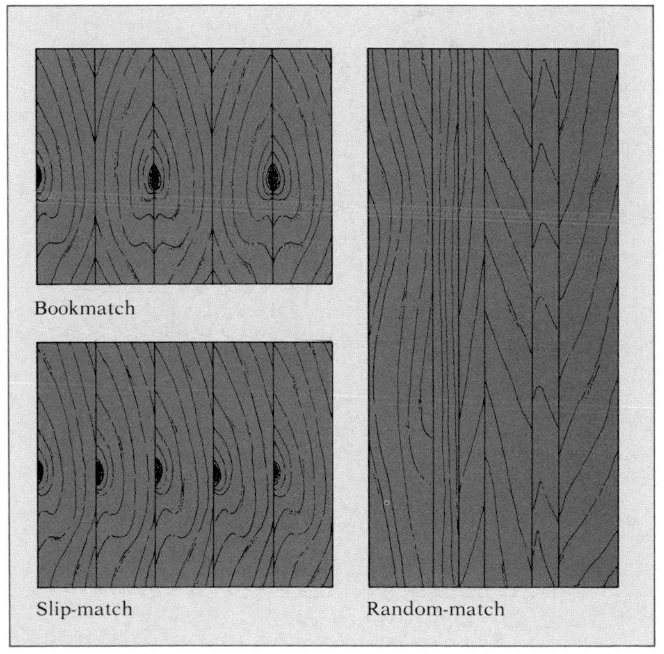

3—Some veneer matches for hardwood plywood.

Composite wood materials

If there is a dominant trend that can be recognized in the modern use of wood, it is the evolution away from the traditional solid-wood or board mentality to the gradual incorporation and acceptance of the growing family of reconstituted or composite wood products. As exemplified by well-established products such as paper and plywood, the development of composites has been inspired by the desire to extend or modify natural wood sizes or properties. Current and future production of composites, however, is mandated by the need to use manufacturing waste and residues, and smaller, lower-grade trees for versatile products of more consistent properties. Future history books probably will recognize our time as a period of drastic transition. Attempting to survey the current picture is somewhat like crystal-ball gazing, for it is likely that despite the many recent advances in technology, the most important breakthroughs lie just ahead. We shall nevertheless review wood composites in a general way and broadly group them into two principal categories, fiber products and solid-wood composites.

In fiber products, wood is broken down to its individual constituent wood cells or fibers, and then reformed into panel, sheet or shape, with or without pressure, by re-establishing chemical bonding among the elements with little or no added resin binder. This category includes cardboard or pasteboard, low and medium-density fiberboard, such as Celotex and Homasote, and hardboards like Masonite.

In solid-wood composites, layers, chips, flakes, splinters, fragments or particles of wood are rebonded using a resin adhesive. An important form is laminated wood, including laminated beams and the more recently developed laminated veneers such as Micro=lam. Laminating randomly redistributes defects and also allows products to be "engineered" by distributing higher and lower-quality elements according to the anticipated distribution of stress. Another important feature of laminating is that by end-to-end scarf or finger-jointing of laminations, members of any length can be produced.

To most, the term solid-wood composite suggests the family of materials loosely termed particle board. In the future, particle board doubtless will be among the most important primary wood products.

Particle board

Although particle-board manufacture is the youngest major segment of the wood-products industry, it has been the fastest growing. By today's standards, the first particle boards manufactured in the early 1950s left much to be desired, and misuse of these early prototypes, often with unsatisfactory results, gave particle board a negative reputation in the minds of many. The greatest dissatisfaction has resulted from attempts to use particle board as a substitute for solid wood or plywood. When used intelligently according to their respective properties, today's particle boards constitute a versatile array of superior products for many facets of building and woodworking.

Because particle products are now being engineered for specific properties and uses, the general term particle board may soon become obsolete. For example, the particles used can be manufacturing residues such as planer shavings, veneer scraps or sawmill chips, or they may be specially machined strands, flakes or splinters. The board may have controlled layering, with the placement of coarse particles in the center and with flakes or finer particles at the surfaces for superior strength and to enhance uniformity and smoothness of the finished board. Materials with directional particles also have been developed to give oriented strength properties, as preferred in structural uses.

Particle boards are classified by particle type and further by adhesive type, density and strength class *(Table 25)*. **Type 1** particle board utilizes a urea-formaldehyde resin suitable for most routine interior uses. However, for applications where greater heat and moisture resistance is required, **Type 2** boards bonded with waterproof adhesive are appropriate. Low, medium and high-density boards, ranging in density from 25 to 70 lb./cu. ft., are produced by varying the percentage of resin and the compaction of the particle mat. Within each density group, boards are produced in two strength classes: **Class 2** is stronger than **Class 1.** Strength properties of some particle boards equal solid-wood strengths of low-density species.

Mat-formed particle board has two outstanding features: Because of the random distribution of particles, properties are uniform across the board faces. Since the particles have their grain direction generally parallel to the board surface, the random arrangement gives cross-ply effects. Particles are mutually stabilized, like veneers in plywood, thus giving particle boards uniformly reduced dimensional properties across the panel faces. The effective shrinkage percentage of particle boards is generally in the same range as that of plywood, averaging about 0.5%. At the same time, because of side-grain compaction and crushing perpendicular to the board surface, long-term delayed recovery, especially under cyclic moisture conditions, reveals that the boards have the least stability perpendicular to the surface.

Mechanical fasteners driven perpendicular to the surface hold well in particle board, especially when appropriate pilot holes are driven. The random particle orientation virtually eliminates splitting. However, because particle boards are usually less dense at the center than

Table 25—Standards for particle board developed by the National Bureau of Standards (Commercial Standard CS 236-66).

Type	Density (Grade) Min. Avg.	Class	Modulus of Rupture, Min. Avg. (psi)	Modulus of Elasticity, Min. Avg. (psi)	Internal Bond, Min. Avg. (psi)	Linear Expansion, Max. Avg. (%)	Screw Holding, Min. Avg. Face (lb.)	Screw Holding, Min. Avg. Edge (lb.)
1	A (High density, 50 lb./cu. ft. and over)	1	2,400	350,000	200	0.55	450	—
		2	3,400	350,000	140	0.55	—	—
	B (Medium density, between 37 and 50 lb./cu. ft.)	1	1,600	250,000	70	0.35	225	160
		2	2,400	400,000	60	0.30	225	200
	C (Low density, 37 lb./cu. ft. and under)	1	800	150,000	20	0.30	125	—
		2	1,400	250,000	30	0.30	175	—
2	A (High density, 50 lb./cu. ft. and over)	1	2,400	350,000	125	0.55	450	—
		2	3,400	500,000	400	0.55	500	350
	B (Medium density, less than 50 lb./cu. ft.)	1	1,800	250,000	65	0.35	225	160
		2	2,500	450,000	60	0.25	250	200

near the surface (by surface compaction in manufacturing or by intentional layering), fasteners driven into the center of panel edges hold poorly. Where it is necessary to fasten into panel edges, edge-banding with solid wood or the use of special fasteners may be preferable.

Most particle board is used as floor underlayment for carpeting or resilient floor covering and as core material for hardwood plywood, wall paneling, doors, furniture and casegoods. It is well suited for facing with veneer to produce panels of various thicknesses. Special grades of particle board are manufactured whose surfaces are especially suited for painting. In manufacturing these boards, fine particles are used to form the surfaces, which are then filled and sealed so that subsequent telegraphing through the paint is minimal. In summary, smoothness of surface is associated with higher density and finer particles. On the other hand, lower-density boards having larger particles, such as flakes, have less smooth surfaces but greater overall dimensional stability.

Particle boards are also being developed for use as structural materials, both as extruded shapes and as components of built-up members. Com-Ply, *(1)* for example, uses particle beams faced with double thicknesses of veneer. Other panel types are comprised of large flakes or oriented strands in homogenous or layered organization to produce boards having properties equivalent to plywood for sheathing applications.

Under sustained bending loads, particle board will creep, so when used as shelving, supports should be more closely spaced than for shelf boards of equal thickness. Particle board (even when bonded with waterproof resins) will not maintain surface integrity under exterior weathering conditions and should therefore be used only where protected from the elements.

1—Com-Ply was developed as a substitute for solid-wood structural members. High-quality veneer cut from the outer portion of logs edges a core of particle board made from the lower-quality inner parts of logs or from logging and mill residues.

Hardboard

Hardboard is a general term referring to fiber-type board interfelted and consolidated under heat and pressure to a density of at least 31 lb./cu. ft. This type of board was first produced in 1924 by W. H. Mason, founder of the Masonite Corp. Although hardboard is now produced by many firms, the term Masonite is used somewhat generically for all types and brands of hardboard.

Fiberboards in the 50 to 80-p.c.f. (lb./cu. ft.) density range are called **high-density hardboards** to distinguish them from what are now more commonly referred to as **medium-density fiberboard (MDF)** in the 33 to 50-p.c.f. class *(Table 26)*. The most commonly available hardboards have a density of 60 to 65 p.c.f. and are termed **standard**. For purposes where lower strength and hardness are acceptable, **service** hardboards having a density of 50 to 55 p.c.f. can be used. **Tempered** hardboards are produced by impregnating standard high-density hardboard with resin and heat-curing. Tempering appreciably improves water resistance, hardness and strength, and results in boards having a density of 60 to 80 p.c.f.

As with particle board, hardboard is unstable in thickness but has dimensional properties similar to particle board in the plane of its surface. However, since it is commonly produced in thicknesses of only ⅛ in. or ¼ in., uneven moisture exchange between the two surfaces may result in bulging or warping.

Depending on the manufacturing process, hardboard is either smooth on both sides or has a screen pattern imprinted on one face. A wide variety of specialty hardboards is produced with simulated wood figure or decorative surfaces, or with surfaces embossed with simulated raised grain. Some hardboard is perforated in pegboard pattern or with geometric designs to form decorative screens. Hardboard has a host of uses in woodworking, ranging from furniture backs and drawer bottoms to curved surfaces and panel inserts.

Table 26—Strength and mechanical properties of fiberboards.

PROPERTY	VALUE FOR STRUCTURAL INSULATING BOARD	VALUE FOR MEDIUM-DENSITY FIBERBOARD	VALUE FOR HIGH-DENSITY HARDBOARD	VALUE FOR TEMPERED HARDBOARD	UNIT
DENSITY	10-30	33-50	50-80	60-80	P.C.F.
SPECIFIC GRAVITY	0.16-0.42	0.53-0.80	0.80-1.28	0.93-1.28	----
MODULUS OF ELASTICITY (BENDING)	25-125	325-700	400-800	650-1,100	1,000 PSI
MODULUS OF RUPTURE	200-800	1,900-6,000	3,000-7,000	5,600-10,000	PSI
TENSILE STRENGTH PARALLEL TO SURFACE	200-500	1,000-4,000	3,000-6,000	3,600-7,800	PSI
TENSILE STRENGTH PERPENDICULAR TO SURFACE	10-25	40-200	75-400	160-450	PSI
COMPRESSIVE STRENGTH PARALLEL TO SURFACE	----	1,000-3,500	1,800-6,000	3,700-6,000	PSI
SHEAR STRENGTH (IN PLANE OF BOARD)	----	100-475	300-600	430-850	PSI
SHEAR STRENGTH (ACROSS PLANE OF BOARD	----	600-2,500	2,000-3,000	2,800-3,400	PSI
24-HOUR WATER ABSORPTION	1-10	----	----	----	PCT. BY VOLUME
24-HOUR WATER ABSORPTION	----	5-20	3-30	3-20	PCT. BY WEIGHT
THICKNESS SWELLING, 24-HOUR SOAKING	----	2-10	10-25	8-15	PCT.
LINEAR EXPANSION FROM 50 TO 90 PERCENT RELATIVE HUMIDITY	0.2-0.5	0.2-0.4	0.15-0.45	0.15-0.45	PCT.
THERMAL CONDUCTIVITY AT MEAN TEMPERATURE OF 75° F	0.27-0.45	0.54-0.75	0.75-1.40	0.75-1.50	BTU IN./H FT. DEG F°*

*British thermal units per inch of thickness per hour per square foot of surface per degree Fahrenheit.

Afterword: Forests Past and Future

CHAPTER 13

Most woodworkers are understandably concerned about our timber supply, both in America and worldwide. We often hear alarming forecasts that continued increases in wood consumption will deplete our forests and eventually result in a timber famine. However, in the face of dire predictions I remain optimistic. I do not suggest that we can continue using our forest resources the way we have in the past and get away with it. Rather, I have faith that we will be able to adopt new ways of managing our forests and of using the timber we grow.

To understand the present situation and to realize where we must go from here, it is important to appreciate the extent to which our civilization, and American society in particular, has evolved with a dependence upon timber. Exactly where the story begins is not clear, for the use of trees and wood as a source of food, shelter, weapons and tools predates recorded history. Since the first civilizations wood has been a primary engineering material. It was also used in sophisticated craft and art, as evidenced by artifacts such as the remarkable wooden objects preserved in Egyptian tombs. Scholars have probably reconstructed prehistory with disproportionate emphasis on the surviving stone and metal artifacts. We grossly underestimate the role wood played in early cultures, simply because the biodegradable evidence no longer remains.

We all know how well our American forefathers built with wood, made furniture and cabinetry, tools, wagons and ships. But perhaps we overlook the fact that in the early days of our country, wood was also the principal source of energy. As late as 1859, 90% of the national heating requirement was met by burning wood. Annual per-capita consumption was about 4½ cords, mostly as fuel in open fireplaces. Further, 80% of the people lived on farms that had been cleared from the continental forest. Many of life's necessities were taken, through home industry, directly from the local woodlot—the farmer who needed a new shed for his animals, or a handle for his hoe, started making it by felling a tree. Our forefathers relied upon a detailed "wood lore" developed through experience, hearsay and handed-down recipes, with emphasis on how or when to do a thing, not on why it would work.

The average American a century or more ago was—by necessity—knowledgeable about wood from local trees to a degree far surpassing what most of us know today. This education doubtless began in childhood, fetching and carrying from woodpile to hearth. The ability to identify both standing trees and converted wood was as routine to children then as learning ABCs or telling time is today. The child could not help learning which species was best for kindling a fire and which would make the best handles, as surely as our offspring learn to adjust vertical roll on a television set.

In the cities, on the other hand, work was likely to be both more specialized and more skilled. Still, many of the tradesmen were primarily woodworkers: the wheelwright, the cabinetmaker, the cooper, the shipwright, the carpenter, the carver. And as the nation grew, depending on wood for its very survival, the local supply of easily available fuelwood became threatened. The hunger for fuel caused woodlots to be recut as fast as sprout growth reached cordwood size. But always, out there just beyond, was the original forest, an inexhaustible supply of old-growth trees silently holding their wide, clear boards and cords of fuel.

As a flourishing America crossed into the 20th century, the most rewarding and respected trades still dealt with woodworking. Though more and more handwork was replaced by powered machines, their operation was largely manual, and a well-versed apprenticeship system continued to pass the traditions of the various trades from generation to generation. Manufacturing specialization and division of labor superseded "do-it-all-yourself" as the American way of life, so commodities were increasingly made by "someone else." Though technologies of metal, stone and glass also expanded, the rising standard of living was paced by wooden homes and furniture, and ever more paper goods, especially for packaging.

The consequences were inevitable. By the late 1800s the prime forests of New England had been lumbered off, and then across the Northeast to the Great Lakes and on to the South and finally to the great Western reserves. The last of the virgin stands began to fall before the loggers' saws.

It took a while for Americans to realize that our great forest resource was not inexhaustible. Forest management came too little, too late. In one area after another, the old growth was logged and relogged, and although most of our forests still exist, their nature has been changed, perhaps forever.

The present century has brought two important changes in the way we use our forests. First, fossil fuels replaced wood as a source of heat and power. As we learned how to use the trees of prehistoric times, the woodbox was replaced first by the coalbin and then by the oil barrel. Although we now face the consequences of depending on petroleum, it may have prevented catastrophe in the forests. For although our population has risen from about 76 million in 1900 to about 220 million today, our annual consumption of wood has risen only from 11 to 14 billion cubic feet. The effect of large numbers of individuals turning back from fossil fuels to wood heat has yet to become apparent.

Second, in this century we have developed the technology of composite products. Made possible by progress in adhesive technology, first plywood and then particle board and fiberboard have expanded the horizons of the previous tradition of using wood mostly in solid form.

Today, we find ourselves in an interesting but complex situation. For although we are considered to be in the space age, our national appetite for the oldest of materials—wood—has never been greater. We in the United States consume wood at an annual rate of about one ton per person. This total tonnage exceeds the combined tonnage of all other materials. About half of that tonnage of wood is made into paper and related fiber products.

At present our annual harvest is approximately balanced by an equal volume of growth, although in the category of softwood sawlogs, the cut exceeds the regrowth. In the future, however, there will be no more offset in consumption by reduced use of fuelwood, for fuelwood consumption is increasing rapidly. Over the next quarter century, as population increases, our consumption of wood is expected to double.

At this point we might wonder if there are not alternatives to our forest resources, such as substituting other materials or importing the wood we need. In recent decades, sophistication in the manufacture of metals, plastics and other synthetics suggested that manmade substitutes for wood might eventually ease the drain on our forests. However, we now appreciate that these potential substitutes are themselves made from non-renewable resources. In their manufacture they require many times the energy and they pollute the environment significantly more than their counterpart products in wood. It appears that we must face, rather than evade, the challenge of increasing timber production. The question is not, "*Shall* we use our forests?" but, "*How* shall we use our forests?"

Unfortunately, the practices of the forest industry seem to be at odds with the quality of our environment. Thoughtless and wasteful treatment of our country's forests has given a large segment of the population a preservationist, rather than a conservationist, attitude. We proudly accept the transformation of our great prairies by "amber waves of grain," but the cropping of our forests creates a frightful image of futureless destruction. I hope the polarized attitudes toward our forest resource can reach a compromise: As timber producers see the need and the wisdom of intelligent management and harvesting practices, the environment-minded public may come to accept the reality of multiple-use forest management that includes timber production.

Will there be enough wood?

The major question today is whether we can in fact continue to meet our demands for wood, or whether we are heading for a timber famine.

My belief that we can meet future needs is based first of all on the fact that trees grow. Wood is a renewable resource. It is easy and pleasant to be reminded of this fact. My area of central Massachusetts has some magnificent stands of hardwoods—sugar maple, red and white oaks, cherry and more. Walking in such impressive forests, one notices a network of stone walls, reminders that this land was once cleared and used for crops or pasture. What wasn't cleared for agriculture was probably leveled at one time or another for lumber, firewood or charcoal. One nearby town, 37% forested in the 1840s, is 97% forested today. On abandoned land, holding back the development of a forest is a greater problem than making one grow.

Nevertheless, in our forests we can see an unfortunate tradition that prevailed in the past and persists even today: the practice of logging and relogging an area, each time taking the best of what grows there with no provision for the future. This has the tragic effect of selecting poorer and poorer residual stock, the inverse of survival of the fittest, with predictable results. When finally allowed to regrow, the forest must spring from the stunted and malformed trees that the loggers did not want, and a forest of the highest quality cannot result. Although unintentionally, the forest has been genetically degraded. In some parts of New England, it has been systematically degraded for 250 years.

For the future, better forest management is imperative. Silvicultural and genetic improvements can vastly increase productivity. Actually speeding up the rate of cutting on some overmature stands may be justified. Reforestation may offset areas falling to the urban sprawl. Since fire, insects and decay consume as much timber as does man, the value of forest protection is obvious.

In the past, timber prices could not support the cost of intensive management. Today, however, the premium value of sawlog and veneer stumpage changes the picture considerably. The increasing value of low-grade trees and weed species for fuelwood or pulpwood now can pay for stand improvement, which heretofore was not economically feasible.

In one way or another we must stop thinking of our forests as wild lands where we simply scavenge what we need. Such a destructive and unproductive approach must be replaced by systematic and intensive forest management.

The other half of the battle is to increase the yield of the trees we grow. Under traditional logging and milling practices that convert trees into lumber or veneer, far less than half of the tree finds its way into the final product. Residues and waste occur at every step of conversion. Logging slash and stumps were left in the woods to rot. At the mill, bark, slabs, edgings, trimmings and sawdust were just a nuisance to be burned or dumped. Throughout secondary processing, additional waste was associated with virtually every manufacturing step. Today the goal is to use the entire tree. This approach involves both reduction of residues by more efficient technology as well as finding uses for those residues that will always be a part of solid-wood conversion.

Ever greater forest production inevitably means relying on smaller and younger trees. Our annual production will include a smaller proportion of large, solid pieces, and a larger proportion of slabs, edgings and other residues. Therefore the forest products of the future will be—and must be—of a new, different and changing technology. Structures will be framed using components and materials manufactured from poorer trees, but will be characterized by closer functional design rather than wasteful use of choice material. Furniture will be made of parts glued up from smaller pieces or stained to allow lower-value species to imitate those of higher traditional appeal. As in all wood products, intermarriage with other materials such as plastics, metals and fabrics will be increasingly appropriate. Many structural and functional wood products will become less and less recognizable as ever having been part of a tree. Plywood and composite products such as particle board will be the rule rather than the exception. So today we find ourselves on the threshold of a new era of changing technology, based on a timber resource characterized by smaller, younger trees of seemingly lower quality, processed for greater total yield, into routine consumer products that are less and less recognizably wood.

Such a glimpse of the future frightens some people into believing that solid-wood products will disappear completely. Hardly. Instead, wood in its natural form will become more respected and more valuable than ever. The crafting of fine furniture and instruments, the creation of carvings and sculpture, will be elevated just that much higher on the scale of cultural respect.

But there will be some changes to reckon with. The high cost of lumber will encourage the woodworker to look beyond the lumber store and shop the mills directly, or start directly with the tree, the way one did a century or more ago. But this time around the craftsman has the tools and knowledge of modern science and technology: a chain saw to harvest the tree and a chain-saw mill to rip it into lumber or planks; a meter to measure its moisture content as it dries; portable high-speed tools with carbide bits to cut through knots; abrasive papers to work the surface smooth; stains and finishes for every desired surface effect and functional requirement.

Wood has a future at all levels of use, for our forests will give us not only a backbone resource for large-scale commodities, but also the semi-precious jewels for our woodcrafting endeavors. Historians tell us we have passed through the stone and bronze and iron ages. We now hear mention of the space age. But it may well be that yet ahead is a new age of wood. . . for those who understand it.

Appendix 1: Commercial names for lumber

Sometimes lumber is sold under the same name as the tree from which it was cut. Sometimes a commercial lumber name corresponds to several species. This list correlates commercial lumber names with common tree names and with botanical names. It is adapted from the U.S. Department of Agriculture Handbook No. 541, *Checklist of United States Trees.*

Commercial name for lumber	Common name of tree	Scientific (botanical) name of tree
Alder:		
Red alder	Red alder	*Alnus rubra*
Ash:		
Black ash	Black ash	*Fraxinus nigra*
Oregon ash	Oregon ash	*F. latifolia*
Pumpkin ash	Pumpkin ash	*F. profunda*
White ash	Blue ash	*F. quadrangulata*
	Green ash	*F. pennsylvanica*
	White ash	*F. americana*
Aspen	Bigtooth aspen	*Populus grandidentata*
	Quaking aspen	*P. tremuloides*
Balsam poplar	Balsam poplar	*P. balsamifera*
Balsam fir	Balsam fir	*Abies balsamea*
	Fraser fir	*A. fraseri*
Basswood	American basswood	*Tilia americana*
	White basswood	*T. heterophylla*
Beech	American beech	*Fagus grandifolia*
Birch	Grey birch	*Betula populifolia*
	Paper birch	*B. papyrifera*
	River birch	*B. nigra*
	Sweet birch	*B. lenta*
	Yellow birch	*B. alleghaniensis*
Box elder	Boxelder	*Acer negundo*
Buckeye	Ohio buckeye	*Aesculus glabra*
	Yellow buckeye	*Ae. octandra*
Butternut	Butternut	*Juglans cinerea*
Cedar:		
Eastern red-cedar	Eastern red-cedar	*Juniperus virginiana*
	Southern red-cedar	*J. silicicola*
Western cedar	Alaska-cedar	*Chamaecyparis nootkatensis*
	Incense cedar	*Libocedrus decurrens*
	Port-orford-cedar	*Chamaecyparis lawsoniana*
	Western red-cedar	*Thuja plicata*
Eastern white cedar	Northern white-cedar	*Thuja occidentalis*
Northern white cedar	Northern white-cedar	*T. occidentalis*
Southern white cedar	Atlantic white-cedar	*Chamaecyparis thyoides*
Cherry	Black cherry	*Prunus serotina*
Chestnut	American chestnut	*Castanea dentata*

Commercial name for lumber	Common name of tree	Scientific (botanical) name of tree
Cottonwood	Black cottonwood	*Populus trichocarpa*
	Eastern cottonwood	*P. deltoides*
	Swamp cottonwood	*P. heterophylla*
	Balsam poplar	*P. balsamifera*
Cucumber	Cucumbertree	*Magnolia acuminata*
Cypress	Baldcypress	*Taxodium distichum*
	Pondcypress	*T. distichum* var. *nutans*
Dogwood	Flowering dogwood	*Cornus florida*
	Pacific dogwood	*C. nuttallii*
Douglas Fir—larch	Douglas fir	*Pseudotsuga menziesii*
	Western larch	*Larix occidentalis*
Douglas fir (south)	Douglas fir	*Pseudotsuga menziesii*
Eastern hemlock—tamarack	Eastern Hemlock	*Tsuga canadensis*
	Tamarack	*Larix laricina*
Eastern spruce—balsam fir	Black spruce	*Picea mariana*
	Red spruce	*P. rubens*
	White spruce	*P. glauca*
	Balsam fir	*Abies balsamea*
Eastern woods	Bigtooth aspen	*Populus grandidentata*
	Quaking aspen	*P. tremuloides*
	Balsam fir	*Abies balsamea*
	Carolina hemlock	*Tsuga caroliniana*
	Eastern hemlock	*T. canadensis*
	Eastern white pine	*Pinus strobus*
	Jack pine	*P. banksiana*
	Red pine	*P. resinosa*
	Black spruce	*Picea mariana*
	Red spruce	*P. rubens*
	White spruce	*P. glauca*
	Tamarack	*Larix laricina*
Elm:		
Rock elm	Cedar elm	*Ulmus crassifolia*
	Rock elm	*U. thomasii*
	September elm	*U. serotina*
	Winged elm	*U. alata*
Soft elm	American elm	*U. americana*
	Slippery elm	*U. rubra*
Engelmann spruce—alpine fir	Engelmann spruce	*Picea engelmannii*
	Subalpine fir	*Abies lasiocarpa*
Engelmann spruce—lodgepole pine	Engelmann spruce	*Picea engelmannii*
	Lodgepole pine	*Pinus contorta*
Gum	Sweetgum	*Liquidambar styraciflua*
Hackberry	Hackberry	*Celtis occidentalis*
	Sugarberry	*C. laevigata*

Commercial name for lumber	Common name of tree	Scientific (botanical) name of tree
Hem-fir	Western hemlock	*Tsuga heterophylla*
	California red fir	*Abies magnifica*
	Grand fir	*A. grandis*
	Noble fir	*A. procera*
	Pacific silver fir	*A. amabilis*
	White fir	*A. concolor*
Hemlock:		
Eastern hemlock	Carolina hemlock	*Tsuga caroliniana*
	Eastern hemlock	*T. canadensis*
Mountain hemlock	Mountain hemlock	*T. mertensiana*
Western hemlock	Western hemlock	*T. heterophylla*
Hickory	Mockernut hickory	*Carya tomentosa*
	Pignut hickory	*C. glabra*
	Shagbark hickory	*C. ovata*
	Shellbark hickory	*C. laciniosa*
Holly	American holly	*Ilex opaca*
Ironwood	Eastern hophornbeam	*Ostrya virginiana*
Juniper	Alligator juniper	*Juniperus deppeana*
	Rocky Mountain juniper	*J. scopulorum*
	Utah juniper	*J. osteosperma*
	Western juniper	*J. occidentalis*
Larch	Western larch	*Larix occidentalis*
Locust	Black locust	*Robinia pseudoacacia*
	Honeylocust	*Gleditsia triacanthos*
Madrone	Pacific madrone	*Arbutus menziesii*
Maple:		
Hard maple	Black maple	*Acer nigrum*
	Sugar maple	*A. saccharum*
Oregon maple	Bigleaf maple	*A. macrophyllum*
Soft maple	Red maple	*A. rubrum*
	Silver maple	*A. saccharinum*
Mixed species	Alaska cedar	*Chamaecyparis nootkatensis*
	Incense cedar	*Libocedrus decurrens*
	Port-Orford-cedar	*Chamaecyparis lawsoniana*
	Western redcedar	*Thuja plicata*
	Douglas fir	*Pseudotsuga menziesii*
	California red fir	*Abies magnifica*
	Grand fir	*A. grandis*
	Noble fir	*A. procera*
	Pacific silver fir	*A. amabilis*
	Subalpine fir	*A. lasiocarpa*
	White fir	*A. concolor*
	Mountain hemlock	*Tsuga mertensiana*
	Western hemlock	*T. heterophylla*
	Western larch	*Larix occidentalis*
	Blue spruce	*Picea pungens*
	Engelmann spruce	*P. engelmannii*
	Sitka spruce	*P. sitchensis*
	Lodgepole pine	*Pinus contorta*
	Ponderosa pine	*P. ponderosa*
	Jeffrey pine	*P. jeffreyi*
	Sugar pine	*P. lambertiana*
	Western white pine	*P. monticola*
Mountain hemlock–hem-fir	California red fir	*Abies magnifica*
	Grand fir	*A. grandis*
	Noble fir	*A. procera*
	Pacific silver fir	*A. amabilis*
	White fir	*A. concolor*
	Mountain hemlock	*Tsuga mertensiana*
	Western hemlock	*T. heterophylla*
Mulberry	Red mulberry	*Morus rubra*
Oak:		
Red oak	Black oak	*Quercus velutina*
	Blackjack oak	*Q. marilandica*
	Calif. black oak	*Q. kelloggii*
	Cherrybark oak	*Q. falcata* var. *pagodifolia*
	Laurel oak	*Q. laurifolia*
	Northern pin oak	*Q. ellipsoidalis*
	Northern red oak	*Q. rubra*
	Nuttall oak	*Q. nuttallii*
	Pin oak	*Q. palustris*
	Scarlet oak	*Q. coccinea*
	Shumard oak	*Q. shumardii*
	Southern red oak	*Q. falcata*
	Turkey oak	*Q. laevis*
	Water oak	*Q. nigra*
	Willow oak	*Q. phellos*
	Emory oak	*Q. emoryi*
White oak	Arizona white oak	*Quercus arizonica*
	Blue oak	*Q. douglasii*
	Bur oak	*Q. macrocarpa*
	Chestnut oak	*Q. prinus*
	Chinkapin oak	*Q. muehlenbergii*
	Gambel oak	*Q. gambelii*
	Live oak	*Q. virginiana*
	Mexican blue oak	*Q. oblongifolia*
	Oregon white oak	*Q. garryana*
	Overcup oak	*Q. lyrata*
	Post oak	*Q. stellata*
	Swamp chestnut oak	*Q. michauxii*
	Swamp white oak	*Q. bicolor*
	Valley oak	*Q. lobata*
	White oak	*Q. alba*
Oregon myrtle	California laurel	*Umbellularia californica*
Osage-Orange	Osage-orange	*Maclura pomifera*
Pecan	Bitternut hickory	*Carya cordiformis*
	Nutmeg hickory	*C. myristiciformis*
	Water hickory	*C. aquatica*
	Pecan	*C. illinoensis*
Persimmon	Common persimmon	*Diospyros virginiana*

Commercial name for lumber	Common name of tree	Scientific (botanical) name of tree
Pine:		
Eastern white pine	Eastern white pine	*Pinus strobus*
Idaho white pine	Western white pine	*P. monticola*
Lodgepole pine	Lodgepole pine	*P. contorta*
Longleaf pine	Longleaf pine	*P. palustris*
	Slash pine	*P. elliottii*
Northern pine	Jack pine	*P. banksiana*
	Red pine	*P. resinosa*
	Pitch pine	*P. rigida*
Ponderosa pine—sugar pine	Ponderosa pine	*P. ponderosa*
	Sugar pine	*P. lambertiana*
Ponderosa pine—lodgepole pine	Ponderosa pine	*P. ponderosa*
	Lodgepole pine	*P. contorta*
Southern pine	Loblolly pine	*P. taeda*
	Longleaf pine	*P. palustris*
	Pitch pine	*P. rigida*
	Pond pine	*P. serotina*
	Sand pine	*P. clausa*
	Shortleaf pine	*P. echinata*
	Slash pine	*P. elliottii*
	Table Mountain pine	*P. pungens*
	Virginia pine	*P. virginiana*
Southern pine (minor)	Pitch pine	*P. rigida*
	Pond pine	*P. serotina*
	Virginia pine	*P. virginiana*
Popple: See Aspen		
Sassafras	Sassafras	*Sassafras albidum*
Silverbell	Carolina silverbell	*Halesia carolina*
Spruce:		
Eastern spruce	Black spruce	*Picea mariana*
	Red spruce	*P. rubens*
	White spruce	*P. glauca*
Engelmann spruce	Blue spruce	*P. pungens*
Sitka Spruce	Engelmann spruce	*P. engelmannii*
	Sitka spruce	*P. sitchensis*
Sycamore	Sycamore	*Platanus occidentalis*
Tamarack	Tamarack	*Larix laricina*
Tanoak	Tanoak	*Lithocarpus densiflorus*
Tupelo	Black tupelo	*Nyssa sylvatica*
	Ogeechee tupelo	*N. ogeche*
	Water tupelo	*N. aquatica*
Walnut	Black walnut	*Juglans nigra*
Western woods	Incense cedar	*Libocedrus decurrens*
	Western redcedar	*Thuja plicata*
	Douglas fir	*Pseudotsuga menziesii*
	California red fir	*Abies magnifica*
	Grand fir	*A. grandis*
	Noble fir	*A. procera*
	Pacific silver fir	*A. amabilis*
	Subalpine fir	*A. lasiocarpa*
	White fir	*A. concolor*
	Mountain hemlock	*Tsuga mertensiana*
	Western hemlock	*T. heterophylla*
	Western larch	*Larix occidentalis*
	Lodgepole pine	*Pinus contorta*
	Ponderosa pine	*P. ponderosa*
	Sugar pine	*P. lambertiana*
	Western white pine	*P. monticola*
	Engelmann spruce	*Picea engelmannii*
White woods	California red fir	*Abies magnifica*
	Grand fir	*A. grandis*
	Noble fir	*A. procera*
	Pacific silver fir	*A. amabilis*
	Subalpine fir	*A. lasiocarpa*
	White fir	*A. concolor*
	Mountain hemlock	*Tsuga mertensiana*
	Western hemlock	*T. heterophylla*
	Lodgepole pine	*Pinus contorta*
	Ponderosa pine	*P. ponderosa*
	Sugar pine	*P. lambertiana*
	Western white pine	*P. monticola*
	Engelmann spruce	*Picea engelmannii*
Willow	Black willow	*Salix nigra*
	Peachleaf willow	*S. amygdaloides*
Yellow poplar	Yellow-poplar	*Liriodendron tulipifera*
Yew	Pacific yew	*Taxus brevifolia*

Appendix 2: Finding the specific gravity of wood

For some time I've been trying to work out a shortcut for determining the specific gravity of wood. (It's useful in wood identification, as well as in other aspects of woodworking, such as in building decoys.) I finally came up with the following method:

1. Machine the sample to rectangular or cylindrical dimensions that will give a volume equal to one of the numbers of cubic inches given in the table. (I have also included sets of dimensions that will yield the given volumes.)
2. Place the sample in an oven at 212°F for 24 hours.
3. Weigh the sample.
4. Use the chart to determine density, as shown in the examples.

Notes:

1. If possible, weights should be to the nearest 0.1 gram or to the nearest 0.01 ounce. (The chart will show how much a deviation of weight will affect accuracy. As an example, for a sample with a volume of 6.75 cubic inches and a weight of 60 grams, a variation of 1 gram will represent only about 0.01 variation in specific gravity. For a sample with a volume of 1 cubic inch and a weight of 10 grams, however, a 1-gram variation will affect accuracy by about 10%.)

2. Remember that the specific gravity of wood machined to dimension while green, room-dry or air-dry will be based on that original volume. If specific gravity on an oven-dry basis is desired, cut the piece 10% or more oversize, oven dry, cool, and quickly machine to the desired dimensions. Then return the wood to the oven for an hour and weigh it again, to determine oven-dry weight.

Examples:

A. Assume a piece of hickory is machined to ¾ in. by ¾ in. by 6 in., for a volume of 3.375 cubic inches, and that after oven-drying it weighs 1.35 ounces. Its specific gravity will be 0.69.

B. Assume a piece of eastern white pine is cut to 1½ in. by 1½ in. by 6 in. for a volume of 13.5 cubic inches, and that after oven-drying it weighs 78.6 grams. Its specific gravity will be 0.355.

C. Suppose you have a piece of sugar maple turned to a diameter of 1 inch. To determine specific gravity, we might want about 5 cubic inches. Since $V = 5\ in.^3 = \pi r^2 h$*, then* $h = 5/\pi(\frac{1}{2})^2 = 6.37$ *in. Therefore, cut the dowel to a length of 6.37 in. Suppose we oven-dry the dowel and it now weighs 50.88 grams. The specific gravity would be 0.62.*

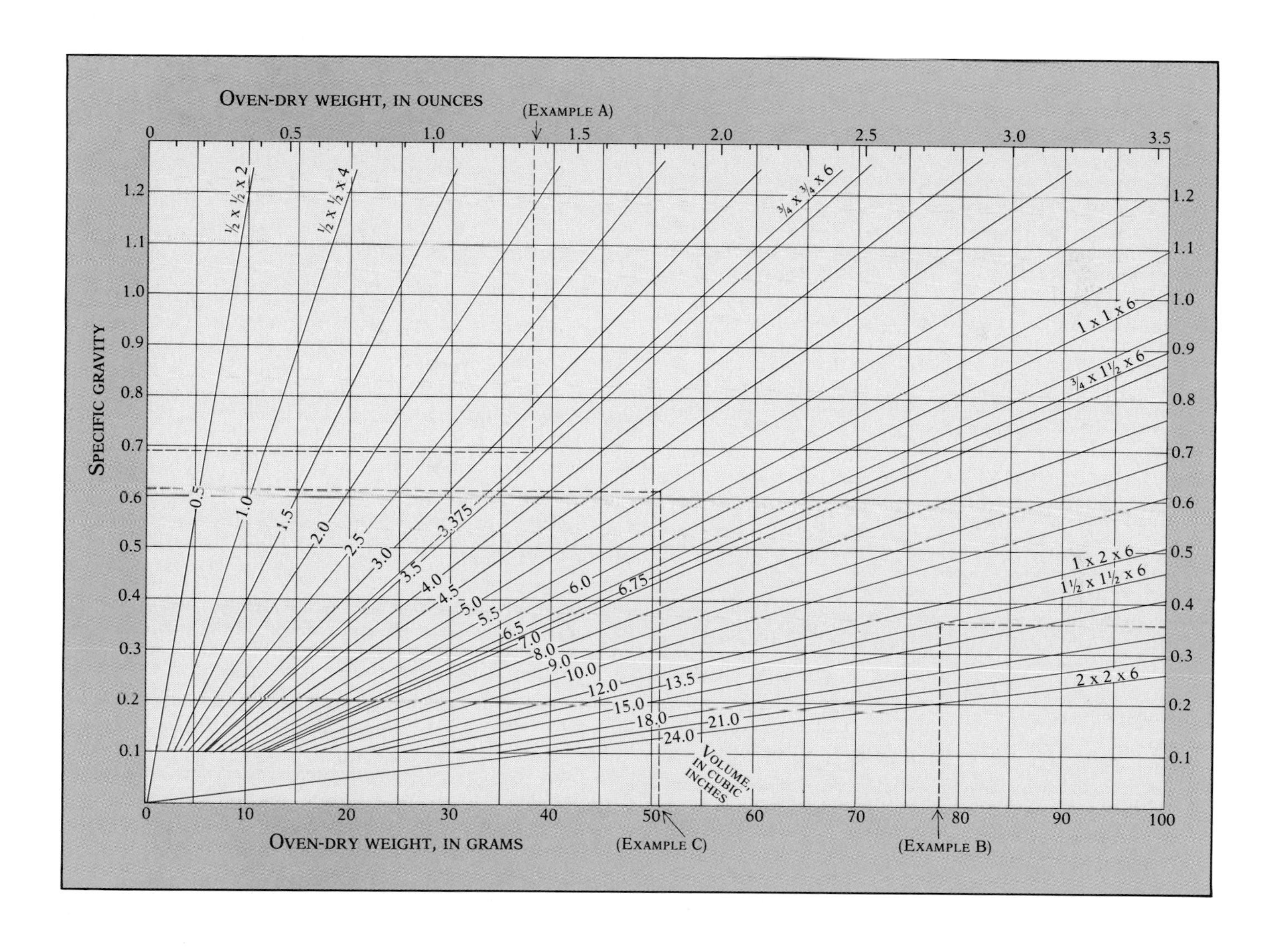

Glossary

ABSOLUTE HUMIDITY
The weight of water vapor per unit volume of air, usually expressed as grains/cu. ft. See also *relative humidity.*

ABSORPTION
The gain of free water by the cell cavities.

ACROSS THE GRAIN
Generally perpendicular to the grain direction.

ADHESIVE
Any substance such as glue, resin, cement or paste capable of holding or joining materials (adherends).

ADSORBED WATER
Same as *bound water.*

ADSORPTION
The gain of bound water by the cell wall from adjacent air.

ADULT WOOD (MATURE WOOD, OUTER WOOD, STEM-FORMED WOOD)
Wood produced after cambial cells have attained maximum dimensions. See also *juvenile wood.*

ADVANCED DECAY (TYPICAL DECAY)
An older stage of decay readily recognized as wood that has become punky, soft and spongy, stringy, ringshaked, pitted or crumbly.

AGAINST THE GRAIN
Reference to cutting direction, as in planing a board surface, such that splitting ahead of the cutter follows the grain direction downward into the wood below the projected cutting surface. Also, generally perpendicular to the grain direction, across the grain. See also *with the grain.*

AIR DRIED (AIR SEASONED)
Having reached equilibrium with outdoor atmospheric humidity. When unspecified, 12% moisture content is the assumed value. See also *kiln dried.*

ALBURNUM
Same as *sapwood.*

ALIFORM PARENCHYMA
As seen in cross section, an arrangement of parenchyma cells grouped closely around pores and forming winglike lateral extensions.

ALONG THE GRAIN
Generally parallel to the grain direction.

ANGIOSPERM
Belonging to the class of plants having seeds enclosed in an ovary. Within this class, the subclass dicotyledons includes all hardwood trees.

ANGLE OF ATTACK
Same as *cutting angle.*

ANISOTROPIC
Not having the same properties in all directions.

ANNUAL RING (ANNUAL GROWTH RING, ANNUAL INCREMENT)
See *growth ring.*

ASSEMBLY TIME
The time elapsed between spreading adhesive on surfaces to be joined and application of pressure to the joint. *Open assembly time* is from the beginning of spreading to joint closure. *Closed assembly time* is from joint closure to application of full pressure.

BACK
The poorer of the two faces of a panel of plywood or other composite.

BALANCED CONSTRUCTION
Symmetrical construction of plywood or other composites having matching layers on both sides of the central plane so that changes in moisture content will not cause warp.

BANDED PARENCHYMA
As viewed in cross section, parenchyma cells that collectively appear as thin, lighter-colored tangential lines (as occur in hickory).

BARK
The tree tissue outside of the cambium, including inner (living) bark and outer (dead) bark.

BASTARD GRAIN (BASTARD SAWN)
Lumber having growth rings that form angles of 30° to 60° with the faces.

BEAM
An elongated member usually supported horizontally and loaded perpendicular to its length.

BIRD'S EYE
Small circular or elliptical areas resembling birds' eyes on the tangential surface of wood, formed by indented fibers. Common in sugar maple and used for decorative purposes; rare in other species.

BLACK KNOT
Same as *encased knot.*

BLEMISH
A defect or anything that mars the appearance of wood.

BLUE STAIN (SAPSTAIN)
A bluish or greyish discoloration of sapwood caused by the growth of certain dark-colored fungi on the surface and inside the wood.

BOARD FOOT
A unit of lumber measurement equivalent in volume to a piece having nominal dimensions of 1 foot (length) by 12 in. (width) by 1 in. (thickness).

BOARDS
Lumber 2 in. or more wide that is nominally less than 2 in. thick. Boards less than 6 in. wide are also called strips.

BOLE
A tree stem or trunk large enough for conversion into lumber or veneer.

BOLT
A short section of a tree trunk.

BOUND WATER (ADSORBED WATER, HYGROSCOPIC WATER)
Water held hygroscopically in the cell wall; water in wood below the fiber saturation point.

BOW
A form of warp; deviation from lengthwise flatness in a board.

BRASHNESS
: Brittleness in wood, characterized by abrupt failure rather than splintering. Causes include reaction wood, juvenile wood, compression failure, high temperature and extremes of growth rate.

BROKEN STRIPE
: Ribbon figure in which the stripe effect is intermittent.

BROWN ROT
: Decay caused by a type of fungus that attacks cellulose rather than lignin, leaving a brownish residue.

BULK PILING
: Same as *solid piling.*

BURL (BURR)
: A hard, woody outgrowth on a tree, more or less rounded in form, usually resulting from the entwined growth of a cluster of adventitious buds. Burls are a source of highly figured veneers.

CAMBIUM
: The thin layer of living, meristematic (reproductive) cells between bark and wood which, by cell division, forms new bark and wood cells.

CASEHARDENING
: A condition in dry lumber wherein residual drying stresses leave the outer layers under compression but the inner core in tension.

CAVITY (CELL CAVITY, LUMEN)
: The void space of a cell enclosed by the cell wall.

CELL (ELEMENT)
: The basic structural unit of wood (and other plant) tissue consisting of an outer cell wall surrounding a central cavity or lumen. Wood cell types include tracheids, vessel elements, fibers, longitudinal parenchyma and ray cells.

CELLULOSE
: A polymer chain carbohydrate, $(C_6H_{10}O_5)_n$, the major constituent of wood cell walls.

CHECKS
: Separations of wood cells along the grain as a result of uneven shrinkage, most common on end-grain surfaces of lumber.

CHIP ANGLE
: Same as *cutting angle.*

CHIPPED GRAIN (TORN GRAIN)
: A machining defect in which small chips are torn from the surface below the intended plane of cut, usually as the result of cutting against the grain.

CLEAR
: In reference to lumber, free of defects or blemishes.

CLEARANCE ANGLE
: The angle between the back of the knife and the path of the cutting edge. It is usually designated by the Greek letter γ.

CLOSE GRAIN (FINE GRAIN, DENSE GRAIN, NARROW GRAIN)
: Slowly grown wood having narrow, usually inconspicuous growth rings in contrast to coarse grain or open grain.

CLOSED ASSEMBLY TIME
: See *assembly time.*

CLOSED FACE
: Same as *closed side.*

CLOSED GRAIN
: See *fine texture.*

CLOSED SIDE (CLOSED FACE, TIGHT SIDE)
: The veneer surface not touching the veneer knife during peeling or slicing, which is free of knife checks.

COARSE GRAIN
: Descriptive of wood having wide and conspicuous growth rings in contrast to close grain. Sometimes used synonymously with "coarse texture" to designate woods with relatively large cell size.

COARSE TEXTURE
: Descriptive of wood having relatively large pores, especially in reference to finishing. Preferred to "open grain."

COLLAPSE
: A shriveled or irregular appearance of wood due to flattened or caved-in cell structure, usually caused by capillary tension during early stages of drying wet wood.

COMB GRAIN
: See *rift grain.*

COMPRESSION FAILURES
: Irregular planes of buckled cells, caused by excessive compressive stress parallel to the grain, appearing as fine cross-wrinkles on longitudinal planed surfaces.

COMPRESSION WOOD
: See *reaction wood.*

CONDITIONING
: Exposure under controlled relative humidity to bring wood to a desired moisture content. Also, the final stage of a kiln schedule designed to relieve residual casehardening stress.

CONFLUENT PARENCHYMA
: As seen in a cross section, parenchyma cells so grouped as to form a more or less tangential band connecting two or more pores.

CONIFEROUS WOOD
: Same as *softwood.*

CORD
: A unit of measure for round-wood, such as firewood and pulpwood. The standard cord is a pile of 4-ft. long logs, 4 ft. high, 8 ft. across.

CORE
: The inner portion of a board, equivalent to half its thickness. See also *shell.* In plywood, the center ply, or collectively, all layers between the face and back plies.

CORE WOOD
: Same as *juvenile wood.*

CRACK
: A large radial check resulting from greater tangential than radial shrinkage.

CROOK
: A form of warp; deviation from end-to-end straightness along the edge of a board.

CROSSBANDS
In plywood with more than three plies, those veneers immediately beneath the faces, having grain direction perpendicular to that of the faces. See also *straight bands.*

CROSS BREAKS
Transverse planes of failure in tension parallel to the grain, caused by localized abnormal longitudinal shrinkage (as in reaction wood) restrained by adjacent normal wood.

CROSS GRAIN
Deviation of grain direction from the longitudinal axis of a piece of wood or from the stem axis in a tree. Pronounced deviation from the surface, especially in veneer, is termed *short grain.* See also *spiral grain, diagonal grain.*

CROSS SECTION (TRANSVERSE SECTION)
A section cut perpendicular to the grain, or the surface exposed by such a cut.

CROTCH GRAIN
Figure produced by cutting centrally through a tree crotch in the common plane of both branches.

CROWN-FORMED WOOD
Same as *juvenile wood.*

CUP
A form of warp; deviation from flatness across the width of a board.

CURING
The setting of an adhesive by chemical reaction. Also, the drying of wood, though this is not the preferred usage.

CURLY FIGURE (FIDDLEBACK GRAIN, TIGER GRAIN)
The figure produced on surfaces, particularly radial, of wood having wavy grain.

CURLY GRAIN
Same as *wavy grain.* Also, sometimes the distorted grain around bird's eyes is called curly-grained wood.

CUTTING ANGLE (ANGLE OF ATTACK; CHIP, HOOK OR RAKE ANGLE)
The angle between the face of a cutting edge and a plane perpendicular to its cutting direction, usually designated by the Greek letter α.

DECAY (ROT)
The decomposition of wood by fungi.

DECIDUOUS TREES
Trees whose leaves normally drop after the yearly period of growth is over.

DEFECTS
Irregularities or abnormalities in wood that lower its strength, grade, value or utility.

DELAMINATION
The separation of layers in laminated wood or plywood caused by failure of the adhesive itself or of the interface between adhesive and adherend.

DENSE GRAIN
Same as *close grain.*

DENSITY
The weight of a body or substance per unit volume.

DESORPTION
The loss of bound (adsorbed) water from the cell wall.

DEWPOINT
The temperature at which atmospheric water vapor condenses out as a liquid.

DIAGONAL GRAIN
Cross grain exhibiting deviation of the growth-ring plane from the longitudinal axis, commonly the result of sawing boards other than parallel to the bark of the log.

DIAMONDING
A form of warp resulting from greater tangential than radial shrinkage, which causes square sections to become diamond-shaped or round sections to become oval.

DICOTS (DICOTYLEDONS)
A class of plants (within the angiosperms) characterized by having two cotyledons or seed leaves. All hardwood tree species fall within the dicots.

DIFFUSE-POROUS WOOD
A hardwood in which the pores are of approximately uniform size and are distributed evenly throughout each growth ring. See also *ring-porous wood.*

DIMENSIONAL STABILIZATION
Treatment of wood to minimize shrinkage and swelling.

DIMPLES
Numerous small depressions in growth rings, especially obvious on split tangential surfaces. Occasionally occurs among certain conifers, notably ponderosa and lodgepole pines.

DIP GRAINED
Having a single wave or undulation of fiber direction, such as occurs in wood along either side of a knot.

DRESSED SIZE
The dimensions of lumber after being surfaced with a planing machine.

DRY-BULB TEMPERATURE
The temperature of the air as indicated by a standard thermometer. See also *wet-bulb temperature.*

DRY KILN
Same as *kiln.*

DRYING DEFECTS
Irregularities resulting from drying that may lower the strength, durability or utility of wood, such as checks or casehardening.

DRY ROT
A term loosely applied to any dry, crumbly rot, but especially to that which, when in an advanced state, permits the wood to be crushed easily to a dry powder. The term is actually a misnomer for any decay, since all fungi require considerable moisture for growth.

DURABILITY
A general term referring to resistance to deterioration; frequently refers specifically to decay resistance of wood, but also to resistance of adhesive bonds and finishes to deterioration.

DURAMEN
Same as *heartwood.*

EARLYWOOD (SPRINGWOOD)
The first-formed portion of the growth ring, often characterized by larger cells and lower density.

EDGE GRAIN (QUARTERSAWN, VERTICAL GRAIN, RADIAL CUT)
Referring to pieces in which the growth rings form an angle of 45° or more (ideally 90°) with the wood surface or lumber face; approaching or coinciding with a radial surface. See also *rift grain.*

ELASTICITY
A property of a material that causes it to return to its original dimensions after being deformed by loading.

ELECTRODES
Components of moisture meters that contact or penetrate the wood when measuring moisture content.

ELEMENT
Same as *cell.*

ENCASED KNOT (LOOSE KNOT, BLACK KNOT)
The dead portion of a branch embedded in the stem by subsequent growth of the tree.

END GRAIN
A cross-sectional surface or the appearance of such a surface.

EQUILIBRIUM MOISTURE CONTENT (EMC)
The moisture content eventually attained in wood exposed to a given level of relative humidity and temperature.

EVEN GRAIN
Wood having uniform or nearly uniform structure throughout the growth ring and little or no earlywood/latewood distinction, as in basswood.

EXTRACTIVES
Substances deposited in wood in the transition from sapwood to heartwood, often imparting significant color and decay resistance.

FACE
Either side or surface of a plywood panel. Also, the surface of plywood having the higher quality, in which case the opposite side is called the *back.*

FACE GRAIN
The figure or pattern on the face side of a plywood panel or board.

FEATHER CROTCH (FEATHER GRAIN)
The figure produced by a longitudinal section through a tree crotch, characterized by a featherlike appearance.

FIBER
A specific hardwood cell type, typically elongated with pointed ends, having thick walls and contributing notably to the strength of wood. Also, in the plural, used as a general term for separated wood cells collectively, as in papermaking.

FIBER SATURATION POINT (FSP)
The condition of moisture content where cell walls are fully saturated but cell cavities are empty of free water.

FIDDLEBACK
Same as *curly figure.*

FIGURE
Any distinctive appearance on a longitudinal wood surface resulting from anatomical structure, irregular coloration or defects.

FINE GRAIN
See *close grain.* Also, sometimes used synonymously with "fine texture" to designate woods with relatively small cells.

FINE TEXTURE
Descriptive of hardwoods having small and closely spaced pores, or softwoods with small-diameter tracheids. Preferred to "closed grain" in reference to finishing.

FLAME GRAIN
Applied to figure produced on flatsawn boards or rotary-cut veneer.

FLATSAWN (FLAT GRAINED, PLAINSAWN, SLASH GRAINED, SIDE GRAINED, TANGENTIAL CUT)
Indicating wood machined along an approximately tangential plane, such that growth rings intersect the surface at an angle of less than 45°.

FLECKS
See *pith flecks, ray flecks.*

FLITCH
A portion of a log sawn on two or more faces, commonly on opposite faces, leaving two waney edges. When intended for resawing into lumber, it is resawn parallel to its original wide faces. Or, it may be sliced or sawn into veneer, in which case the resulting sheets of veneer laid together in the sequence of cutting are called a flitch.

FLUORESCENCE
The absorption of invisible ultraviolet (black) light by a material that transforms the energy and emits it as visible light of a particular color.

FREE WATER
Moisture held in the cell cavities of the wood, not bound in the cell wall.

FUNGI
Simple forms of microscopic plants, whose parasitic development in wood may cause mold, stain or decay.

FUZZY GRAIN
Same as *woolly grain.*

GRADE
A designation of the quality of a log, or of a wood product such as lumber, veneer or panels.

GRAIN
A confusingly versatile term whose specific meaning must be made apparent by context or by associated adjectives. Among its uses are direction of cells (e.g., *along the grain, spiral grain*), surface appearance or figure (e.g., *ribbon grain*), growth-ring placement (e.g., *vertical grain*), plane of cut (e.g., *end grain*), growth rate (e.g., *narrow grain*), early/latewood contrast (e.g., *uneven grain*), relative cell size (e.g., *open grain,* meaning coarse-textured), machining defects (e.g., *chipped grain*) or artificial decorative effects (e.g., *graining*). See specific compound terms.

GRAIN DIRECTION
The direction of the long axes of the dominant longitudinal cells or fibers in a piece of wood.

GRAINING
Painting or otherwise imitating the figure of wood on a surface.

GRAIN SLOPE
Same as *slope of grain.*

GREEN
Freshly cut and unseasoned. Also, having moisture content above fiber saturation point.

GROSS FEATURES
Physical features of wood that can be perceived with the unaided eye, sometimes including macroscopic features.

GROWTH RING (GROWTH LAYER, GROWTH INCREMENT)
The layer of wood (or bark) added to the stem in a given growth period; in the temperate zones, one layer is added per yearly growth period and is often termed annual ring.

GYMNOSPERM
The class of plants having naked seeds (not enclosed in an ovary). Within this group are all trees yielding softwood lumber.

HAND LENS
A magnifying lens, hand-held, used to examine wood; 10x is the standard magnification used.

HARDWOOD (POROUS WOOD)
Woods produced by broad-leaved trees in the botanical group referred to as angiosperms. Since these woods have vessels, they are also termed porous woods, because a vessel in cross section is termed a *pore*. (The term hardwood is not a designation of actual hardness in wood.)

HEART SHAKE (HEART CHECK, RIFT CRACK)
Radial crack in the vicinity of the pith.

HEARTWOOD (DURAMEN)
The central core of wood in mature stems. At one time heartwood was sapwood but it no longer conducts sap or has living cells. In most species, extractives impart a darker color to heartwood.

HONEYCOMBING
Checks that occur in the interior of a piece of wood, usually in the plane of the rays, as a result of casehardening stresses developed in drying.

HOOK ANGLE
Same as *cutting angle*.

HOOKE'S LAW
A law that states for elastic materials, strain is proportional to stress within the elastic range. See *modulus of elasticity*.

HYDROMETER
An elongated, weighted glass instrument that measures the specific gravity of liquid by the depth to which it sinks when floated in the liquid. Used by woodworkers to determine the strength of PEG solutions.

HYGROMETER
An instrument that measures the relative humidity of the atmosphere. See also *psychrometer*.

HYGROSCOPICITY
The ability of a substance to adsorb and desorb water.

HYGROSCOPIC WATER
Same as *bound water*.

HYPHAE
The microscopic filaments of a fungus, which digest and adsorb material from its host. See also *incipient decay, mycelium*.

INCIPIENT DECAY
An early stage of decay in which hyphae have invaded the cell structure, sometimes discoloring the wood, but have not perceptibly reduced the hardness of the wood.

INCLUDED SAPWOOD
Areas of light-colored wood, apparently sapwood, found within the portion of stem that has become heartwood.

INCREMENT
The growth added in a given period. See also *growth ring*.

INNER BARK
See *bark*.

INTERGROWN KNOT (TIGHT KNOT, RED KNOT)
A portion of a branch that was alive when intergrown with the surrounding stem.

INTERLOCKED GRAIN
Repeated alternation of left and right-hand spiral grain, each reversal usually distributed over several growth rings.

ISOTROPIC
Having equal properties in all structural directions.

JOIST
One of a series of parallel beams used to support floor and ceiling loads, usually installed with its wide dimension vertical. See also *plank*.

JUVENILE WOOD (CORE WOOD, CROWN-FORMED WOOD, PITH WOOD)
Wood formed near the pith of the tree, often characterized by wide growth rings of lower density and abnormal properties. See also *adult wood*.

KILN (DRY KILN)
A heated chamber for drying lumber, veneer and other wood products, in which temperature, humidity and air circulation are controlled.

KILN DRIED (HOT-AIR DRIED)
Having been dried in a kiln to a specified moisture content; for cabinet woods, usually implies dryness below that attainable by air drying. See also *air dried*.

KILN SCHEDULE
A sequence of dry-bulb and wet-bulb temperatures specified for successive stages of drying lumber in a kiln, designed to attain desired dryness without defects.

KINK
A form of warp characterized by abrupt deviation from straightness or flatness due to either localized grain distortion (as around knots) or to deformation by misplaced stickers.

KNIFE CHECKS (LATHE CHECKS)
Parallel-to-grain failures developed cyclically in one side of knife-cut veneer during its manufacture; depth of checks varies with species and cutting conditions. Also called lathe checks, especially in peeled veneer. See also *closed side, open side*.

KNOT
A portion of a branch overgrown by the expanding girth of the bole or a larger branch. See also *intergrown knot, encased knot, spike knot, pin knot*.

LATEWOOD (SUMMERWOOD)
The portion of the growth ring formed after earlywood, often characterized by smaller cells or higher density.

LATHE CHECKS
Same as *knife checks*.

LEAF GRAIN
Another term for flat-grain figure.

LIGNIN
A complex chemical substance making up approximately 25% of wood substance, interspersed with cellulose in forming the cell wall. Lignin stiffens the cell and functions as a bonding agent between cells.

LONGITUDINAL
Parallel to the stem axis of the tree or branches, therefore describing the axial direction of the dominant cell structure; along the grain.

LONGITUDINAL GRAIN
Any plane cut parallel to the grain direction of wood. It may be radial, tangential or an intermediate plane.

LOOSE KNOT
Same as *encased knot.*

LOOSENED GRAIN (SHELLED GRAIN)
Separation of latewood layers from a planed surface, usually accompanying pronounced raised grain and typically occurring on the pith side of flatsawn boards.

LOOSE SIDE
Same as *open side.*

LUMBER
Pieces of wood no further manufactured than by sawing, planing, crosscutting to length and perhaps edge-matching.

LUMEN (CELL LUMEN)
Same as *cavity.*

MACROSCOPIC
Referring to features visible with low-power magnification (e.g., a 10x hand lens), as distinguished from microscopic features.

MATURE WOOD
Same as *adult wood.*

MEDULLA
Same as *pith.*

MEDULLARY RAYS
Rays connected with the pith. Often used (loosely) to refer to all rays.

MERISTEM
Reproductive tissue. Apical meristems are located in twig tips and produce elongation. The cambium is a lateral meristem producing girth.

MICROBEVEL
An extremely narrow bevel along a cutting edge, which increases the sharpness angle for greater edge durability.

MICRON
1/1000 of a millimeter.

MIL
1/1000 of an inch.

MIXED GRAIN
Referring to a quantity of lumber containing both edge-grain and flat-grain pieces.

MODULUS OF ELASTICITY
The ratio of stress to strain, within the elastic range of a material.

MODULUS OF RUPTURE
In reference to wood, the stress in bending sustained at failure.

MOISTURE CONTENT
The weight of water in the cell walls and cavities of wood, expressed as a percentage of oven-dry weight.

MOISTURE GRADIENT
The variation of moisture content in wood, such as the gradation from wetter core to drier surface in a drying board.

MOISTURE METER
An instrument used for the rapid determination of moisture content in wood by electrical means.

MOLD
A fungal growth on wood taking place at or near the surface, usually greenish to black in color.

MONOCOTS (MONOCOTYLEDONS)
A class of plants (within the angiosperms) characterized by a single cotyledon or seed leaf. See also *dicots.*

MOTTLED FIGURE
A type of broken stripe figure having irregular interruptions of curly figure.

MYCELIUM
The mass of hyphae (microscopic elements) of a fungus, often visible as a cottony mat or layer on the surface of wood with advanced decay.

NARROW GRAIN
Same as *close grain.*

NEEDLE-POINT GRAIN
Same as *rift grain.*

NOMINAL DIMENSION
The dimension by which lumber is known and sold in the market (the actual dimension after drying and dressing may be somewhat less).

NONPOROUS WOODS
See *softwood.*

OLD GROWTH (VIRGIN TIMBER)
Trees in a mature, naturally established forest, whose timber is characterized by large size, straight boles and freedom from knots.

OPEN ASSEMBLY TIME
See *assembly time.*

OPEN FACE
Same as *open side.*

OPEN GRAIN
See *coarse texture.* Also, descriptive of wood having widely spaced growth rings, in contrast to *close* or *dense grain.*

OPEN PILING
Stacking wood products in layers separated by stickers to permit air circulation.

OPEN SIDE (LOOSE SIDE, OPEN FACE)
The surface of veneer against the knife during peeling or slicing; may contain knife checks.

OUTER BARK
See *bark.*

OUTER WOOD
Same as *adult wood.*

OVEN-DRY WOOD
Wood dried to constant weight in an oven maintained at temperatures of 101°C to 105°C (214°F to 221°F).

OVERLAY
Sheet materials other than veneer (plastics, paper, metal) glued to the surface of wood panels.

PARENCHYMA
Thin-walled wood cells (living when part of sapwood) involved mainly with food storage and distribution. With a hand lens, groupings of parenchyma may appear as light-colored areas on cross sections. See also *prosenchyma.*

PECK (PECKINESS, PECKY DRY ROT)
Advanced decay in living trees that occurs in the form of elongated pockets of rot; most familiar in bald cypress and incense cedar.

PEELED VENEER
Same as *rotary-cut veneer.*

PHLOEM
The tissue of the inner bark, which conducts food in the tree. Also used loosely in reference to bark in general.

PIGMENT FIGURE
Figure in wood resulting from irregular deposits of colored extractives.

PINHOLES
Small round holes in wood caused by insects.

PIN KNOT
A knot less than ¼ in. in diameter.

PITCH
Material formed in the resin canals of softwood; also, accumulation of resin, as in pockets, streaks or seams, or as is exuded from wounds.

PITCH POCKET
A flattened round or oval tangential separation in the wood of conifers, which contains (or did contain) solid or liquid resin (pitch).

PITH (MEDULLA)
The small core of soft, spongy tissue located at the center of tree stems, branches and twigs.

PITH FLECKS
Longitudinal streaks of wound tissue caused by the vertical tunneling of fly larvae belonging to the genus *Agromyza*.

PITH WOOD
Same as *juvenile wood.*

PITS
Recesses or unthickened portions of the secondary cell walls through which fluids pass from cell to cell.

PLAINSAWN (PLAIN GRAIN)
Same as *flatsawn.*

PLANK
A piece of structural lumber installed with its wide dimension horizontal, usually intended as a bearing surface. See also *joist.*

PLYWOOD
A composite board of veneers glued together with the grain directions of adjacent layers mutually perpendicular.

POCKET ROT
A localized, sharply delineated volume or pocket of advanced decay surrounded by apparently sound wood.

PORE
The cross section of a hardwood vessel.

PORE MULTIPLE
Two or more pores arranged radially and in close contact.

POROUS WOOD
Same as *hardwood.*

POT LIFE
Same as *working life.*

POWDER-POST BEETLES
Small beetles, especially of the genus *Lyctus,* that attack mainly sapwood of large-pored hardwoods, reducing the tunneled wood to fine powder.

PRESERVATIVE
Any substance used to treat wood for protection against fungi, insects or marine borers.

PROSENCHYMA
Nonliving wood cells that function in conduction and support. Includes tracheids, vessels and fibers and accounts for most of the volume of wood structure. See also *parenchyma.*

PSYCHROMETER
A type of hygrometer for measuring atmospheric humidity by dry-bulb and wet-bulb thermometers.

QUARTERSAWN (QUARTERED, QUARTER GRAIN)
Same as *edge grain.*

QUILTED FIGURE
Figure sometimes found in bigleaf maple, characterized by crowded bulges in the grain direction.

R
Symbol for radial section or surface.

RADIAL
The horizontal direction in a tree between pith and bark. A radial section is along a plane that would pass lengthwise through the pith.

RADIAL CUT (RADIAL GRAIN)
Same as *edge grain.*

RAISED GRAIN
A condition developed in planing causing the elevation of latewood above earlywood, without separation, typically on the pith side of flatsawn boards. Also, the severed cells caused to rise above a surface by intentional wetting, as done in preparation for final sanding.

RAKE ANGLE
Same as *cutting angle.*

RATE OF GROWTH
The relative rate of increase in tree girth, usually expressed as rings per inch.

RAYS
Flattened bands of tissue composed of ray cells, extending horizontally in a radial plane through the tree stem. See also *medullary ray.*

RAY FLECKS
The conspicuous appearance of rays on an edge-grain surface.

REACTION WOOD
Abnormal wood formed in leaning stems and branches in trees. In softwood trees, it forms on the lower side of the stem and is called *compression wood*; it is denser but more brittle, and has greater than normal longitudinal shrinkage. In hardwood trees, reaction wood forms typically on the upper side of the stem and is termed *tension wood,* characterized by woolly surfaces when machined and greater than normal longitudinal shrinkage.

RED KNOT
Same as *intergrown knot.*

RELATIVE HUMIDITY
The ratio of the amount of water vapor present in the air to that which the air would hold at saturation at the same temperature. Usually expressed as a percent. See also *absolute humidity.*

RESIN
Material secreted into resin canals of softwood trees. Also, a term applied to certain synthetic organic products (as used in glues and finishes) similar to natural resins. See also *pitch.*

RESIN CANAL
Tubular passageways containing resin in the wood of certain softwood trees.

RIBBON FIGURE (RIBBON GRAIN, STRIPE FIGURE)
Figure apparent on an edge-grain surface of wood with interlocked grain, characterized by vertical bands of varying luster and vessel markings. See also *broken stripe*.

RIFT CRACK
See *heart shake*.

RIFT GRAIN (COMB GRAIN, NEEDLE-POINT GRAIN)
The surface or figure produced by a longitudinal plane of cut which is at approximately 45° to both rays and growth rings. The term is used especially for white oak with its large rays. The term comb grain is used where the vessel lines are parallel to the board edge and the rays produce a uniform pencil stripe.

RING-POROUS WOOD
Hardwood having relatively large pores concentrated in earlywood and distinctly smaller pores in latewood. See also *diffuse-porous wood*.

RING SHAKE (RING FAILURE, SHELL SHAKE)
A separation of wood structure parallel to the growth rings, often in the first layer(s) of earlywood, usually occurring in the standing tree.

RIPPLE MARKS
Fine striations perpendicular to the grain, most apparent on a tangential surface, produced in wood with storied rays.

ROE FIGURE (ROEY GRAIN)
The appearance of a radial surface when stripes less than 1 ft. long are formed by irregular interlocked grain.

ROT
Same as *decay*.

ROTARY-CUT VENEER (PEELED VENEER)
Veneer cut on a lathe by rotating a log against a fixed knife, which produces a continuous veneer sheet.

ROUND-EDGE LUMBER
Lumber having bark along both edges.

ROUND KNOT
The round or oval exposed section of a knot cut more or less crosswise to the limb axis. See also *spike knot*.

SAP
The water in a tree, including any dissolved nutrients and extractives.

SAPSTAIN
Same as *blue stain*.

SAPWOOD (ALBURNUM)
The physiologically active wood comprising one to many outermost growth rings, usually lighter in color than heartwood.

SEASONING (CURING)
The process of drying wood.

SECOND GROWTH
Timber that has grown in an area following harvest or destruction of previous timber.

SELECT
In softwood lumber, the highest appearance grades are Select grades, usually separated as B and better, C, and D Select grades. In hardwood factory lumber, Selects is one specific grade, placing in quality below Firsts and Seconds, but higher than Common grades.

SEMI-DIFFUSE POROUS WOOD (SEMI-RING-POROUS WOOD)
Hardwood having fairly evenly distributed pores of gradually decreasing size from earlywood to latewood portions of the growth ring.

SHAKE
See *ring shake, heart shake*. Also, a hand-split shingle.

SHARPNESS ANGLE
The angle between the face and back of a cutting edge, usually designated by the Greek letter β.

SHEAR
A condition of stress (and resulting strain) acting to cause portions of an object to move or slide in parallel but opposite direction from one another.

SHELL
The outer portion of a board, equivalent to one-quarter the thickness. See also *core*.

SHELLED GRAIN
Same as *loosened grain*.

SHELL SHAKE
Same as *ring shake*.

SHORT GRAIN
See *cross grain*.

SHORT IN THE GRAIN
Term sometimes used to describe brittle fractures in wood.

SHRINKAGE
Change in dimension due to loss of moisture below the fiber saturation point, expressed numerically as a percentage of green dimension.

SIDE GRAIN (SIDE CUT)
Same as *flatsawn*. Also, any longitudinal surface, as opposed to end grain.

SILVER GRAIN
Figure produced by showy or lustrous ray fleck on a quartered surface.

SLAB
A broad, flat, thick piece of wood.

SLASH GRAIN (SLASH SAWN)
Same as *flatsawn*.

SLICED VENEER
Veneer produced by moving a log or flitch vertically against a fixed veneer knife.

SLOPE OF GRAIN (GRAIN SLOPE)
A measurement of cross grain taken as the amount of grain deviation across a board in a measured distance along its length, expressed as a ratio such as 1 in. in 12 in., or 1 in 12, or simply 1:12.

SOFTWOOD (CONIFEROUS WOOD)
Wood produced by coniferous trees in the botanical group referred to as gymnosperms. Since these woods lack vessels, they are sometimes referred to as nonporous woods. (The term softwood does not necessarily refer to the actual hardness of the wood.)

SOLID PILING (BULK PILING)
Close stacking of lumber or other products, without separation of layers with stickers, as in *open piling*.

SOLITARY PORE
Pores that do not touch other pores but are surrounded completely by other types of cells as seen in cross section.

SOUND KNOT
A knot that is solid throughout and shows no sign of decay.

SOUND WOOD
Wood having no decay.

SPALTED WOOD
Partially decayed wood characterized by irregular discolorations appearing as dark *zone lines* on the surface.

SPECIFIC GRAVITY
The ratio of the weight of a body to the weight of an equal volume of water; relative density.

SPIKE KNOT
A knot cut more or less parallel to its long axis so that the exposed section is definitely elongated. See also *round knot.*

SPIRAL GRAIN
Cross grain indicated by grain deviation from the edge of a tangential surface, resulting naturally from helical grain direction in a tree or artificially by misaligned sawing.

SPRINGWOOD
Same as *earlywood.*

STAIN
A discoloration in wood caused by stain fungi, metals or chemicals. Also, a finishing material used intentionally to change the color of wood.

STAR SHAKE (STAR CHECK)
Multiple heart shake, having a more or less star effect.

STARVED JOINT
A poorly bonded glue joint, due to insufficient glue.

STEEP GRAIN
Rather severe cross grain.

STEM-FORMED WOOD
Same as *adult wood.*

STICKERS
Wooden strips used to separate layers of lumber to permit air circulation.

STORIED RAYS
Rays whose arrangement (as viewed on a tangential section) is in horizontal rows. See also *ripple marks.*

STRAIGHT BANDS
Internal plies of plywood, other than the core ply, having grain direction parallel to that of the face plies.

STRAIGHT GRAIN
Indicating grain direction parallel to the axis or edges of a piece.

STRAIN
Unit deformation resulting from applied stress.

STRENGTH
The ability of wood to resist applied load.

STRESS
The force or load per unit area resulting from external loads as in a structure, or internal conditions as in drying.

STRIPE FIGURE (STRIPE GRAIN)
Same as *ribbon figure.*

SUMMERWOOD
Same as *latewood.*

SUNKEN JOINT
A depression at a glue joint resulting from surfacing edge-glued material too soon after gluing.

SURFACE CHECKS
Checks that develop on a side-grain surface and penetrate the interior to some extent.

SWELLING
Increase in the dimensions of wood due to increase in moisture content.

T
Symbol for tangential section or surface.

TANGENTIAL
Describing surfaces and sections of wood perpendicular to the rays and more or less parallel to the growth ring.

TANGENTIAL CUT
Same as *flatsawn.*

TENSION WOOD
See *reaction wood.*

TERMINAL PARENCHYMA
Parenchyma cells located at the end of the growth ring, sometimes forming a conspicuous light line delineating the growth ring, as in yellow poplar.

TEXTURE
Relative cell size indicated by adjectives from fine to coarse; in softwoods, determined by relative tracheid diameter; in hardwoods, determined by relative pore diameter. Also, sometimes used to indicate evenness of grain, e.g., "uniform texture" and uniformity in size and distribution of pores, e.g., even texture. See also *coarse texture, fine texture.*

THERMAL CONDUCTIVITY
The transfer of heat through a material by conduction; the K factor indicates the relative rate of thermal conductivity. The lower the K factor, the better the insulating properties and the poorer the conducting properties.

THERMAL EXPANSION
The increase in dimension of a material in response to increase in temperature.

TIGER GRAIN
Same as *curly figure.*

TIGHT KNOT
Same as *intergrown knot.*

TIGHT SIDE
Same as *closed side.*

TIMBER
Wood in standing trees having potential for lumber.

TISSUE
A group or mass of cells having similar function or a common origin.

TORN GRAIN
Same as *chipped grain.* Also, in plywood or veneer, a growth ring separation.

TRACHEIDS
Elongated conductive cells comprising over 90% of softwood tissue; also found in some hardwoods.

TRANSVERSE SECTION (TRANSVERSE SURFACE)
Same as *cross section.*

TRUNK
The main stem of a tree. See also *bole.*

TWIST
A form of warp in which the four corners of a flat face are no longer in the same plane.

TYLOSES
Bubblelike structures that form in the vessels of certain hardwoods, usually in conjunction with heartwood formation.

TYLOSOIDS
Bubblelike structures that form in the resin canals of certain softwoods.

TYPICAL DECAY
Same as *advanced decay.*

UNEVEN GRAIN
Wood with growth rings exhibiting pronounced difference in appearance between earlywood and latewood, as in southern yellow pine or white ash.

VENEER
Wood cut by slicing, peeling or sawing into sheets ¼ in. or less in thickness.

VERTICAL GRAIN
Same as *edge grain.*

VESSEL
A conductive tube in hardwoods formed by end-to-end arrangement of cells whose end walls are open. The cross section of a vessel is called a pore.

VESSEL LINES
In hardwoods with fairly large diameter vessels, the visible lines produced on longitudinal surfaces wherever the plane of cut opens vessels lengthwise.

VIRGIN TIMBER (VIRGIN GROWTH)
Same as *old growth.*

WANE
Bark, or lack of wood from any cause, on the edge or corner of a piece of lumber.

WARP
Distortion of the intended shape of a piece of wood. See also *bow, crook, cup, diamonding, kink* and *twist.*

WAVY BANDS OF PORES
Pores arranged in undulating bands approximately parallel to the growth rings, as in the latewood of elm and hackberry.

WAVY GRAIN (CURLY GRAIN)
Undulations of the grain direction creating horizontal corrugations on a radially (and sometimes tangentially) split surface.

WEATHERING
Discoloration and disintegration of wood surfaces due to environmental influences such as wind and dust, abrasion, light and variations in precipitation and humidity.

WET-BULB DEPRESSION
The difference between dry-bulb and wet-bulb temperatures.

WET-BULB TEMPERATURE
The temperature as measured by a thermometer whose bulb is covered by water-saturated cloth, the evaporation from which lowers the temperature in relation to the relative humidity of the air.

WHITE ROT
Decay caused by a type of fungus that leaves a whitish spongy or stringy residue. See also *brown rot.*

WHORL
In coniferous trees, a group of branches that occurs at regular intervals or nodes along the main stem.

WITH THE GRAIN
Reference to cutting direction, as in planing a board surface, such that splitting ahead of the cutter follows the grain direction upward and out of the projected surface. See also *against the grain.*

WOOD (XYLEM)
The cellular tissue of the tree (exclusive of pith) inside the cambium.

WOOD SUBSTANCE
The solid material of which wood is composed, principally cellulose and lignin, exclusive of extractives and sap.

WOOLLY GRAIN (WOOLLINESS, FUZZY GRAIN)
Wood surfaces having wood fibers frayed loose, rather than severed cleanly, at the surface; commonly encountered in machining tension wood.

WORKING LIFE (POT LIFE)
The period of time after mixing during which an adhesive remains usable.

X
Symbol for cross section or cross-sectional surface.

XYLEM
Same as *wood.*

ZONE LINES
See *spalted wood.*

Bibliography

Butler, Robert L. 1974. *Wood For Wood Carvers and Craftsmen,* 122 pp. Cranbury, N.J., and London: A. S. Barnes and Company. A craftsman's introduction to do-it-yourself operations from tree to finished product. The author discusses wood and where to find it, explains log defects and gives detailed instructions on felling trees, cutting, drying and storing flitches. Suggestions for carving are outlined.

Coleman, Donald G. 1966. *Woodworking Factbook,* 240 pp. New York: Robert Speller & Sons. Reviews anatomical and visual features of wood and summarizes identification characteristics of various domestic and imported woods. Quantitative data in simplified tables accompany discussions of properties, selection and suitability of woods for woodworking, painting, finishing, gluing and carving.

Constantine, Albert, Jr. 1975. *Know Your Woods,* 384 pp. New York: Charles Scribner's Sons. Describes more than 300 species of wood, including source, tree and wood features, workability and uses. Gives information on general tree structure, wood anatomy and characteristics. Other chapters discuss edible tree products, poisonous trees, drugs from trees, official state trees and woods of the Bible.

Core, H. A., W. A. Cote and A. C. Day. 1979. *Wood Structure and Identification,* second edition, 168 pp. Syracuse, N.Y.: Syracuse University Press. Wood structure presented at gross, microscopic and ultrastructural levels. Features outstanding photomicrographs and electron micrographs. Gives individual anatomical analysis for commercially important American species. Identification keys are based on hand-lens and microscopic features. Useful to wood identification students and to woodworkers as well.

Edlin, Herbert L. 1969. *What Wood is That—A Manual of Wood Identification,* 160 pp. New York: The Viking Press. Discusses timber conversion, past and present, to lumber, veneer and other wood products. Tree structure and anatomical features are the basis for a systematic key for recognizing 40 woods. Includes a fold-out chart containing 40 veneer specimens (7/8 in. by 3 in.).

Hand, Jackson. 1976. *How to Do Your Own Wood Finishing,* 170 pp. New York: Popular Science/Harper & Row. This practical guide to wood finishing and refinishing discusses stripping old finishes and surface preparation of wood by sanding, staining, bleaching and filling. Compares the features and application methods of penetrating resin finish, shellac, varnish, lacquer and enamel.

Harlow, William M. 1970. *Inside Wood...Masterpiece of Nature,* 120 pp. Washington, D.C.: The American Forestry Association. An exploration of the beautiful structure of wood as seen through a hand lens, light microscope and electron microscope. Wood properties, decay and dendrochronology are also presented. This thorough and technically accurate book is easily understood by non-specialists.

Harrar, Ellwood S. 1957. *Hough's Encyclopaedia of American Woods.* Vols. I-XIII, 25 species per vol. New York: Robert Speller & Sons. A treasured classic, in limited supply. Each two-part volume covers 25 species. Part One is a ring binder containing transverse, radial and tangential slices (1.6 in. by 3.7 in. by 0.01 in.) of actual wood tissue. Part Two is a bound edition of species descriptions.

James, William L. 1975. *Electric Moisture Meters for Wood,* 27 pp. U.S.D.A., Forest Products Laboratory Gen. Tech. Rpt. FPL-6. Washington, D.C.: U.S. Govt. Printing Office. Reviews pertinent electrical properties of wood and the design, accuracy, operation and maintenance of resistance and dielectric moisture meters. A list of commercial suppliers is also included. Electrical resistance values for wood are tabulated.

Koch, Peter. 1964. *Wood Machining Processes,* 530 pp. New York: The Ronald Press Company. This highly technical approach to all aspects of wood machining includes analyses of orthogonal cutting and peripheral milling. Woodworkers will be interested in the detailed discussions of sawing, jointing, planing, molding, shaping, turning, boring, routing, carving, mortising and tenoning, and abrasive machining.

Koch, Peter. 1977. *Utilization of the Southern Pines. Volumes I and II,* 1,663 pp. U.S.D.A. Agr. Handb. No. 420. Washington, D.C.: U.S. Govt. Printing Office. An exhaustive compilation on the ten species of southern pine. Volume I, *The Raw Material,* discusses the trees and the anatomical, physical and mechanical characteristics of their woods. Volume II, *Processing,* covers all phases of processing technology from machining to finishing.

Kollmann, Franz F. P., and Wilfred A. Cote, Jr. 1968. *Principles of Wood Science and Technology. Vol. I. Solid Wood,* 592 pp. New York: Springer-Verlag. Perhaps the most comprehensive work on wood science and technology ever written. Although much of the material is highly technical, fundamentals are also thoroughly covered. Includes chapters on anatomical structure, chemical composition, defects, biological deterioration, wood preservation, physical properties, strength properties, seasoning and machining.

Kollman, Franz F. P., Edward W. Kuenzi, and Alfred J. Stamm. 1975. *Principles of Wood Science and Technology. Vol. II, Wood Base Materials,* 703 pp. New York: Springer-Verlag. A companion to Volume I, this equally thorough treatise covers modified woods in all forms. Although many fundamentals are presented, the emphasis is on commercial processing technology. The six chapters discuss adhesion and adhesives for wood, solid modified wood, veneer and plywood, sandwich composites, particle board and fiberboard.

Kribs, David A. 1968. *Commercial Foreign Woods on the American Market,* 241 pp. New York: Dover Publications, Inc. A most useful guide to foreign woods. Descriptions of more than 350 species (including general properties, anatomy and uses) are accompanied by photographs of transverse (10x and 80x) and tangential (100x) sections. Contains an illustrated glossary, bibliographies and identification keys.

Little, Elbert L., Jr. 1980. *Checklist of United States Trees (native and naturalized).* Revised, 375 pp. U.S.D.A. Agr. Handb. No. 541. Washington, D.C.: U.S. Govt. Printing Office. Contains a compilation of scientific names and current synonyms, approved common names and others in use, and the geographic ranges of native (679 species) and naturalized (69 species) trees of the United States (excluding Hawaii). A principal authority on scientific and common names.

Mitchell, H. L. 1972. *How PEG Helps the Hobbyist Who Works With Wood,* 20 pp. U.S.D.A. Forest Products Laboratory. Washington, D.C.: U.S. Govt. Printing Office. Polyethylene glycol-1000 (PEG) is introduced as an agent for dimensional stabilization of wood. This booklet gives instructions for mixing PEG solutions, preparing treating vats, and drying and gluing wood. This booklet also explains how to treat turnings, green wood carvings, gunstocks and statuary, and how to preserve archeological specimens.

Murphey, Wayne K., and Richard N. Jorgensen. 1974. *Wood As An Industrial Arts Material,* 164 pp. New York: Pergamon Press, Inc. Intended to provide students, teachers and interested non-specialists with an understanding of the nature and properties of wood. Full chapters are devoted to machining, joint construction, grading, gluing, bending, finishing and other fabrication processes. The appendix suggests easy experiments to demonstrate wood properties.

Panshin, A. J. and Carl deZeeuw. 1980. *Textbook of Wood Technology,* fourth edition, 722 pp. New York: McGraw-Hill Book Company. Probably the most widely used text on wood anatomy and identification; essentially two books in one. Part One discusses the anatomy of commercially important wood and defects; Part Two includes identification features and keys. For each of 75 woods there are photomicrographs with descriptions of gross and minute features.

Princes Risborough Laboratory. 1972. *Handbook of Hardwoods,* second edition, revised by R. H. Farmer. 243 pp. London: Her Majesty's Stationery Office. Describes hardwoods from around the world to assist users in selecting timber. Information is conveniently arranged under headings such as weight, strength, working properties, veneer, defects, durability, preservation and uses. Full descriptions of 117 species, brief descriptions of 103, no illustrations.

Rasmussen, Edmund F. 1961. *Dry Kiln Operator's Handbook,* 197 pp. U.S.D.A., Forest Products Laboratory Agr. Handb. No. 188. Washington, D.C.: U.S. Govt. Printing Office. A comprehensive, easily understood text on dry-kiln technology. In addition to design, construction and operation of kilns, information is presented on other topics related to drying. Kiln schedules are given for numerous lumber species and thicknesses.

Record, Samuel J., and Robert W. Hess. 1943. *Timbers of the New World,* 640 pp. New Haven, Conn.: Yale University Press. Includes most of the important timbers of North and South America, with descriptions of geographic range, site type and tree characteristics. The woods are described mainly in terms of gross features and physical properties with notes on workability, uses and unique features. Scientific and common names are correlated.

Rendle, B. J. 1969, 1970. *World Timbers: Volume 1, Europe & Africa. Volume 2, North & South America. Volume 3, Asia & Australia & New Zealand,* 516 pp. London: Ernest Benn Ltd., and Toronto: University of Toronto Press. Re-issued from a series that originally appeared in the journal *Wood,* this comprehensive work describes more than 200 timbers of the world, each illustrated by a color photograph. Properties and characteristics are presented with a minimum of technical terms.

Rietz, Raymond C. and Rufus A. Page. 1971. *Air Drying of Lumber: A Guide to Industry Practices,* 110 pp. U.S.D.A., Forest Products Laboratory Agr. Handb. No. 402. Washington, D.C.: U.S. Govt. Printing Office. Describes how lumber can be dried most effectively under outdoor conditions, and illustrates air-drying principles and procedures. Discusses why lumber is air-dried, properties of wood in relation to drying, piling methods, air-drying defects, drying costs and protection of air-dried stock.

Selbo, M. L. 1975. *Adhesive Bonding of Wood,* 122 pp. U.S.D.A., Forest Products Laboratory Tech. Bull. No. 1512. Washington, D.C.: U.S. Govt. Printing Office. Summarizes the current information on the gluing of wood, and discusses various natural and synthetic adhesives. The preparation of wood for gluing and gluing procedures are explained. Although written primarily in reference to commercial production, the principles are applicable to small operations.

Shelton, Jay, and Andrew B. Shapiro. 1976. *The Woodburners Encyclopedia,* 155 pp. Waitsfield, Vt.: Vermont Crossroads Press. An informative, readable source of theory, practice and equipment for anyone interested in wood for heating. Includes comprehensive and technical information on fuelwood, stoves, chimneys, installation and safety, plus data on commercial stoves and other products.

Spencer, Albert G. and Jack A. Luy. 1975. *Wood and Wood Products,* 246 pp. Columbus, Ohio: Charles E. Merrill Publishing Co. A meeting ground between the very general and the highly technical, this book provides a comprehensive view of wood from identification and properties to wood-related products, processes, materials and industries.

Stevens, W. C., and N. Turner. 1970. *Wood Bending Handbook,* 109 pp. London: Her Majesty's Stationery Office. Covers hand and machine-bending of solid wood, laminated bending and plywood bending. Also discusses selection of material, preparation, softening and setting of bends, and explores the theory of optimum bending restraint. Tables of bending properties for numerous species are invaluable.

Symonds, George W. D. 1958. *The Tree Identification Book,* 266 pp. New York: William Morrow & Company, Inc. Simplifies the identification of principal American trees by allowing comparison of reader-collected or observed material to the book's photographs. Leaves, fruits, twigs, bark and needles are shown separately, then combined on master pages for each species.

Tsoumis, George. 1968. *Wood as Raw Material,* 276 pp. New York: Pergamon Press. A well-written text describing wood in macroscopic, microscopic, physical, chemical and ultrastructural terms. Wood formation and degradation are reviewed. Identification keys for North American and European woods are accompanied by useful photographs and instructions for microscopic study of wood.

U.S. Forest Products Laboratory. 1956. *Wood—Colors and Kinds,* 36 pp. U.S.D.A Agr. Handb. No. 101. Washington, D.C.: U.S. Govt. Printing Office. A pictorial identification guide (now out of print) for the species most commonly found in retail lumber markets. For each of 18 hardwoods and 14 softwoods, full-scale color photographs of wood surfaces and descriptions of the range, properties, uses and gross features.

U.S. Forest Products Laboratory. 1974. *Wood Handbook: Wood As An Engineering Material,* revised, U.S.D.A. Agr. Handb. No. 72. 446 pp. Washington, D.C: U.S. Govt. Printing Office. An important source for engineers, architects, designers and woodworkers. Includes basic concepts plus valuable tables of numerical data. Subjects range from wood anatomy and properties, lumber grades and composites to fastening, gluing, bending, finishing and preservative treatment.

Walton, Harry. 1976. *Home and Workshop Guide to Sharpening,* 160 pp. New York: Popular Science/Harper & Row. This comprehensive and clearly written book covers virtually every sharpening need. Principles of cutting edges are explained and the basic sharpening materials are surveyed. Specific instructions are given for sharpening everything from knives, chisels and scrapers to saws, drills, jointer knives and router bits.

Williston, Ed M. 1976. *Lumber Manufacturing,* 512 pp. San Francisco: Miller Freeman Publications, Inc. A discussion of sawmilling, in both practice and principle, from the raw material to sawing, sorting, drying and planing. Also covers design and operation of all types and sizes of mills, as well as saw characteristics, filing and maintenance.

The WoodBook 80, 372 pp. 1980. San Francisco: Avery-Phares, Inc. A compilation of product specification and use manuals from leading wood product associations, issued annually. Covers softwood building products (lumber, laminated timber, framing, millwork, plywood, particle board, etc.) and related building materials (stains, paints, etc.) Includes design details and construction recommendations.

Youngquist, W. G., and H. O. Fleischer. 1977. *Wood in American Life, 1776-2076,* 192 pp. Madison, Wis.: Forest Products Research Society. Written by a former director of the U.S. Forest Products Laboratory, this book looks at the role of wood and wood products in shaping our nation's past, present and future, combining an experienced scientific and technical viewpoint with exhaustive historical research.

Index